The Death of Argument

APPLIED LOGIC SERIES

VOLUME 32

SCOPE OF THE SERIES

Logic is applied in an increasingly wide variety of disciplines, from the traditional subjects of philosophy and mathematics to the more recent disciplines of cognitive science, computer science, artificial intelligence, and linguistics, leading to new vigor in this ancient subject. Kluwer, through its Applied Logic Series, seeks to provide a home for outstanding books and research monographs in applied logic, and in doing so demonstrates the underlying unity and applicability of logic.

The titles published in this series are listed at the end of this volume.

The Death of Argument

Fallacies in Agent Based Reasoning

by

JOHN WOODS

The Abductive Systems Group,
University of British Columbia, Vancouver, Canada
and
Department of Computer Science,
King's College, London, England
and
Department of Philosophy,
University of Lethbridge, Canada

Springer-Science+Business Media, B.V.

A C.I.P. Catalogue record for this book is available from the Library of Congress.

ISBN 978-90-481-6700-5 ISBN 978-1-4020-2712-3 (eBook)
DOI 10.1007/978-1-4020-2712-3

Printed on acid-free paper

For their unstinting support, their love and endless good humour,
I dedicate this work to
Carol Woods
Catherine Armstrong
Kelly Woods
Michael Woods.

Contents

Dedication v
Preface xiii
Acknowledgments xvii
Prologue xix

Part I Metatheoretical Questions

1. WHO CARES ABOUT THE FALLACIES? 3
   1 Questions About Fallacies. 3
   2 The Current State of Fallacy Theory. 5
   3 Light At The End Of The Tunnel? 7
   4 Their Importance. 12
   5 Is That All There Is? 15
   6 Critical Thinking 19
   7 So What? 20

2. THE NECESSITY OF FORMALISM 25
   1 Methodological Pluralism 25
   2 Formal Logic 26
   3 Nomic Systematicity 30
   4 Technical Information 31
   5 Theory and Practice 32
      5.1 Massey 33
      5.2 McPeck 34
      5.3 McP**k 35
      5.4 Scriven 35

6 Pedagogy 36
7 Conclusion 38
8 Appendix A 39
9 Appendix B 39
10 Appendix C 41

3. THE INFORMAL CORE OF FORMAL LOGIC 43
1 Artificial Languages 43
2 The Language of PC 45
3 The Utility of PC 48
4 The Irreducible Informality 49
5 Quantum Logic 53
6 Further Work for Informal Logic 55
7 Informal Theories of Implication and Consistency 60

Part II Threats and Intimidation

4. *AD BACULUM* AND PASCAL'S WAGER 65
1 The Standard Treatment 65
2 Prudential Argument 66
3 Case One: The Heist 67
4 Case Two: The Anti-smoking Commercial 67
5 Case Three: Pascal's Wager 68
6 Doxastic Surrender 69
7 Inaccessibility to Reason 70
8 The State, the Party and the Wager 71

5. APPEAL TO FORCE 75
1 Arguments From the Stick 75
2 Case One 76
3 Case Two 76
4 Case Three 76
5 Case Four: Risk Aversion Strategies 77
6 Case Five: The Stick-Up 78
7 Case Six: Negotiations 80
8 Case Seven: Veiled Threats 82

9 Practical Arguments 84
10 *Ad Baculum* Reasoning 86
10.1 Collective Bargaining 87
10.2 The Mugger 89
10.3 Anti-smoking Arguments 89
10.4 Utilities 90
11 Description of a Basic Model 91
11.1 Example 92
11.2 Example 92
11.3 Propositions 93
11.4 Actions 93

Part III Arguments Involving Reference to Persons

6. DIALECTICAL BLINDSPOTS 97
1 Not Practising What You Preach 97
2 Insincerity and Irrationality 99
3 Explanatory Opacity 104
4 Blindspots 106
5 Trust 107

7. *AD HOMINEM* *111*
1 Johnstone to Locke 111
2 Aristotelian Refutations 112
3 Falsifying Refutations 118
4 Peirastic Refutations 120
5 Proof *Ad Hominem* 122

8. AND SO INDEED ARE PERFECT CHEAT 125
1 "Fallacies" that aren't Fallacies 125
2 Bias 126
3 Popular Appeals and Common Knowledge 131
4 Preconception and Dogmatism 137
4.1 *Ad Verecundiam* 138
4.2 Conservatism 139
4.3 Inconsistency 141
5 Doxastic Loyalty 144

Part IV Pragma-Dialectics

9. PRAGMA-DIALECTICS 151
1 The Pragma-dialectical Approach 151
2 Difficulties 153
3 Unification 155
4 Prospects for Pragma-dialectics 156

10. BUTTERCUPS, GNP'S AND QUARKS 161
1 Theory-dependency 161
2 Fallacies 163
3 Moderate Dependency 165
4 Theory and Analysis 168

11. UNIFYING THE FALLACIES? 171
1 Theory Reductions 171
2 Exemplar Theory 174
3 *VEG* as Theoretical Stipulation 177
4 Reconciling *WW* and *VEG* 179

Part V Intractable Disagreement

12. STANDOFFS IN PUBLIC POLICY 185
1 Degrees of Standoff, Force I-III 185
2 Extremism 187
3 Dealing with Standoffs of Force Five 194
4 Education as Strategy: Persuasion and Thought Control 197

13. STANDOFFS AND DEMORALIZATION 201
1 Demoralization 201
2 Spinning 208
3 Witnessing 211
4 Privatizing the Good 212
5 Indoctrination 214

Part VI How to Interpret Arguments

14. CHARITY: CAN WE FIND IT A HOME? 219
   1 Slogans 219
   2 Radical Charity 222
   3 Dialogical Charity 226
   4 Radical Charity Again 232
   5 How Reliable is Charity? 233

15. MISSING PREMISSES 239
   1 Cooperative Conditions 239
   2 Validity 242
   3 The Logical Minimum and Pragmatic Optimum 243
   4 Genericity 246
   5 Find the Premisses 249

Part VII Analogy

16. BY PARITY OF REASONING 253
   1 Parity of Reason 253
   2 Standoffs 257
   3 The Structure of Analogy 260
   4 Counterexample 262
   5 Relevance 265
   6 Analogical Predication 268
   7 Deep Structure 269

17. VERDI IS THE PUCCINI OF MUSIC 273
   1 Analogies 273
      1.1 The Method of Cases 275
      1.2 Analogical Prediction 276
      1.3 What the Cases Suggest 279
      1.4 Numerical Analogical Prediction 283
      1.5 More Theory 285
   2 Analogical Semantics 290
   3 Concluding Reflections 296

Part VIII Induction

18. *SECUNDUM QUID* *301*
1 Omitting a Qualification 301
2 DeMorgan's Approach 304
3 Extending the Reach of the Fallacy 306
4 *Petitio Principii* 309

19. HASTY GENERALIZATION 311
1 Hasty Generalization: The Basic Idea 311
2 Primitive Induction 314
3 Strategic Rationality 315
4 Practical Reasoning 317
5 Notional and Behavioural Belief 319
6 The Raven Paradox 320
7 The Problem of Induction Again 324
8 Strategic Rationality Again 331
9 Conclusion 333

20. THE PROBLEM OF ABDUCTION 335
1 Abduction 335
2 Induction Has a Justification 338
3 Objections 341
4 Skepticism 345

Epilogue: The Way Ahead? 349

References 359

Index 373

Preface

The present work is a fair record of work I've done on the fallacies and related matters in the fifteen years since 1986. The book may be seen as a sequel to *Fallacies: Selected papers 1972–1982*, which I wrote with Douglas Walton, and which appeared in 1989 with Foris. This time I am on my own. Douglas Walton has, long since, found his own voice, as the saying has it; and so have I. Both of us greatly value the time we spent performing duets, but we also recognize the attractions of solo work. If I had to characterize the difference that has manifested itself in our later work, I would venture that Walton has strayed more, and I less, from what has come to be called the Woods-Walton Approach to the study of fallacies. Perhaps, on reflection "stray" is not the word for it, inasmuch as Walton's deviation from and my fidelity to the WWA are serious matters of methodological principle.

The WWA was always conceived of as a way of handling the analysis of various kinds of fallacious argument or reasoning. It was a response to a particular challenge [Hamblin, 1970]. The challenge was that since logicians had allowed the investigation of fallacious reasoning to fall into disgraceful disarray, it was up to them to put things right. Accordingly, the WWA sought these repairs amidst the rich pluralisms of logic in the 1970s and beyond. The WWA was never intended as a general theory of argument or a comprehensive articulation of an informal logic. It was a dominantly logical approach to the study of fallacies, an approach that emphasized the utility of nonstandard systems of logic, including dialogue logic. Neither was it supposed by the WWA that fallacies were *intrinsically* dialogical, as is now supposed by Douglas Walton in company with the pragma-dialecticians of the Amsterdam School. Nor was the WWA much interested in finding a unified theoretical structure within which everything bearing the name of fallacy would have a well-elucidated home. This was not a decision borne by failed attempts at federal amity, but rather one that grew out of the conviction that any such unification was destined to end up as theoretically thin. Here, too, as with our friends in Amsterdam, Walton's present inclination is to see in unification prospects of greater theoretical robustness than do I; and I on the other hand remain unconvinced, and so an unrepentant pluralist, at least for now, about the structure of fallacious reasoning.

There remains a point on which Walton and I remain in perfect accord. It is that getting the fallacies right is an extremely important task for logic, indeed a central part of its mandate, and that it is a task much more avowed than performed. I hope that I might be forgiven a brief shudder of impatience over the slapdash work that still makes its way into basic textbooks in logic and critical thinking.

In one way or another, everything in this book has been influenced by the Netherlands, my intellectual home away from home. Some chapters first took shape at the Netherlands Institute for Advanced Study, where under the direction of Frans van Eemeren I was part of the Research Group on Fallacies, together with Rob Grootendorst, Sally Jackson, Scott Jacobs, Agnès van Rees, Agnes Verbiest, Douglas Walton and Charles Willard. The remaining chapters were all written in whole or part when for portions of most of the years in this period I was Adjunct Professor in Frans van Eemeren's Leerstoelgroep Taalbeheersing, Argumentatietheorie en Rhetorica at the University of Amsterdam. I greatly value my intellectual association with Holland, and am gratified that over the years it has been the occasion of enduring friendships. For their interest and support of my work I warmly thank my Dutch colleagues and friends, Frans van Eemeren, the late Rob Grootendorst, Else Barth, Erik Krabbe, Agnès van Rees, Eveline Feteris, Francesca Snoeck Henkemans, Peter Houtlosser, Bart Garsen, Johan van Benthem, Jeanne Peijnenburg, and Frank Veltman. I am no less grateful to my non-Dutch supporters and critics: In Canada, J.A. Blair, Ralph Johnson, Robert Pinto, Hans Hansen, David Hitchcock, Christopher Tindale, Trudy Govier, Andrew Irvine, Leslie Burkholder, Leo Groarke, Mark Vorobej, Thomas, Hurka, George Englebretsen, J.J. MacIntosh, Michael Stingl, Ian Hacking, and Douglas Walton; in the United States, the late Henry Johnstone, Jonathan Adler, Harvey Siegel, Michael Wreen, Lawrence Powers, Mark Weinstein, Michael Scriven, James Freeman, Jaakko Hintikka, William Lycan, Alvin Goldman, Jonathan Bennett, Jonathan Bennett, Julius Moravcsik, Patrick Suppes, Bas van Fraassen, Sally Jackson, Scott Jacobs, Joseph Wenzel, Charles Willard, Isaac Levi, Dale Jacquette, and John Hoaglund; and Dov Gabbay, Jonathan Cohen, Chris Reed, Peter McBurney, Timothy Williamson, Jim Cunningham and Alec Fisher (U.K.); Harald Wohlrapp and Hans Jürgen

Ohlbach (Germany); Jean-Paul van Bendegem and Paul Gochet (Belgium); Gabriella Pigozzi (Italy); Lorenzo Peña (Spain); Graham Priest, Greg Restall, the late Richard Sylvan and Robert Meyer (Australia); and Roderic Girle (New Zealand).

Research for these chapters has been supported financially by, in addition to the Netherlands Institute for Advanced Study and the Leerstoelgroep Taalbeheersing, Argumentatietheorie en Rhetorica, the Social Science and Humanities Research Council of Canada, the University of Lethbridge Research Fund, the Dean of Arts and Science, University of Lethbridge, and King's College, London. My thanks to all.

My research assistants in Lethbridge, Dawn Collins, Ethan Toombs and David Graham, have put the book into camera-ready form and prepared the index. I am most grateful for this excellent help. My thanks also to the publisher, Charles Erkelens and his assistant, Lucy Fleet as well as Deborah Doherty, from Author Support, for technical assistance and direction.

Acknowledgments

I would like to thank the following co-author, editors, and publishers for their permissions to reprint material in this book.

- Co-author **Brent Hudak** for "Verdi is the Puccini of Music", *Synthese*, 92 (1992), 189–220; and "By Parity of Reasoning", *Informal Logic*, 11 (1990), 135–139.

- **Kluwer Academic Publishers** for "Verdi is the Puccini of Music", *Synthese*, 92 (1992), 189–220; "And So Indeed Are Perfect Cheat", *Argumentation* 9 (1995), 645–668; and "The Necessity of Formalism in Informal Logic", *Argumentation* 3 (1989) 149–167.

- **Royal Netherlands Academy of Arts and Sciences** for "Public Policy and Standoffs of Force Five", in E.M. Barth and E.C.W. Krabbe (eds.), *Logic and Politics of Culture*, 1992, 9–108.

- **Informal Logic** for "By Parity of Reasoning", *Informal Logic*, XI (1990), 125–139; "Buttercups, GNP's and Quarks: Are Fallacies Theoretical Entities?", *Informal Logic*, X(1989), 67–77; and "Is the Theoretical Unity of the Fallacies Possible?", *Informal Logic*, XVI (1994), 77–85.

- **Springer-Verlag GmbH & Co. KG** for "Deep Disagreements and Public Demoralization", in Dov M. Gabbay and Hans Jürgen Ohlbach (eds.), *Practical Reasoning: Springer Notes on Artificial Intelligence*, 1996, 650–662.

- **Logique et Analyse** for "Missing Premisses in Pragma-Dialectics", *Logique et Analyse*, 129–130 (1993), 155–168.

- **Uitgever Koninklijke Van Gorcum BV, Assen** for "The Problem of Abduction", *Algemeen Nederlands Tijdschrift voor Wijsbegeerte*, jaargang 93 nr. 4, pp. 265–273.

- **Sic Sat Press** for "Who Cares About the Fallacies?", in F.H. van Eemeren, Rob Grootendorst, J. Anthony Blair, Charles A. Willard (eds.), *Argumentation Illuminated*, Amsterdam: Sic Sat Press, 1992, 22–48; "*Ad Baculum* is Not a Fallacy", *Proceedings of the Fourth International Conference of the International Society for the Study of Argumentation*, in Frans H. van Eemeren, Rob Grootendorst, J. Anthony Blair and Charles A. Willard (eds.), Amsterdam: Sic Sat Press, 1999, 221–224; and "*Ad Hominem*: From Johnstone to Locke to Aristotle", in Frans H. van Eemeren, *et al.* (eds.), *Analysis and Evaluation*, Amsterdam: Sic Sat Press, 1995, 295–408.

- **Editions Rodopi BV** for "*Secundum Quid* as a Research Programme", in E.C.W. Krabbe, *et al.* (eds.), *Empirical Logic and Public Debate*, Amsterdam: Rodopi, 1993, 27–36.

- **Penn State University Press** for "Dialectical Blindspots" *Philosophy and Rhetoric*, 26 (1993), 251–265; and "Appeal to Force" in Hans Hansen and Robert Pinto (eds.), *Fallacies: Classical and Contemporary Readings*, University Park, PA: Penn State University Press, 1999, 181–196.

- **Communication and Cognition** for "Pragma-Dialectics: A Radical Departure in Fallacy Theory", *Communication and Cognition*, 24 (1991), 43–53.

- **Walter de Gruyter GmbH & Co. KG** for "*Ad Baculum*, Self-Interest and Pascal's Wager", in F.H. van Emeeren *et al.* (eds.), *Argumentation: Across the Lines of Discipline*, 1987, 343–349.

Prologue

In 1970, the Australian logician C.L. Hamblin published a book entitled *Fallacies.* It was a bombshell. Hamblin marshalled some very strong criticisms against what he called the Standard Treatment in fallacy theory. Here is part of that attack:

> We have no *theory* of fallacy.... In some respects... we are in the position of the medieval logicians before the twelfth century: we have lost the doctrine of fallacy, and need to rediscover it [Hamblin, 1970, p. 12].

Speaking of the treatment of fallacies in logic textbooks at the time of writing, Hamblin insisted that

> what we find in most cases... is as debased, worn-out and dogmatic treatment as could be imagined — incredibly tradition-bound, yet lacking in logic and in historical sense alike, and almost without connection to anything else in modern Logic at all [Hamblin, 1970, p. 12].

Hamblin's criticisms proved to be a fruitful provocation. To a large extent, three intellectual developments can claim some degree of origination in Hamblin's goading. These three are informal logic, fallacy theory and argumentation theory. A good sense of the development of informal logic can be got from Ralph Johnson's *The Rise of Informal Logic* [Johnson, 1996], as well as *Fundamentals of argumentation theory* [van Eemeren et al., 1996]. *Fallacies: Selected Papers, 1972–1982* by John Woods and Douglas Walton [Woods and Walton, 1989] renders a similar service for fallacy theory, as does Hansen and Pinto [Hansen and Pinto, 1995]. A useful *entré* into argument theory is the collection of essays, *Studies in Pragma Dialectics*, edited by Frans van Eemeren and Rob Grootendorst [van Eemeren and Grootendorst, 1994]. It is hard to know what Hamblin would have thought of the work done on fallacies during the past thirty years, but there can be little doubt that he would not in the year 2003 see occasion to lodge the same criticisms that he levelled in 1970. Fallacy theory may not yet have achieved the theoretical maturity that Hamblin called for, but it has come a long way from the despised Standard Treatment. The present book is a further attempt to take the fallacies research programme in a direction which, as I hope, Hamblin might have approved of.

I am a traditionalist about fallacy theory. I accept the traditional notion of fallacy as an argument or piece of reasoning which seems to be good in a certain way but isn't in fact good in that way. I also tend to think of fallacy theory as a branch of logic. I am moreover something of a traditionalist about logic. I yield to no one in the respect I have for modern mathematical logic in its four princely domains of model theory, proof theory, set theory and recursion theory. But no one seriously believes that modern mathematical logic is much good for the analysis of real-life argument and reasoning, nor that it is the best place in which to transact the business of fallacy theory.

Mathematical logic has achieved a hegemony that is scarcely a hundred years old, whereas logic itself has been a dominant intellectual force in the tradition that ensues from Athens for two and half thousand years. What logic has always been until comparatively recently is the theoretical core of general theories of argument and inference. It was so with the first logician and remained so, with little variation, until Frege. One thing that the old logic could not do is capture the structure of mathematics, not even Greek mathematics. Centuries later Descartes would celebrate algebra as a universal method of mathematical problem solving. After 1879, adequate logic would inherit this hegemonic character. It would purport to be wholly for the representation of the structure of mathematics. In this fateful transformation, logic lost its customary tie to the *agora*.

There is some evidence that the mathematical hold on logic is relaxing its grip. Informal logic, investigations into critical thinking, argumentation theory and fallacy theory itself are underwriting research programmes of considerable power. Computer scientists have led the way in re-jigging logic in ways that cater to more realistic models of actual human practice. Default logic [Reiter, 1980], nonmonotonic logic [Sandewall, 1972] and [McCarthy, 1980] and autoepistemic logic [Moore, 1985] are all products of work done by computer scientists. Even mainstream logicians have been involved in making logic more realistic, what with developments in dynamic logic [van Benthem, 1994], labelled deductive systems [Gabbay, 1996], situation logic [Barwise and Etchemendy, 1984], paraconsistent logic [Batens et al., 2000], among numbers of other such developments. Nor should we overlook the burgeoning literature on dialogue logic ([MacKenzie, 1990; Girle, 1993; Girle, 1996; Walton and Krabbe, 1995; Lorenzen and Lorenz, 1978; Gabbay and Woods, 2001c; Barth and Krabbe, 1982], to name just a few).

Mine therefore is both a more traditional and a more liberal view of logic. (A more detailed account of this conception than I have space for here can be found in Woods, Johnson, Gabbay and Ohlbach [Woods et al., 2002].) One of the virtues of this liberalism is that it makes wholly unnecessary (in fact, wholly impossible) fruitless wrangles about whether fallacy theory is logic or something else (dialectic, for example). And if we heed Hamblin's wish that fallacy theory achieve at least some sort of connection with modern logic,[1] we have taken a stand against those who think that ordinary reasoning and everyday argument should not be subjected to accounts that are technically theoretical.

Though I favour the more traditional conception of logic, I am not inclined to be greedy about it. I am far from believing that everything good in argumentation theory and fallacy theory is logic. Though my notion of logic is already something of an interdisciplinary concept, the disciplines required for the production of deep and mature theories of human cognitive and argumentative practice significantly outreach the stretch of anything deserving of the name of logic. I welcome the need for interdisciplinary cooperation, linking up all branches of logic with computer science, cognitive psychology, neurobiology, forensic science, linguistics (including conversation analysis and discourse analysis) and argumentation theory.

It is important that we be aware that there has been a considerable conceptual shift in fallacy theory since Aristotle's original classification in *On Sophistical Refutations.* In some cases the name persists while what it denotes has changed. Aristotle defined a fallacy as an argument that appears to be a *syllogism*, but isn't a syllogism in fact. But no one today, even those who take the traditional approach, would define a fallacy in just this way. This is because hardly anyone is much inclined to think the theory of syllogisms as a serious rival of modern systems of logic. (Perhaps the dismissal is too abrupt. The logic of the syllogism was the first intuitionistic, relevant, nonmonotonic and paraconsistent logic ever. See below). Similarly, present-day accounts of the *ad hominem* fallacy have nothing to do with Aristotle's discussion (interestingly the *ad hominem* did not make Aristotle's original list of thirteen) or with Locke's treatment of it [Locke, 1961]. Nor do our accounts of begging the question have the same breadth as Aristotle's, or our ac-

[1] A case in point is Woods and Walton's modelling of question-begging in the Kripke-semantics for intuitionistic modal logic [Woods and Walton, 1989, ch. 10].

counts of the complex question fallacy bear the slightest resemblance to what Aristotle said about it. Then, too, the inductive and statistical fallacies are comparatively new, arising in the 17th century with the birth of probability theory and the emergence of inductive methods in science. (See here [Woods, 1999b; Woods, 2002a; Woods, 1999d; Woods, 1999a; Woods, 1999c].)

'Dialectic' is also a word with a history. In Aristotle's *On Sophistical Refutations*, a dialectical argument is a question-and-answer exercise with respect to some received view or dialectical proposition. A dialectical proposition is one that expresses an *endoxon*. An *endoxon* is an opinion held by everyone, or by most people, or by experts. Today a dialectic argument is generally understood as one in which an issue is in *dispute*, with the parties attempting to prevail on each other with regard to that issue. However, in some cases, modern theorists understand a dialectical argument as little more than a *dialogue* in which the parties are having a rational discussion. (See here [van Eemeren and Grootendorst, 1984] and [Hamblin, 1970, p. 320].)

When names don't change, underlying conceptual change is likely to go unnoticed, except for investigators who have tried to master the history of fallacy theory. Since these are patently in the minority, bloopers are bound to be made. One is that John Stuart Mill charged that every deductively valid argument committed the fallacy of begging the question. Aristotle considered the possibility, and then rejected it; and Sextus Empiricus made the complaint and stuck by it. But not only did Mill not make the claim, he actually and emphatically denied it. Another blooper is one in which Locke is credited with originating the terms *ad verecundiam* and *ad ignorantiam* (and borrowing the term *ad hominem*), which he then used to denote the fallacies that have borne those names ever since. But Locke in fact does not think that arguments of these kinds are fallacious at all. Conceptual shifts and historical inaccuracies can mar a theorist's best efforts. I have tried to avoid these difficulties here, but I can't claim a perfect result. Conceptual shifts and historical mistakes are a little like fallacies. They don't in the general case announce themselves.

A word or two would be in order about the title of this book. At $176^a10 - 12$ of *Sophistical Refutations* Aristotle writes as follows:

> For it is possible for it to be true to say 'Yes' or 'No' without qualification to countless different questions; but still one should not answer them with a single answer for that is *the death of argument* ([Barnes, 1984, p. 299]; emphasis added).

Aristotle is here discussing the fallacy of several questions or, as it has come to be called, the fallacy of many questions or complex question. In its modern sense, the many question fallacy is exemplified by queries such as, "Have you stopped beating your dog?" in the context of interrogation arguments in which questions must always be answered by 'Yes' or 'No'. On some accounts, the fallacy involves the question's presupposition, "You have beaten or used to beat your dog". If the question is allowed to stand, then since it must be answered 'Yes' or 'No', each answer presupposes that the answerer has beaten his dog. In this way, the questioner commits the respondent to a proposition which the respondent has not conceded (and in most real-life cases would emphatically reject if he could). This is not, however, Aristotle's analysis. His view is that

> [t]hose fallacies that depend upon the making of several questions into one consist in our failure to articulate the account of a proposition. For a proposition predicates a single thing of a single thing .... Now since a deduction [= syllogism] starts from propositions and a refutation is a deduction, a refutation, too, will start from propositions [Barnes, 1984, pp. 286;169^{a}6 – 14].

Aristotle's example is the question "Are Coriscus and Callias at home or not at home?" But what goes wrong with this question? To understand Aristotle's position it is necessary to emphasize that he is discussing the thirteen fallacies in the context of a kind of argument called *refutation*. In a refutation there are two participants, a questioner and an answerer. The answerer puts forward a thesis T (e.g., that all virtues can be taught), which he then defends by answering critical queries put by the questioner. Aristotle requires that the answerer's responses be either 'Yes' or 'No' and therefore that the questioner's questions be answerable in this way. The further task of the questioner is to take the concessions made by the answerer as premisses and to construct from those premisses a syllogism whose conclusion is not-T, i.e., the contradictory of the thesis that the questioner was attempting to defend. Syllogisms for Aristotle are a special kind of valid argument. They are valid arguments that satisfy additional constraints. One is that the argument contain no redundant premisses (which make syllogistic logic a kind of nonomontonic relevance logic); another is that no premiss may occur as the conclusion, and derivatively, that premiss-sets must always be consistent (which makes syllogistic logic a paraconsistent logic); a further requirement is that syllogisms not have multiple conclusions (which makes the system a kind of intuitionistic logic). (See [Woods, 2001].) A refutation, then, is an argument that satisfies these conditions (among others) and whose conclusion is the contradictory of the questioner's original thesis. A further requirement imposed by Aristotle on syllogisms is that they be constructed from a special kind of statement, for which Aristotle reserves the technical term "proposition". As we have seen, a proposition is a statement that predicates a single thing of a single thing. Propositions therefore cannot be molecular constructions of pairs of statements. If we examine the question that Aristotle discusses, "Are Coriscus and Callias at home or not at home?", it is clear that it admits of a 'Yes' or 'No' answer. The 'Yes' answer concedes to the questioner the statement, "Coriscus and Callias are at home or are not at home". A negative answer concedes to the questioner the statement, "It is not so that Coriscus and Callias are at home or not at home". Neither of these statements is of any use to the questioner. They are not propositions. The first is equivalent to "Coriscus is at home and Callias is not at home, or Coriscus is at home and Callias is at home, or Coriscus is not at home and Callias is at home, or Coriscus is not at home and Callais is not at home". And the second is equivalent to the negation of the first. Since they are not propositions in Aristotle's technical sense, they are ineligible for use in constructing syllogisms; hence they cannot be used in constructing a réfutation.

This leaves us with two rather pressing questions. Why did Aristotle think that he was justified in requiring that syllogisms be constructed exclusively out of propositions? And what is it about the concession of non-propositions in the context of refutation-arguments that justifies Aristotle's passionate denunciation? ("the death of argument"). The first question is more easily answered than the second. Aristotle imposes the proposition-only condition on syllogisms because he holds the thesis of *propositional simplification.* According to this thesis, anything stateable in any natural language is also stateable without relevant loss in the sublanguage of propositions. Propositions in turn are statements of the form "All A are B", "No A are B", "Some A are B" and "Some A are not-B". If the thesis of propositional simplification is true, this works a substantial economy into Aristotle's logic. For Aristotle can now claim that he can capture a complete theory of deduction in this very economical sublanguage of propositions. Aristotle states the theory at *On Interpretation* $17^a13, 18^a19$ff. and 18^a24 [Barnes, 1984, pp. 26–28]. There is no known extant proof.

Our second question asks for what justifies Aristotle's claim that allowing non-propositions into refutation-arguments would be the death of argument. Aristotle does not answer this question, or anyhow does not answer it in any direct way. Even so, an answer can be conjectured. Take as a second example the rather more interesting (and modern-sounding) "Have you stopped beating your dog?" It is equivalent to "Do you have a dog you have beaten and, if you have a dog you have beaten, have you stopped beating it?" As Aristotle correctly notes, the question is answerable 'Yes' or 'No'. If the answer is affirmative it concedes the statement "I have a dog that I have beaten and I still beat it". But now consider what a negative reply gives us: "It is not the case that I have a dog that I've beaten and that I still beat it", or symbolically,

not-((D and B) and S).

Let me say again that Aristotle rightly recognizes that this statement could be true, that it could be a *fact* that not-((D and B) and S). But it is not a fact that the questioner can make use of as long as he is interested in determining the questioner's status as a (potential) past and present dog-beater. Imagine a dialogue in which every question were complex in this sort of way, and every answer were (as might well be the case in real-life interrogation-argument) in the negative. Then it would be impossible for the questioner to get to the *particular* facts of the case at hand. This indeed would be the death of argument with regard to any thesis which itself is a proposition (e.g.,"The questioner is not a violent person"). If every fact of the case is complex in the kind of way that the fact not-((D and B) and S) is, then no fact will imply any statement in the form "The questioner is a violent person" which is the contradictory of the thesis we are presently considering. Hence no refutation can be constructed.

It is arguable that Aristotle would have been prepared to extend the death-of-argument metaphor to all fallacies. There is no doubt that Aristotle thinks that fallacies are serious mistakes. This is also my own view, but it is subject to a tautologous seeming qualification: Fallacies are serious mistakes *when they are indeed fallacies*! Whether we are thinking of Aristotle's original thirteen or what I call the gang of eighteen (see chapter 1), it is apparent that rarely are arguments of these various types fallacies just because they are of that type. So, for example, when an *ad hominem* argument or an *ad verecundiam* argument is a fallacy, it is not so merely because it has the form of an *ad hominem* or an *ad verecundiam* argument. That is to say, fallaciousness is not intrinsic to arguments of these kinds; and the same is true for nearly them all, whether Aristotle's thirteen or the gang of eighteen. This is both a setback and an attraction. The setback is that the fallacy theorist must be able to specify the varying conditions under which an argument of a given kind is and is not fallacious. This is more easily said than done; but doing it, even so, is filled with interesting challenges, which are the potential occasion of theoretical depth. And that is the attraction of fallacy theory.

A word or two about my subtitle would also be in order. Since all reasoning is agent-based, *Fallacies in Agent-Based Reasoning* might strike the reader as rather pleonastic. Of course. But not all *theories* of reasoning make express provision for agents in their working vocabularies. In some circles, it is still widely held that the predicate calculus is the general theory of reasoning, but not even the proof theory of this approach has the means to establish theorems about what reasoning agents do. What the subtitle of this book is intended to signal is that an account of fallacies needs to be set in a more general theory of *cognitive agency*, a theory of how agents prosecute their alethic and strategic agendas.[2] I do not here have the space to propose a complete theory of agent-based reasoning. For the most part, it looms in the background, with occasional partial articulations as may be needed, to fill out a point in the emerging account of fallacies.

Some of the standard fallacies have something to do with real-life dialogues. To the extent that this is so, it falls to the fallacy theorist to have a clear-eyed view of what real-life dialogues are and of the conditions under which they are properly transacted. One of the earliest (and best) of such contributions is Aristotle's account of refutations, in *On Sophistical Refutations*. Mediaeval writers were logicians of considerable sophistication, and they made important headway with precise analyses of disputation games such as *Obligation* [Sherwood, 1963]. In the present-day dialogue logic is a growth industry.

At the heart of these developments is a conception of dialogical propriety captured in what has been called the Goody Two-Shoes Model of Argument [Gabbay and Woods, 2001c]. As applied here the name is a somewhat light-hearted reflection of the idea of a person who makes too much of being good. (The ancient word "goody" is a contraction of "goodwife".) In appropriating this name, I intend no mere levity, but wish rather to dramatize the model's distinctive features. In its most basic sense, Goody Two-Shoes argument is a project of open, objective enquiry in which the parties behave (or are expected to behave) reasonably, honestly, helpfully and equably. On one reading, Grice's logic of conversation is an instance of the Goody Two-Shoes model, as is the pragma-dialectical approach to critical discussions. Much, too, of modern dialogue logic approximates to this model. In the chapters that follow, I shall explain to the unacquainted reader details of Grice's and the pragma-dialectical approaches. For the present, it suffices to say that in a number of places in this book I shall attempt to demonstrate that a robust and imaginative fallacy theory requires that we should not accord hegemonic standing to the Goody Two-Shoes model; and that there are types of argument that deviate significantly from Goody Two-Shoes and yet are objects of serious theoretical attention.

---

[2] "Alethic" derives from the Greek and here means "pertaining to what is true" or "truth-oriented".

Strange as it might initially strike us, such arguments, though they are in one straightforward sense of the term *non-cooperative*, are not, just so, irrational or stupid. Part of their importance lies in the emphasis they lend to a distinction between what might be called *alethic rationality* and *strategic rationality*, the one an intellectual virtue and the other a practical virtue. This is a distinction which we shall repeatedly find it profitable to engage.

PART I

METATHEORETICAL QUESTIONS

In the interests of inter-chapter coherence, I have tried to give the book a kind of thematic organization. The three chapters that make up Part 1 are devoted to the following questions, among others. What is a fallacy? What is it that we want fallacy theory to do (what are its targets)? What resources are needed for fallacy theory and what are its principal methods? What is the connection between fallacy theory and informal logic? What is the connection, or lack of it, between informal logic and formal methods generally? What, in addition to its possible contributions to fallacy theory, is the proper business of informal logic?

Chapter 1

WHO CARES ABOUT THE FALLACIES?

The main business of this chapter is to motivate the systematic study of fallacious reasoning and argument. In its most usual meaning, a fallacy is a common misconception; that is, a false statement that is widely believed. Examples of such statements are, "Handling frogs causes warts", "You'll catch a cold if you sit in a draft", and — at a more academic level — "John Stuart Mill thought that every valid argument begs the question". (Recall the Prologue!)

This is not, however, the meaning of "fallacy" that has occupied logicians and argumentation theorists since the time of Aristotle. In this second meaning, a fallacy is something that characterizes the *reasoning* or the line of *argument* that gets one to assert or assent to or believe a given statement. And there is, in this second sense, no requirement that the statement that the reasoning has led to be a fallacy in the *first* sense.

If the initial task of the chapter is to say why studying fallacies in this second sense is a valuable thing to do, I also wish to call attention to the importance of the distinction between alethic and strategic rationality, mentioned briefly in the Prologue.

1. Questions About Fallacies.

There are mistakes. And there are mistakes that people actually make. The difference between them is that mistakes that are actually committed are mistakes that seem not to be mistakes to those who commit them. They are errors, blunders, and missteps whose mistakenness is unapparent. Some theorists of reasoning reserve their particular attention for a particular group of mistakes that are actually committed — or anyhow are committed with a notable frequency. They call these particular mistakes "fallacies", a term that has been used since antiquity. In the traditional meaning originating with Aristotle in the fourth century B.C., a fallacy is a bit of reasoning which seems to be good in a certain respect but is not in fact good in that respect. The deficiency being unapparent, a fallacy then would be a mistake of some sort that people actually make, or anyhow are disposed to make. Three questions immediately arise.

1 How does a fallacy differ from other mistakes that people are inclined to make? I call this the *boundary question*.

2 Do fallacies have something in common, apart from the fact that they are mistakes that people make or are inclined to make with a notable frequency? I call this the *common factor question*.

The third question is the most important of the three.

3. Apart from the fact that they are mistakes that people make or are inclined to make, do fallacies have any particular importance for human life? Are fallacies mistakes whose commission matters, and if so how? This I call the *who-cares question*.

My aim in the present chapter is to make some headway in answering these questions. The very fact that I have set myself this task is enough to indicate to the reader my conviction that the three questions are important, and that the last is especially important. This suggests an answer to the query embedded in the title of the chapter, "Who Cares About the Fallacies?" The answer is "Everyone does (or should)". The burden of the chapter will be to explain and justify this reply.

The fallacies I am thinking of in this chapter are collectively referred to as "the gang of eighteen". The term is a loose convenience of course. [Copi, 1986] treats seventeen fallacies, [Carney and Scheer, 1980] eighteen, [Schipper and Schuh, 1959] twenty-eight, [Black, 1946] eleven. Woods, Irvine and Walton discusses the following fallacies.

- *ad baculum*
- *ad hominem*
- *ad misericordiam*
- *ad populum*
- *ad verecundiam*
- affirming the consequent
- denying the antecedent
- amphiboly
- begging the question

- biased statistics
- complex question
- composition and division
- faulty analogy
- equivocation
- hasty generalization
- *secundum quid*
- gambler's
- *ignoratio elenchi*

[Woods et al., 2000]

It is worth noting that though these lists are obviously not identical, they do display a rather striking overlap.

2. The Current State of Fallacy Theory.

The fallacies are as usual up to their ears in definitional and analytical dissensus. People who are distrustful of anything so portentous as fallacy "theory" often wonder why it is that the more or less standard list of them, eighteen or so, usually of Latin name and dubious origins in a tired tradition, should still command a place in contemporary effort. After all, the scope for error, misstep and misconception is vast,[1] and who is to say that many a more deserving blunder does not call out at least equally for the celebrity of a name? So why these? (The boundary question).

A further objection is that for all the endless attention that the fallacies attract, there seems to be no settled and credible idea of what fallaciousness is. "Even when one learns to recognize alleged examples of the various 'fallacies', it is difficult to see what common factor makes them [so]" [Lambert and Ulrich, 1980, p. 25]. (The common factor question.)

Is it distinctive of them that the fallacies are bad *arguments* (or argument instances)[2] which occur commonly or with a notable or striking frequency?[3] Are the fallacies always arguments involving incorrect *inferences*, or is such a view, common though it is, seriously misconceived

---

[1]See, for example, [Massey, 1975a]. Shades of DeMorgan: "There is no such thing as a classification of the ways in which men arrive at an error: it is much to be doubted whether there ever *can* be", [DeMorgan, 1926, p. 276].

[2]"We should retain the historical nucleus of the idea of a fallacy as a logically bad argument" [Johnson, 1987a, p. 245].

[3]Johnson "we shall introduce the notion of frequency; because a fallacy is not just any mistake in an argument, but one that occurs with some frequency" [Johnson, 1987a, p. 245]. Cf. Scriven "Fallacies are the attractive nuisances of argumentation, the ideal types of improper inference. They acquire labels because they are thought to be common enough or important enough or to make the cost of labels worthwhile" [Scriven, 1987a, p. 333].

and itself a kind of fallacy?[4] Are the fallacies always arguments (or strategic mismanoeuvres, dialectical infelicities, or whatever else they might be) which exhibit a peculiar disposition to trick or deceive?[5] If so, whatever are we doing in exposing first-year students the wide world over to purported paradigms of the fallacies which would dupe no one (thus "There is an afterlife, hence there is an afterlife")?

This is not asked frivolously. If a fallacy is a bad argument which appears not to be bad, then how is it possible to identify an argument as a fallacy or to give an example of a fallacy? For the example to work, the person for whom the example is intended must recognize that the argument is bad, but it must also be an argument whose badness is not apparent to him or her. So, on the face of it, to see something as a fallacy is to see it as bad and not to see it as bad. But this makes it impossible to recognize fallacies. This is the *Fallacy Dilemma.* It imposes on the theorist and the student of fallacies a burden of some subtlety. It is the burden of getting people to see the badness of arguments they are not inclined to *see* as bad.

Undoubtedly among the strongest contemporary criticisms that have been launched against the Standard Treatment of the fallacies are, as we saw in the Prologue, those of Hamblin in his influential book, *Fallacies.*

> The truth is that nobody, these days, is particularly satisfied with this corner of logic.... We have no theory of fallacy at all.... In some respects... we are in the position of the medieval logicians before the 12th century: we have lost the doctrine of fallacy, and we need to rediscover it [Hamblin, 1970, pp. 11–12].

And

> ...what we find in most cases, I think it should be admitted, is as debased, worn-out and dogmatic a treatment as could be imagined — incredibly tradition-found, yet lacking in logic and historical sense alike, and almost without connection to anything else in modern logic at all. This is the part of his book in which a writer throws away logic and keeps his reader's attention, if at all, only by retailing traditional puns, anecdotes, and written examples of his forbears [Hamblin, 1970, p. 12].

Hamblin is making two complaints. One is, in effect, that informal logicians, and other people who were writing about fallacies in the late

[4]Treating fallacies as inference-mistakes is "a colossal mistake, a super-fallacy..." [Hintikka, 1987, p. 211].

[5]See Govier, "By definition, a fallacy is a mistake in reasoning, a mistake that occurs with some frequency in real arguments and which is characteristically deceptive" [Govier, 1987, p. 177], emphasis added. Cf. Woods and Walton, "a fallacy is an argument that is a tricky deception, because it is incorrect even while it has a tendency to seem correct" [Woods and Walton, 1982, p. 2]. And, more carefully, Fogelin and Duggan, "the term 'fallacy' is our most general term for criticizing any general procedure used for the fixation of beliefs that has an unacceptably high tendency to generate false or unfounded beliefs" [Fogelin and Duggan, 1987, p. 262].

1960s, had not done a good job in solving the Fallacies Dilemma. The other is, in effect, that these same people had not done a good job in answering the three questions that I set out in the introduction to this chapter: the boundary question; the common factor question; and the who-cares question.

Since, as I have already said, my aim is to answer these questions, and also to deal with the Fallacies Dilemma, it should be evident that I think that Hamblin's objections can be met.

3. Light At The End Of The Tunnel?

My answer to the third question, "Why should we care about the fallacies, why are they so important?", will have something to do with rationality. Rationality for me is a down-and-dirty affair, the more or less efficient management of life's dynamic turnings with the aid of low-finite cognitive resources in low-finite time against persisting invitations to err. It seems to me quite clear that part of the main business of human rationality lies in the contribution it makes to our survival and our prosperity. The human agent is genetically and culturally endowed with skills, some of which may have been naturally selected for, and these are the basic equipment in what we might with a certain looseness call his *rational survival kit.* It is, of course, a good question how far to press the idiom of natural selection. The human agent is a global maximizer, that is, a being who knows how to endure local setbacks in return for longer term advantages. It is debatable whether "the mechanism of natural selection... can [always] simulate rational human action directed toward a purpose or aim" [Popper, 1972, p. 267]. This is not to say that human agents wait passively by for new mutations to occur; in fact "one can talk of many, if not most, new cultural elements occurring by virtue of the fact that they are needed" [Ruse, 1974, p. 435]. When I speak here of our evolutionary endowments, I intend reference to rational skills that either have been naturally selected for or have come to us by way of cultural evolution, that is, because they "were needed".

What has this to do with fallacies? I am struck by something that Michael Scriven said, that the fallacies are a kind of ideal type of attractive nuisance or impropriety [Scriven, 1987a]. They are, I think, at least that. And as they have come down to us in the tradition they are also a kind of caricature of their associated improprieties, which lie deeply embedded in human practice. Certainly, "There is an afterlife; therefore, there is an afterlife" is nothing if not a caricature, and it is a perfect illustration of the Fallacy Dilemma. Good caricatures have a certain memorable vividness, and this is a helpful feature of them when it comes to what I think is the most distinctive aspect of the fallacies that the tradition has preserved for us at mid-century. For I think that the fallacies are idealized caricatures that are also symptoms of the sorts of things that can go wrong with those skills that make up our rational survival kits. So my contention will be that the fallacies that the tradition has thrown up are a highly idealized, vivid symptomatology of basic rational misperformance. If this is right, then it won't suffice to say that the fallacies are worthless as they stand — overdrawn, unrealistic, and unanalyzed (though often they are indeed these things). Thus construed, the fallacies would fare, well or badly, with other idealized symptomatologies, such as the one-month and two-month commercial bill discount rates that are loosely symptomatic of Japan's money supply rising too quickly.

It follows from this that the importance of the fallacies is derivative. They owe their importance to the importance of survival and to the contribution thereto made by those skills with which they are associated by caricatured misperformance. What are these skills? Without any pretence to completeness, it seems that they will include at least the following (however imperfectly they may be realized in practice):

- The ability to see the significance of what goes on and, in particular the ability to draw consequences.
- The ability to generalize from positive instances.
- The ability to make primitive inductions.
- The various routines of epistemic cooperation and, in particular, knowing when to rely on the say-so of others.
- The capacity to generate (not just receive) information.
- The means of discerning type-ambiguities in nature.
- The rudiments of cost-benefit protections of self-interest, including measures for conflict resolution.

No doubt other candidates could be furnished for inclusion in this list; nor are all the skills there displayed strictly disjoint. I do think, however, that these are basic in the sense that they are necessary, that human survival would be deranged without them. Supposing this to be so, the question recurs: "What has all this to do with the traditional fallacies?" I am going to propose that the traditional fallacies are errors or missteps which involve the misperformance of one or other of these basic skills; and, correspondingly, that the importance of fallacies is in part at least parasitic on the importance of these skills. Of course, it would be ludicrous to conceive of fallacies as mistakes that are *intrinsically* fatal or some near thing. (No coroner will give "fallacy" as the cause of death!) Not every misperformance of a fundamental skill has the importance that attaches generically to the skill itself. The connection that I am suggesting is subtler and more indirect. The items that make up our rational survival kits are largely constitutive of what it is for beings like us to be cognitive agents; and they go some way toward defining our rationality. And they are topic-neutral. So, on the present suggestion, *one* thing that fallacies have in common is that they are violations of comparatively primitive, topic-neutral cognitive norms. Fallacies in turn imbibe this topic-neutrality. A fallacy is not a mistake in physics or political science. It is a mistake *period*! When it is committed in physics or political science, it is not an error intrinsic to those disciplines.

I should note in passing that there are other skills whose misperformance lands us in fallacies, and other norms whose violation does so as well. Consider in this regard the important work of Robert Ennis (e.g., in his classic textbooks [Ennis, 1969b; Ennis, 1969a]). My list of skills is not a rival of Ennis', but rather a complement.

Let us turn to some examples.

A. In the business of survival, we will fare badly unless we know how to draw consequences. If we cannot draw consequences at all, we cannot avert disastrous consequences. And associated with this is the fallacy of the *non sequitur*.

B. We will not survive if we cannot generalize from samples, that is, generalize in confirmation-theoretic ways. The inability to project the "right" correlation is a fundamental failure. Equally, an over-zealousness for projecting coincidences can be ruinous, as in the generalization from a single observed case of food-poisoning with attendant epic discouragement of nutrition across the board. Thus, concerning generalizations, the gamut goes from null to hasty, for which latter the tradition offers us the slovenly caricature, "Cynthia is a bad driver; so women are bad drivers".

C. Not every generalization from a single case is a fallacy. In fact it is sometimes a blunder and *worse* not to generalize in this way, such is the role of "primitive inductions". The toddler on touching the element on his mother's kitchen stove learns in one case never to do that again. It is a remarkable feat and, it may seem, a full-blown generalization. 'Never' quantifies and 'again' projects. Here is a situation in which instantial experimentation would be, to the say the least, ill-advised, as if the toddler's confidence in

> the generalization had to await fifty or so repetitions of the encounter (and the early retirement of his poor hand from the pool of his motor resources). A mother noticing any such affection for the experimental method would at once summon the assistance of her pediatrician. And she would, of course, be nothing but right to do so. Here, too, the tradition reminds us of the importance of analogy-recognition in rational inductive practice, and intensifies the reminder with a caricature (oftener than not of misperformance rather than non-performance). Thus the silly but harmless, "It's perfectly all right for moose to go about without clothes, so why not us?" (It may, by the way, be preferable to detach primitive inductions from induction proper and assimilate them outright to analogical inference. Even so, it is doubtful that the link with induction would be altogether severed; for "[a]nalogy thus pictured is an inferential leap, whereof the top of the trajectory is a *slurred-over* induction") ([Quine and Ullian, 1970, p. 61], emphasis added).

Before going on with examples, it might be helpful to pause and to say again what I am trying to establish. We notice, first of all, that in virtually every treatment of the fallacies in the last hundred and fifty years or so, there is substantial evidence of a consensus that the traditional eighteen are possessed of sufficient wherewithal to grant them a non-trivial place in our speculations. Why, given that there are, as DeMorgan has said, many more ways of going wrong than any taxonomy can capture, do these recur in the literature, generation-in and generation-out? Is it simply a matter of intellectual conventions which we find lazily congenial?

It is also noticeable that, though there are some exceptions, the traditional lore about the gang of eighteen is dominated by brief and next-to-empty descriptions, together with examples of striking simple-mindedness. These I have been calling caricatures. Caricatures are meant to be simple-minded, and my point is that this is their principal function in the traditional accounts of the fallacies, and that in its most recurring aspects traditional fallacy "theory" is exemplary rather than analytic. It has largely been the case that traditional fallacy theory teaches whatever it does teach by example; and simple examples are more easily remembered than complex ones. So much for fallacious caricature. But what explains the importance and persistence of the gang of eighteen? Again, this is explicable as follows: the standard examples caricature (because they over-simplify) improprieties of rational performance of which they themselves can be considered a kind of idealized symptom. Those misperformances in turn are failures of one or another of our basic rational survival skills.

If this is a correct picture, it should be discernible in further examples. Let us see.

D. It is imperative that we make causal inferences. Crossing the streets of Amsterdam is a guaranteed disaster if the physics of traffic is construed in accidental, so to speak, correlations. Thus, *post hoc, ergo propter hoc.*

E. We need to make judgements of relevance. If the gunman has fired his pistol six times, should Dirty Harry try to infer whether the gun is empty (is it a Luger, perhaps?), or is he free to infer the completeness of the propositional calculus? The failure of relevance would be lunatic (though the implication, if he lived long enough to draw it, would be just fine). Well, it would be *truth-preserving.* Thus, *fallacies of relevance.*

F. We need to engage in routines of cooperative rational acceptance, of learning when to accept and act on the say-so of others. Was it Peirce who said that I know when my birthday is and who my parents are by hearsay? Who can pretend to any knowledge of current affairs without newspapers, or of science without scientific textbooks, or of the whereabouts of Central Station without asking Suzanne? So we have the *ad verecundiam*, the *ad hominem*, and the *ad populum* (that version of it in the form, "People say ... ", or "It is common knowledge that ..."). And we have caricatures: "Fifty million Frenchmen can't be wrong, so policies in Micronesia are fine".

G. It is necessary to know how to command or summon up information in a timely way. An efficiently managed life cannot be one of ungoaded informational passivity. We need to engage in techniques of information-seeking, including the asking of questions. Bleeding from an artery, it is insufficient to wait and see whether one's perceptual field will eventually register a sign reading "Hospital fifty metres ahead". One needs to *ask.* And so we have asking-fallacies such as *many questions*: "Have you stopped beating your dog?"

H. We must be able to recognize and to adjust to type ambiguities in nature. If our toddler, on burning his finger on the kitchen stove, quits the kitchen forever, and the house, and the neighbourhood, and the county ("Sussex is too rough for me!"), he is defeated by part-whole asymmetries. The point is that even a house can't be figured out at all until the basic part-whole asymmetries are cottoned onto. Our toddler needs to see that though the house is safe and cozy, the kitchen is differentially tricky. And so, we speak of the fallacies of *composition and division.*

I. Elementary routines of cost-benefit analysis are needed for the protection of self-interest. This is prudence, seriously underestimated by Oscar Wilde as "that bitch-goddess of incapability", the recognition of where one's interests lie and of what will and won't conduce to their realization. Whereupon the *ad baculum* enters the picture: "Your money or your life!"

4. Their Importance.

The fallacies that the tradition has given us have been victimized and traduced for failing every manner of task that it was not theirs to perform. They are not decision procedures, and not complete analyses of the actual reasoning practice. It is perfectly all right if someone wants to pursue such analyses — encouraged by the idealizations to muck in empirically and find out how induction, for example, actually does work, and in what ways it can systematically misfire (see below). But this has not been the mandate of the traditional typological accounts of the fallacies. The propositional calculus allows us to identify certain invalidity-schemata which might interest us for one reason or another. Some of these have names. So, "If A then B, not-A, not-B" is the fallacy of denying the antecedent; and "Not (A and B), therefore not-A", though it hasn't a name, so far as I know, does have a description: "the fallacy of distributing negation through conjunction". Though structurally more lucid than many of the fallacy schemata we have been discussing, these too are idealizations, caricatures of the real thing. They are no more an account or a theory of actual deductive malpractice than a traditional description of the *petitio* is a determination of the admissibility of certain patterns of explanation in paleontology.[6] So I conclude that the traditional fallacies have been unfairly knocked, though sometimes, as we have seen, their associated descriptions are over-stated or just wrong.

We also now have an answer to the boundary question, "Why *these*?" Because they are symptoms of errors that afflict the basic rational skills needed for survival. And, in so saying, the common factor question can likewise be put to rest.

[6]See Harper in reaction to the Woods-Walton treatment of the *petitio* in Kripke-structures [Harper, 1980].

The idea of rational misperformance covers at least two different kinds of cases, what we might call misfirings and abuses. Misfirings indicate malfunctions of the cognitive apparatus: not being able to manage the distinction between necessary and sufficient conditions; processing difficulties with nested negations; confusions (sometimes) over conjunctive probability; confusions of primitive inductions with instantial inductions; and so on. Abuses lie closer to the plane of conscious manipulation. Thus resorting to an irrelevant *ad hominem* rather than being taken in by one; exploiting another's confusion of necessary and sufficient conditions; tripping up an adversary with incorrect conjunctive probability determinations; and so on.[7] With the distinction at hand, we can now capture another intuition about the fallacies, the idea that they are tricks or deceptions. So they are, dialogically speaking, when one speaker commits a fallacy knowingly with a view to having the other speaker accept it innocently, with intended contamination of his commitment store.

I have also been assuming without argument what some readers may not be prepared to grant, namely, that the cognitive skills required for human survival involve rules and strategies which unfailingly qualify as rational. It may be, however, that sometimes such manoeuvres turn out to be in some interesting sense *incorrect.*

For example, hasty generalization is sometimes a *prudent* strategy, especially when the risks are high, even though the haste of the generalization might attract a charge of fallaciousness. If fallacies are violations or failures of rationality, then we could expect to have it that survival skills are sometimes exercised successfully but *not rationally.* Robert Fogelin seems to be making a similar point when he says: "human beings seem endowed with innate capacities for messing things up above the simplest levels of complexity. (This is particularly true of inductive inferences where, for example, the tendency for hasty and unfounded generalization seems to be hardwired into the human brain)" [Fogelin and Duggan, 1987, p. 4]. But, as Thomas Hurka has suggested, in conversation, sometimes it is the better strategy to conjoin probabilities non-multiplicatively, even though it violates the multiplicative axiom of the probability calculus.

The conjunction axiom of the probability calculus provides that the probability of a conjunctive proposition "A & B" is the probability of A *multiplied* by the probability of B (provided that A and B are probabilistically independent). But consider the following problem. Suppose that Lisa graduated in sociology from an Ivy League university, and that

---

[7] I am grateful to Scott Jacobs for discussion of these matters.

she is an active feminist and environmentalist. Which of the following two statements is more probable?

- Lisa is a Republican
- Lisa is a socially moderate Republican

If you selected the second instead of the first, your choice agrees with that of test subjects in psychological experiments. But it violates the conjunction axiom of the probability calculus. To see how, suppose that the probability of Lisa's being a Republican is 0.3 and of being a social moderate is 0.6. Then the probability that she is both is $0.3 \times 0.6 = 0.18$. Jonathan Cohen has argued in several places that although non-multiplicative conjunctive probability behaviour does violence to a concept of probability with which we associate the name of Pascal, it nevertheless honours a notion of probability with which we associate the name of Bacon and that sometimes this latter is the concept of probability which it is more appropriate for people to act upon.[8]

As for hasty generalization, it may be that it is not always fallacious (see item C, Section 3), and so it is likely that a condition on rational generalization will be something like this: generalizations may be made more or less hastily depending, among other things, on the risk that attaches to inductive caution.

Stephen Stich points out that rats are notorious hasty generalizers about toxicity [Stich, 1985]. See also, [Garcia et al., 1972]. They will refuse food tasting like any food ingested shortly prior to any sickness. Stich concludes that doing so probably equips that rat with more false beliefs than true, yet nevertheless the haste and breadth of the generalization will be indulged by evolution. It seems clear that there are two ways of looking at such situations. We could say, as Stich himself does, that there is, in evolutionary terms, nothing particularly crucial about truth. Or we could say that there are situations in which it is best to risk or make certain sorts of errors so as to avoid even more critical errors.

[8]See, for example, Cohen. Pascalian probability (also called aleatory) is the sort of probability embedded in games of chance, such as roulette. One of the founders of the mathematical theory of this concept was Blaise Pascal (1623-1662), hence the name. "Baconian" probability is named after Francis Bacon (1561-1626), one of the earliest, and staunchest, inductivists of modern science. Baconian probability is the degree of likelihood conferred on general propositions by their true positive instances. Cohen's submission is that these are different concepts of probability, and that each has (a contextually qualified) place in our cognitive lives [Cohen, 1989].

I also note that among the alternative formalisms for representing uncertainty is what some AI theorists call possibility theory (and which makes essential use of fuzzy set theory) as well as Dempster–Shafer theory (which permits the probabilities of disjoint sets to sum to more than 1.)

The second seems to me the better lesson to draw; and if this is right, it is some encouragement for thinking that in certain cases broad and hasty generalization are the more rational strategies, never mind that in some perfectly allowable ways they are (or might be) errors. Whether we decide to align the idea of fallacy with such errors is basically a question of whether to interpret fallacies as prudential or strategic failures rather than cognitive ones, and when to do so.

In any case, whatever the better option might be, it leaves undisturbed the fact that sometimes — often in fact — the strategic error and the cognitive error coincide, that the strategic error inheres in the falsehood of the generalization. And so we can continue to say that the hasty generalizations of the Standard Treatment are reminders that things can go wrong with inductions and that sometimes what goes wrong is the embrace of cognitive (or epistemic) error for the sake of misplaced strategic caution.

We shall note, however, that what we have just said does not determine that our toddler's one-shot generalization, though correct strategically (as it certainly is), is incorrect, or anyhow not as correct, cognitively or epistemically. Our description of the case does not require that we postulate for the toddler any more epistemic confidence in the generalization than that it *might* be true. This suffices to ground his determination not to act on its contradictory.

Before leaving this matter, there is a related issue raised by Isaac Levi, also in conversation. Fallacies, I say, are idealized symptoms of *misperformance* of rational skills necessary for human survival. Levi's concern is that I leave the account underdeveloped. For how are we to understand the idea of misperformance without a background theory of rationality which is constructable independently of any particular account of the fallacies? Fair enough, but it is not strictly to the point of the present endeavour. I mean to explain the importance and persistence of the gang of eighteen. For this I need the concept of misperformance. For present purposes, I am prepared to take it as primitive. This would be a bad choice were misperformance, unanalyzed, obscure to the point of explanatory impotence. I don't think that such is the case, nor do I think that Levi thinks so either. So for the task at hand I shall let things be. (Hasty generalization is also the topic of chapter 19.)

5. Is That All There Is?

I have been suggesting that the importance of the gang of eighteen derives from the importance of the cognitive skills whose failure or misperformance they caricature. If this were the whole story, it could fairly be said that it wouldn't make the fallacies interesting enough to justify the attribution of importance to *them.* I think that this is right. The case for the importance of the fallacies needs to be strengthened. It would be strengthened — in fact made — if we were able to justify a further claim, namely, that the traditional fallacies are *snares and delusions.* Let me turn to this now.

It is surely no stretch of the term to say that people are often *deluded* by complexity. There is evidence galore, some of it contentious, that human beings handle complexity badly. Nested negations, quantifier iterations,[9] the compounding of probabilities and the routine manoueuvres of induction all tend to go wrong in interesting and experimentally discernible ways.[10] What is striking about these errors is that everybody makes them, they are confidently made and they are hard to correct. Fogelin posits a genetic defect, a default in the cognitive hard-wiring. The idea of deludedness, thus explained, catches in its net a goodly selection from the gang of eighteen: all the inductive and statistical fallacies, affirming the consequent and denying the antecedent. In the Standard Treatment we find caricatures of this kind of cognitive malpractice and it is interesting and important because it is delusional. We now have the resources to hand to cash Scriven's engaging metaphor, in which fallacies are "attractive nuisances of argumentation" and "ideal types of improper inference". Such misperformances are attractive because they are natural to commit and they are confidently committed; they are nuisances because they are errors. They are ideal types of error exemplified in their caricatures; and they are improper because, again, they are errors. So much for delusions.

Snares are another matter. If some of the traditional fallacies are delusional because, and in the sense that, they caricature delusional performance, others are snares because, and in the sense that, they caricature cognitive effort of a kind that entraps the theorist, not the quotidien reasoner. All of the inductive fallacies saddle the theorist with the necessity to say something helpful about the problem of induction. The causal fallacy theorist must stand ready with solutions to various issues in the theory of causality. Unless these things are done, the theorist is all but guaranteed to have little of theoretical interest to say about

[9]Those familiar with the basic AI literature will be aware that von Neumann has a celebrated argument about the complexity of functions, an impediment to information-processing by humans which computers manage to overcome. Nested functions are the problem. The problem extends to existential quantifiers by way of Skolem functions with which they can be replaced [von Neumann, 1958].

[10]See the references herein to [Fogelin and Duggan, 1987; Cohen, 1989; Stich, 1985; Garcia et al., 1972]. See also [Nisbett et al., 1983; Nisbett and Ross, 1980; Tversky and Kahneman, 1983; Cohen, 1981; Holland et al., 1986].

his fallacies. The theorist is thus entrapped by difficult and intractable philosophical questions. It is not for nothing that the inductive and causal fallacies did not have much of a place in the tradition until after Hume's ransacking of induction and causality. For example, not much was made of inductive fallacies by Aristotle. Aristotle does discuss *secundum quid*, but in an desultory way, as the mistake of not qualifying a generalization. It was probably not until Antoine Arnauld and Pierre Nicole in their *Port Royal Logic* [Arnauld and Nicole, 1996][11] that faulty inductions were categorized as a species of fallacy, and it was not until Mill's *A System of Logic* (Book V) that these faults acquired their central importance in the evolving tradition.

Composition and Division give the theorist no particular worry about whether to bet on the Calgary Flames just because all of its members are good players. Rather, the theorist is faced with the necessity of getting clear the part-whole relationship [Woods and Walton, 1989, ch. 8]. He might also be held to the task of helping with, "Is my nose the mereological sum of its atoms?" [Putnam, 1988, p. 12]. Putnam imagines Kripke's reply as "since your nose could have consisted of different atoms, it has a modal property the group of atoms does not. So your nose is not identical with the group of atoms" [Putnam, 1988, p. 12]. And more inexorably still, when does the term "thing" compose and when does it divide? To ask this is to be well on the way to asking for a theory of (Aristolelian) substances.

The *ad ignorantiam* appears as the problem of inference to the best explanation. We might agree that Max Planck's postulation of quanta produces an entirely good derivation of blackbody radiation; and we might further agree that this is a reason to infer the truth of Planck's conjecture. But, then, why not also infer the existence of those traditionally despised ghosts, for which there is also no evidence, as theoretical entities which explain poltergeist phenomena? And if inference to the best explanation is all right, the theorist can be expected to explain why

(a) P explains Q
(b) Q
(c) ∴ P

is fine, while

(d) If P then Q
(e) Q
(f) ∴ P

[11]Similarly, the statistical fallacies came to light with the emergence of probability theory. We find the gambler's fallacy, for example, in [Laplace, 1951, p. 163].

is a notorious fallacy. Mind you, lots of people think that inference to the best explanation isn't all right, after all [Duhem, 1969; Laudan, 1981; van Fraassen, 1980].[12] Even so, I hold that the fallacy theorist cannot pretend to a good account of the *ad ignorantiam* without taking a principled stand on inference to the best explanation.

The *ad verecundiam* faces the theorist with the task of explaining the legitimacy of expertise in our communal epistemic practices. He or she will need, among other things, a theory of rational co-operation [Axelrod, 1984; Woods, 1989a]. He or she might also be expected to pronounce upon the Principle of Charity (co-operation theorists having appropriated the term from the precincts of radical translation). (See chapter 14 below.)

All these cases, and more, answer well to the idea of entrapment. It is the theorist who is trapped, not the everyday practical reasoner, and in every case the theorist will have been snared by a hard problem bequeathed to him by the philosophical tradition which oftener than not is only guide to such matters. In so saying we have the basis for resisting Arnauld's words which are a kind of rallying cry for legions of writers for whom the notion of fallacy "theory" is anathema. If what I've been saying is true, then not only is the gang of eighteen important, but Arnauld misses the point, and those who follow him are deeply misled by his having done so.

> But reason does not find its principle use in science [=theory]; to err in science is not a grave matter, for science has but little bearing on the conduct of life [Arnauld and Nicole, 1996, pp. 265–266].

Let me sum up the argument to date. The traditional fallacies caricature epistemic (and sometimes strategic) misperformance of skills necessary for our survival. Thus far, they are important because those skills are important. Moreover, many of the gang of eighteen portray defects which apparently are so embedded in human nature as to make such malpractice delusional, natural, universal, confident, wrong and incorrigible. Further still, several of the traditional lot turn on philosophical issues which one finds entrenched in even their most casual traditional descriptions, never mind that these descriptions are often wrong. The fallacy theorist is snared by these problems, for they stand menacingly in the way of his saying something sensible and deep about his target fallacies. He cannot make his way without presenting himself on in-

---

[12] See also [Cartwright, 1983, 87 ff] and [Hintikka, 1989, p. 4]. See [Gabbay and Woods, 2004] for further discussion.

duction, causality, rational co-operation, charity, inference to the best explanation, Aristotelian substances and much more besides. Of course, this is not to say that theoretical headway with, say, the *post hoc* fallacy encumbers the researcher with the need to produce a complete and comprehensive theory of causality. But at a minimum, he must (1) attend to Hume's contention that there is no objective difference between and no fact of the matter about whether a correlation is also a causal correlation; and (2) integrate his remarks about the *post hoc* with his contribution to the prior question. So the requirement is that the fallacy theorist must make such contributions to central philosophical problems as make possible their integration with and reinforcement of his assertions about the fallacies in question. Another way of saying this (and only slightly overstated), is that each of the traditional fallacies is an entire *research-programme* in its own right.

The further importance of the fallacies, then, is that they reflect in their caricatures performance-limitations which may be congenital in all reasoners, and/or they collect in their traditional casual descriptions philosophical problems of great difficulty, *some* success with which is an inescapable condition of advancing the theory of the fallacies in question. For fallacy theory is continuous with philosophy itself, and yet it branches into all the disciplines.

6. Critical Thinking

One of the more interesting contentions among contemporary critical thinking theorists has to do with the question of field-specificity. Some hold with McPeck that there are few, if any, interesting canons of successful critical thought shorn of disciplinary attachment and, in any event, differential disciplinary attachment will always make for differential critical procedures.[13] This sounds right, though it is an interesting question whether it forecloses upon all domain-neutral principles. (Cf. [Paul, 1985].) It is quite true that we shouldn't appraise paleontology in just the way that we do quantum field theory, but that may not entirely settle the hash of the generalization that denying the antecedent is always and everywhere a critical error.

Some people (for example, [Weinstein, 1990]) have also come to the opinion that philosophy — "pure epistemology" — is conspiciously disabled for the analysis of critical thinking. It is thought, apparently, to

[13]The modern *locus classicus* of this view is [Toulmin, 1969]. McPeck may have taken Toulmin's insight somewhat further than originally intended [McPeck, 1981]. An excellent response to McPeck is [Ennis, 1996].

be too general and too abstract, leaving unattended the peculiarities of domain-specificity. It also stints on the supply of application models.

For all its allure, the domain-specificity thesis produces for its enthusiasts a huge encumbrance. If the critical thinking theorist is now expected to plow the fields of all of human knowledge, how is he to do this with low finite cognitive resources and within his biblical allotment of three score and ten years (or twenty, if he doesn't smoke and watches his diet)? Even our most robust and tenacious polymath will stagger under the cruel burdens of such learnedness. Of course, it is always open to the critical thinking theorist to make his task manageable by leaving most of it undone. Thus, "You do physics; Sarah law; Harry art history; and I'll take a crack at policy-formation theory". When you, Sarah, Harry and I reconvene in Amsterdam at the ISSA Conference in 2006, whether or not we will be able to exchange our results will be a direct function of how deeply true the domain-specificity thesis was in the first place. If the respective investigative canons of those disciplines *do* inhere in their dark centres, we will have nothing to say to one another.

Even if subject depth doesn't completely foreclose upon chitchat across the lines of discipline, the proposed division of labour is not really what my critical thinker wants. He wants his theories to be, if not general, at least fairly comprehensive. He wants to outdo even Gerald Ford, who is said to be able to do two things at once. So he must be prepared to spread himself around. At this juncture the fallacy theorist has something useful to propose. He can invite the critical thinker to appropriate the gang of eighteen and let it take him where it will. If what I have been saying is so, the gang of eighteen will take him very many places and will make of the theorist a formidable cosmopolite. It is difficult to imagine that its tendrils would fail to reach any significant suburb of critical effort, as I hope the earlier discussion will have made plain.

The particular virtue of the suggestion is that it gives to our polymath some useful procedural structure. He will have some settled idea of where to begin, and of which issues have been settled and of which are outstanding. The gang of eighteen furnishes the theorist with a rich and challenging research programme in *medias res*. It is a programme with an ancient philosophical pedigree, but it is also a programme that insists upon application models and tough conditions of empiricial adequacy. The truth in the contemporary complaint against the Standard Treatment is not that the research programme is no good, but rather that too many a textbook has not taken it seriously. So there is a rapprochement between the traditional fallacies and the critical thinking agenda begging to get underway. I hope that it will get underway, and soon, since such is the farther importance of the fallacies. The present proposal supplies us with an interesting, in fact a crucial, byproduct. It tests in an orderly way the suggestion that the differential critical canon lies deep in the disciplinary heart of each subject. If we did meet with this radical eccentricity, my sunny optimism of just lines above would have been called into question. But now the doubts would be principled ones. We would have demonstrated a Limitation Theorem, to speak tendentiously. And that too would be important.

7. So What?

Part of what answers the boundary question is that fallacies are misperformances of topic-neutral cognitive skills of a very basic kind. Another part of the answer is that fallacies are snares and delusions. They are snares because they embed theoretically elusive concepts. They are delusions in the sense that they are confusions that *appear* to have a genetic origin (we appear to be, in Fogelin's words "hard-wired" for these errors). They are attractive mistakes, they are made by everyone, and they are hard to correct. The two answers together also go some way towards answering the common factor problem, for not every type of mistake carries the appearance of a genetic etiology, not every type of mistake is attractive, universal and incorrigible; not every type of mistake embeds a research programme; and so on.

This leaves the who-cares question. What is the importance of studying the fallacies? I have suggested that their importance is derivative upon the importance of the skills of which they are misperformances. I haven't yet discussed what may strike some people an the most obvious evidence of the importance of fallacies and fallacy theory: Fallacies damage cognitive performance, (both personally and interpersonally); and fallacy theory gives instruction in how to identify and *avert* fallacious reasoning and fallacious argument.

The reason I have not yet cited this presumed fact is that it is not a fact. Books and articles on fallacy theory are not manuals of self-help for the cognitively (or argumentatively) insecure. I won't take the time to review the empirical and experimental record. Interested readers could consult, with profit, the combined e-mail lists, the OSSA list for the study of argumentation and ARGTHRY list at York University, from December 10, 1999 to February 10, 2000, for a fascinating discussion of the real-life efficacy of courses in critical thinking and informal logic. For something hard-printed see also [Manktelow, 1999].

On the face of it, this is a horrific admission. It is tantamount to conceding the uselessness of fallacy theory, and of converting our title into a merely rhetorical question.

Upon reflection, we see that the expectation that a theory of fallacies would make us adept at avoiding their commission was both naive and careless. It was naive because, like most investigations of the human condition, there is little by way of a direct connection between disclosure and efficacy (in this respect, the traditional classification of logic as a *humanities* discipline is nothing but appropriate). It was careless in not taking seriously how fallacies have been characterized. They are snares and delusions. They are mistakes that are easy to make, that are widely made, and are difficult to correct, and difficult to detect.

The value of fallacy theory is the value of any of the human sciences generally. A good account of the fallacies will specify the conditions under which certain inclinations to err cash out in human performance. Understanding that is understanding a part of the human condition and a nontrivial part, as has been the burden of this chapter to show.

Bearing on the question of the utilitarian value of fallacy theory is a principle which I shall dub the *Easy-Easy Principle.* The principle does not strictly derive from the utilitarian expectation we have just been discussing, but it resembles it in spirit. To explain the principle, we need the concept of *easiness* as applied to human tasks. A task is easy when a human being can perform it competently without formal tutelage, and without noticeable effort. Having a conversation is easy for any pair of co-linguists, but producing a grammar of their language is not. Reading a timetable at the airport and inferring whether there is a flight from Vancouver to London on Saturday is easy; generating a theory of negation-as-failure (on which that inference in part rests) is not. Riding a bicycle is easy; plotting the underlying kinesiology is not.

The Easy-Easy Principle asserts that easy behaviour has in principle an easy theory. An easy theory is similarly one that can be understood by an arbitrarily selected competent individual without tutelage and without noticeable effort. Easy theories include common sense theories, but are not restricted to them.

What is surprising about the Easy-Easy Principle is the attraction it holds for people who should know better (briefly discussed in [Woods, 2000a]). It takes little reflection to see that it is a hopelessly mistaken maxim. Language is easy, but linguistics is hard. Inference is easy, but logic and cognitive science are hard. Arguing is easy, but there is no chance that even some of the simpler theories of argumentation are easy. A case in point is the pragma-dialectical approach to argumentation. For many people it is an attractive and accessible theory. But it is not easy.

Fallacy theory fails the Easy-Easy Principle hands down. Getting a deep and principled description of a fallacy turns out to be a far more complex thing than is superficially evident in the ease with which it is committed in real life. Determining, for example, whether it is fallacious to judge that the probability of "Lisa is a socially moderate Republican" is greater than "Lisa is a Republican" is, as we saw, something that takes us into the thick of inductive logic and probability theory; and these are far from easy.

The failure of the Easy-Easy Principle for the investigation of fallacious reasoning and argument serves as an indispensable guide to the fallacy theorist. It bids him not to seek for easy answers. This is hugely important guidance. It implies that a few confident intuitions on the part of the would-be theorist will not get the job done. It follows from this that if, for example, formal deductive logic is not the best theoretical instrument for modelling the fallacies, it is because it is the *wrong model* (concerning which see [Woods, 2000a]), not that it is difficult or technical or even that it is formal. It is here, I think, that we can best appreciate Hamblin's hostility to what he calls the Standard Treatment found in many of the logic textbooks of the late 1960s. Hamblin appears to be saying that the brief and largely unanalyzed descriptions of the fallacies that we find in such works cannot possibly be theoretically adequate accounts of them. This matters, according to Hamblin, since at the time he was writing, these textbook treatments were by and large all there was. It is as if the research community had given up on fallacies. It is also likely that Hamblin was responding to another concern. In most disciplines, textbooks have what I call the *trickle-down property.* They give accounts, albeit in simplified form perhaps, of the best theoretical work currently done in that discipline. This is true of textbooks on economics, biochemistry, physics, and banking and finance. Hamblin's point seems to be that there was in the late sixties nothing to trickle-down into textbooks or chapters on fallacies. Hamblin might agree with me that one of the attractions of the Standard Treatment is that it provides an idealized caricature of the fallacies, a kind of shortcut symptomatology. But if that is all there is to fallacy theory, there *is* no fallacy theory, and the caricatures are unavailing.

If what I have been saying in the present chapter is so, we may hope that the would-be fallacy theorist has a reasonable conception of how he or she must proceed, if Hamblin's dream of a regenerated research programme is to be realized. Of course, much water has flowed under many bridges since 1970, the year in which Hamblin's book was published. It is gratifying to see how much of Hamblin's dream has indeed been realized; and, if I may make so bold to say so, realized in ways not obviously incompatible with the account of fallacies sketched in these pages.

Chapter 2

THE NECESSITY OF FORMALISM

In this chapter and the next I want to take notice of the special relationship between fallacy theory and informal logic. Most of fallacy theory has nothing to do with the logical forms investigated by systems such as *FDL* (formal deductive logic). So it is reasonable to see most of fallacy theory as informal logic. At one time, I was inclined to the skeptical view that there was nothing to informal logic but fallacy theory. Now I am pretty sure that this is wrong. As will become clear, the *fundamental* question for informal logic is its connection or lack of it to the idea of logical form.

1. Methodological Pluralism

This chapter is a sequel to an earlier attempt to find and justify uses of formal methods in informal logic and argumentation theory. Here I wish to give greater attention than I was able to previously to various motivations of anti-formalist skepticism in (much of) argumentation theory [Woods, 1980], reprinted in [Woods and Walton, 1989, ch. 17]. I shall also attempt to set these remarks in the context of the complex web of relations between and among theory, pedagogy and classroom practice. I acknowledge that my title is something of a provocation. So, too, was my conclusion in *What is Informal Logic?*: "What is Informal Logic? *Nothing is!*" [Woods, 1980, p. 62]. But the present title's purpose is not merely rhetorical. For I hope that it can be said without offence that fueled by disapproval, a good deal of informal logic is crotchety, suspicious, anti-theoretical, doctrinaire and apocalyptic and thus in many ways wrongheaded and precipitate. There is, I think, a disturbing amount of methodological anxiety and self-scrutiny in argumentation theory, which suggests rather modest levels of conceptual and operational maturity in the subject. This being so, I rather think that what is required is a good deal of pragmatic toleration of methodological pluralism, and it is in this spirit that I offer my thoughts here. I propose to come at the question of the motivation of anti-formalist skepticism under five headings: *formal logic; nomic systematicity; technical information; theory and praxis;* and *pedagogy.*

2. Formal Logic

Here I have in mind the modalized variant of first order logic known as the (generic) system of strict implication. It is this system against which Ralph Johnson blasphemed so mischieviously when he coined the acronym *FDL*, pronounced 'fuddle' [Johnson, 1987a, p. 47]. It is a wonderfully funny turn. But why should it be? The fact is that the disapproval of *FDL* among nonformalist argumentation theorists is immense; and it can be accounted for somewhat along the following lines. Among the leading complaints are these: (1) *FDL* is too narrow to model adequately anything as rich and complex as actual argumentation. *FDL* is still basically the theory of deduction restrictively conceived in simple sentential, first order and modal environments. (2) *FDL* is too abstract, as witness first order model theory in infinite domains. (3) *FDL* squanders too much of its energies on such merely technical questions as the representation of truth-functionality by the Sheffer stroke. (4) *FDL* is too metatheoretical, as witness its affection for such things as the Compactness Metatheorem. And (5) *FDL* is too counterintuitive, sacrificing credibility to algorithmicity, as witness its treatment of implication and equivalence.

Each of these observations has a justification of sorts and each makes a point of greater or less importance. Each however is answerable, more or less convincingly (see Appendix A); and so I am minded to think that such criticisms pass their targets in the dark. (See also the Prologue.) What is truly surprising, however, is that the standard list of anti-formalist complaints does not contain two of the strongest attacks on the relevance of formal logic to the study and modelling of argument.

It has been around awhile. I proposed it in 1974 at the St. Louis International Conference on Relevant Logic [Woods, 1989b], and the same point had been made even earlier and independently by Gilbert Harman; so it is Harman's point by precedence [Harman, 1970]. It is simple to state: It is said that the rules of inference of formal logic are more or less canonical for human reasoning; but this is provably untrue. The so-called "rules of inference" actually fail for inference.

Harman, and I too, take inference to be a kind of reasoning which involves the adjustment of one's beliefs under the dynamic press of new stimuli. Inference can be likened to a function *f* from a set Gamma of belief-expressing statements to a not necessarily distinct set Delta of belief-expressing statements such that for any statement, *B*, it may typically be the case that either Delta=Gamma∪{*B*} or Delta=Gamma-{*B*}, where *B* is some member of Gamma inconsistent with *A*. Where the possessor of Gamma is not sure about *B*, then Delta might be Gamma∪{"I am not sure whether *A*"}; and so on. ('∪ here denotes the operation of set-theoretic sum, or union; '-' denotes set-theoretic subtraction.) The function *f* thus works typically either as an extension or a restriction of Gamma. A possessor of a stock of beliefs who is exposed to life's dynamic turnings and who has access to this function is able to modify his belief-stock by *adding* a new statement or by *deleting* an old statement, or by withholding either or both.[1]

It is a serious error to think (what virtually every elementary textbook in formal logic asserts with such lofty confidence) that the rules that govern *f* are (a subset of) the "inference rules" of classical logic. For, given that a belief-stock Gamma contains *A* and $A \rightarrow B$, Delta may be set in *at least three ways*: either Delta=Gamma ∪ {*B*} or Delta=Gamma-{*A*} or Delta=Gamma-{$A \rightarrow B$}. But the detachment rule for formal logic allows for just one manoeuvre: 'to infer' *B*. Hence the detachment rule is not a valid rule of inference. Similar considerations disqualify the other rules of formal logic. They also, by the way, disqualify (modalized) *ex falso quodlibet*, which is the classical theorem that says that a contradiction entails any statement whatever. But it is important to notice two things about the unseating of this rule. First, there is nothing peculiarly or exotically wrong with *ex falso quodlibet*; it is just one of several rules which fail for inference. Secondly, the demonstrable failure of *ex falso quodlibet* is a failure for inference, not for entailment. C.I. Lewis was, I think, seriously wrong when he said the system of strict implication has a "primary advantage over any present system", namely, "... that its meaning of implication is precisely that of ordinary inference..." [Lewis, 1912, p. 531].

---

[1]Of course not every adjustment of one's belief-stock is accomplished *inferentially*. A change in retinal stimuli may well lead to a change in belief (of what you see), but this is not inference (*pace* Peirce). For more on the type of belief-adjustment that inference is, see [Gabbay and Woods, 2003].

Well, then, the rules of formal logic misregulate inference, and this is a devastating body blow to strict implication as a theory of inference. It does not mean, however, that the system of strict implication is of no imaginable use to the argumentation theorist. For the system of strict implication still stands as a theory of *entailment*. (Most of the more celebrated attacks on disjunctive syllogism and *ex falso quodlibet*, confuse inference with entailment in some or other lethal way!). And so, the answer to our question would seem to be: if a knowledge of entailment theory would be of use to argumentation-analysis, then formal logic, i.e., *FDL*, still has a contemporary *raison d'être* within argumentation theory. And surely *some* knowledge of entailment is of imperative interest to the argumentation analyst. For one thing, he must devote some thought to whether a rational arguer and rational inferer must close their doxastic consequences, that is, believe all propositions entailed by anything he believes. (For a negative answer, see [Woods, 1988].)

The second attack on the relevance of *FDL* to the theory of rational argument is a reaction to something attributed to Quine. It is legitimate and necessary that the theory of discursive rationality take some interest in the nature and limits of human rationality itself. It is appropriate to ask in virtue of what characteristics and capacities is a human able to perform the routines of argument *rationally*. In a famous passage, Quine says that when it comes to interpreting a person's beliefs, the preferable translation "imposes our logic" [Quine, 1960]; for example, "the logical truths, or the simple ones, will go without saying; everyone will unhesitatingly assent to them if asked..." [Quine, 1970, p. 102]. It seems then that Quine is forwarding an important thesis about human rationality: *a rational agent has internalized a complete first order logic.* I do not have the space to evaluate this claim as completely as I would like. (For a fuller treatment see [Cherniak, 1986].) Suffice it to say that there is ample empirical evidence to contradict any notion of a human performance-model of (even) the propositional calculus. For consider the decidability of logical truth in the propositional calculus. It is now known that for this task there would be required the kind of intellectual effort that would swamp the capacities of an ideal computer equipped with boundless resources. This is not merely a theoretical limitation. It is a tremendous practical difficulty, which places the human performance-model in a position of computational derangement. Such a task is practically non-computable in the sense in which, and because, it is not computable in polynomial time. Similarly, the determination of truth functional consistency belongs to a set of problems characterized as nondeterministic polynomial time problems — problems solvable in polynomial time but on a *non*deterministic Turing machine; and also solvable by a *deterministic* computer but only in exponential time [Cook, 1971; Carp, 1972]. Such problems are human-computation impossible. And finite human rationality is no performance-model of a complete first order logic.

A Turing machine is an imaginary calculating device with which to model the concept of *algorithm*. A deterministic Turing machine is one whose output is produced deductively. In a nondeterministic Turing machine, output is governed by statistical considerations. A problem has a practical solution if it can be solved in polynomial time. That is, the time needed to solve the problem depends directly on the size of the input, or on the size of the input multiplied by itself once, or twice, or three times, or so on. Such problems are problems with fast solution. Collectively they are the class *P* (for polynomial time). Note that problems with fast solutions are often only comparatively fast. Intuitively, they can take a very long time. They are fast *in contrast to* problems that are in a class called *NP*.

NP is the class of problems whose solutions can be *checked* (as opposed to *found*) in polynomial time. Here is an example. There is a manual in which 10,000 different numbers are denoted, one by one. Do any of these numbers sum to 16? To find the solution, one must (nearly enough) inspect every possible group of the 10,000 and check its sum. (This is called the *subset sum problem.*) The best that a computer can do is to inspect slightly in excess of one thousand billion billion combinations. But the fastest machine handles only slight millions of instructions per second. So the subset sums problem takes forever and a day to solve. Programs for the subset sum problem have what computer scientists call *exponential time complexity.* When each new element is added to the input, the account of computation time that is needed is multiplied by some number. This produces what is called an *exponential blow-up* or a *computational explosion.* It is not known that there is no way in which *NP*-problems can have *P*-solutions. One of the central questions for computer science is whether *NP=P.*

In the early seventies Stephen Cook proved that all *NP*-problems could be reduced to a sole *NP*-problem in logic. The problem is called 'satisfiability' or *SAT*. The significance of the reduction is that if *SAT* has a fast solution, so does every *NP*-problem; i.e., *NP=P* [Cook, 1971]. A year after Cook's extraordinary theorem, Richard Carp, proved that *SAT* reduces to a variety of other *NP*-problems, including subset-sum [Carp, 1972]. This showed that no *NP*-problem is harder than subset-sum. So we say that, like *SAT* itself, subset-sum is *NP*-complete. *NP*-complete problems are therefore hardest problems in *NP*.

It is still not known whether *NP*=*P*.

Does all this complexity show that (modalized) *FDL* is useless for the argumentation theorist? Not at all, unless, again, he has no interest in entailment and no appreciation of the fact that his favourite theory of entailment is one in which entailment is *truth-preserving*. What is more, though its rules are not inference rules, entailment theory does aid in the task of consistency-management. In the general case in which a belief-stock contains A and $A \rightarrow B$, the appropriate entailment information shapes our inference choices, of which there are seven: (i) assert B; (ii) deny B and deny A; (iii) deny B and deny $A \rightarrow B$; (iv) deny B and withhold A; (v) deny B and withhold $A \rightarrow B$; (vi) withhold B and withhold A; (vii) withhold B and withhold A.

To sum up, so far: It is quite certain that *FDL* cannot be our theory of inference, though it may well be our best theory of entailment, and, just so, of some relevance to the analysis of argumentation. Similarly *FDL* (whether modalized or not) cannot be the internalized logic of actual human rational performance, at least along the lines attributable to Quine. These are important limitations, and they should be honoured. But they will not bear careless generalizations, e.g., that there is *no* place in argumentation for any formal theories or methods of any kind. That would be a fallacy.

3. Nomic Systematicity

Formalism has a tendency to simplify. That is a large part of its appeal. Simplification is achieved by abstraction and suppression, each of which has considerable potential for abuse. If it is forgotten that a formal theory ultimately serves its intended applicational interpretation, then a formal arrangement can end up without an application or it can dismiss or leave out of account features of genuine importance for actual practice. So one is indeed justified in keeping an eye open for potential abuses of formal methods. But it is surely an over-reaction to plunge into 'situational logic' — an enterprise "of merely anecdotal casuistry which over-concentrates upon the non-recurring case in all its ineffable singularity" [Woods, 1980, p. 62]. Tendencies of this 'situational' sort, to think that actual real-life argumentation is neither systematic nor nomic (i.e., subject to laws), are fairly easy to discourage empirically. Discourse, dialogue, argument, inference and memory-retrieval are all in part *structural* undertakings and therefore admit of, if not call out for, systematic treatment. Similarly, if such goings on are nomic, then by their systematic representation this lawlikeness is nicely expressible.

So we have two questions that lie at the very heart of the question of nomic systematicity. First, what is the value of systematic representation of actual argumentation? And, second, is there any reason to think that there is, in argumentation theory, an interesting, informative and non-trivial lawlike generalization of any kind? The answer to the first question is that systematicity, whether in mathematics, physics, musicology or argumentation theory, makes the dense, complex, multivarious flux *more intelligible*. And the answer to the second question is Yes. Here is one example of such a generalization: "No *ad baculum* argument is ever fallacious just because it is *ad baculic*" [Woods, 1987, p. 349], reprinted here as chapter 4.

There are others. The traditional informal fallacies loosely subdivide into two categories. There are the generic fallacies and the episodic. Generic fallacies involve a type of argumentative feature that disposes an argument-episode exhibiting that feature to error. Thus "... the term 'fallacy' is our most general term for criticizing any general procedure used for the fixation of beliefs that has an unacceptably high tendency to generate false or unfounded beliefs, relative to that method of fixing beliefs" [Fogelin and Duggan, 1987, p. 262]. And: "A fallacy is an argument which violates one of the criteria/standards of good argument *and which occurs with sufficient frequency* in discourse to warrant being baptized" [Johnson, 1987a, p. 246], emphasis added. Argument types having such features sometimes provide fallacious instantiations and sometimes not. So among the generic fallacies we might expect to find the *ad populum* and the *ad hominem*. (See Appendix B for an attempt to expose some of the nomic structure of the abusive *ad hominem*.) Furthermore, the weight of evidence suggests that their fallacious instantiations often, if not typically, turn on the irrelevance of what the fallacy's name picks out to the argument's conclusion; for example, in the modern analysis of the fallacious circumstantial *ad hominem*, the personal attribution is irrelevant, even if true.

Episodic fallacies are argument types which, relative to some specification of the cognitive act underway in which the argument is set (e.g., the argument is mimicking an inference) provide only erroneous instantiations. So a question-begging *inference* is always incorrect, and a *deduction* that commits the *post hoc, ergo propter hoc* is always incorrect.

Perhaps these are not yet knock-down generalizations or very deep ones. But they suggest that argumentation has the potential to yield up at least some of its secrets in such ways, and that such ways should therefore be explored.

4. Technical Information

There is much too much gobbledy-gook in contemporary intellectual affairs. Every academic discipline is simply awash in it. There is, accordingly, the perfectly natural disposition on the part of those who love unofficious plain speaking to keep argumentation theory, which in modern form is still in its comparative infancy and is still comparatively innocent, free of a jargony adolescence and an esoteric old age. Laudable though it is to keep argumentation theory unblemished by unnecessary obscurities, it is also going too far to think that its deepest truths are expressible in a vocabulary and conceptual inventory not much different from that which is required to handle a morning's shopping at the *Gap*, or an evening's entertainment of friends at dinner. In work that Douglas Walton and I have done over a number of years, we have been repeatedly struck by the helpfulness for argument theory of imported or adapted intellectual work from elsewhere, some of which is inescapably technical. Some of our work on Composition and Division would have petered out inconclusively except for instructive negative examples from set theory and mereology and positive examples from aggregate theory. So too with our work on dialectic and circularity; there Kripke's semantics for intuitionistic logic proved to be a congenial analytical tool. In some of our more recent efforts, Rescher's plausibility screening has helped with the question of how to adjudicate conflicting expert testimony [Woods et al., 2000, ch. 11]; probabilistic decision theory is helpful for the analysis of the negative consequence arguments, and so on.

It seems to me that there are two sensible rules concerning the admissibility of technical information into argumentation theory: First, it must do real work for argumentation theory. Consider, for example, the empirical utility claimed by Charles W. Harper Jr. for Woods and Walton [Woods and Walton, 1989]. According to Harper for analysis of circularity developed in that somewhat technical paper can be used to acquit paleontological explanation of the charge of vicious circularity [Harper, 1980]. Second, its non-technical paraphrase should not be too costly (i.e., too difficult or long to produce), to bother with. I am perfectly content to have the onus fall on the importer of technical information to establish its legitimacy and its helpfulness. (See here Appendix C.)

5. Theory and Practice

Here I have in mind four recent swipes, real or imagined, at the rôle of theories in the study of argumentation. First is Gerald Massey's celebrated claim, sometimes understood as the claim that fallacious argument is unamenable to theoretical articulation since there can be no such thing as a theory of invalidity. Second, there is John McPeck's charge that argumentation theory is devoid of general principles. Third is an argument that I have heard attributed to McPeck (thus the fruits of notoriety) but I haven't been able to pin it on him. So I shall offer its ownership to the anonymous McP**k. McP**k's criticism is that fallacy theory fails to identify and define the fallacies prior to enshrining them in theory; that is, fallacy theory fails to show that there is something "there" to theorize about. And, finally, there is Michael Scriven's plea, that argumentation researchers forego theory for a while and concentrate more on actual happenings [Scriven, 1987b, pp. 16–17]. Or, in a memorable turn of phrase, that argumentation researchers should concentrate not on the "physics" of their work but rather upon its "engineering". What to make of these challenges?

5.1 Massey

Massey asserts that *FDL* is better at detecting and describing validity than at detecting and describing invalidity. In this he is surely right; but he quickly goes on to say that there is "reason to hope that the gap, or the chasm, that separates our ability to show validity from our ability to show invalidity can be narrowed or even closed through a successful unification of logic and grammar" ([Massey, 1975a, p. 62] and [Massey, 1975b]. Cf. [Johnson, 1996, ch. 10]). Even so, *FDL*, illuminates validity only partially, as Massey has no need to be reminded. Some valid arguments — sequences whose premisses entail their conclusions — are under-paraphrased in *FDL*. Think, for example, of "*X* is red, therefore, *X* is coloured" and "*Y* is square, therefore, *Y* is rectangular", arguments which turn on determinate-determinable or species-genus considerations, and which used to be characterized as holding by "immediate inference, analytically" or "in virtue of the appropriate meaning-postulates" [Woods, 1974b]. Similarly, the counter-valid argument "*X* is orange, therefore, *X* is purple" will, paraphrased in *FDL*, collect the same verdict as, and will be given the same structural analysis as, the merely invalid "*Y* is orange, therefore, *Y* is a fruit". So Massey's point is that though formal analysis is not rich enough for the definitive account of either the validity or invalidity of argumentation, it does a better job for validity and nothing else does as good a job for invalidity. True enough, and important, but it is no counsel of despair, for it leaves open to attempt to track these properties in a natural logic, which is George Lakoff's term for a logical theory of deep structure developed by generative grammar [Lakoff, 1970].

I conclude, then, that Massey's asymmetry argument is not decisive against argumentation theory; it is cautionary, and deservedly so. If this is right, it is not quite so important for argumentation theorists to try to make the point that fallaciousness is not invalidity, and that even if Massey were right in pointing out the formal intractability of invalidity, this would not show the impossibility of a theory of fallacies. "I am inclined to grant the point that there cannot be a theory of invalidity.... The conclusion that I draw is not that there cannot be a theory of fallacy but rather that "fallacy" should not be defined in terms of validity/invalidity" [Johnson, 1987b, p. 242]. The important point is rather, is that you can't do (all of) fallacy theory in *FDL*. But who would have thought otherwise?

5.2 McPeck

McPeck advances the following argument: Each subject or discipline has its own peculiar domain of argumentation, each of which possesses and is governed by its own logic. Principles of argument are domain-relative and so too are the canons governing violations of such principles. There are no such things as fallacies as such, for there are no utterly general principles for them to be violations of [McPeck, 1981]; cf. [Johnson, 1996, ch. 11].

Now if McPeck means that there is no fallacy that is not fallacious in every argumentative domain, then it seems that he has made a basic mistake. For "p, so p" is always a fallacious inference (not to be confused with the correct and unexceptional entailment, "p entails p"). Mind you, not every fallacy is committable in every argumentative domain. Short of attributing causal powers to the large cardinal numbers, *post hoc* is uncommittable in Cantorian arithmetic, and the circumstantial *ad hominem* is uncommittable where there has been no personal reference to an opponent. If this were McPeck's thesis, it would be correct, but it would not be in any sense a discouragement of argumentation theory.

I have heard it said that McPeck thinks that the only mistakes worthy of the term "errors of reasoning" are bits and pieces of bad chemistry, shoddy paleontology, inept jurisprudence and so on. It can be granted that bad science and bad law might be influenced by feeble powers of reasoning (there or anywhere) but I should have thought that we would have lost a useful, if inexact, distinction between empirical inadequacies and logical failures in those very domains. In any event, the conclusion that some mistakes can be made only in chemistry, e.g., mistakes about the ionic structure of inert gases, is rather trivial, and it patently does not generalize. *That* would be a fallacy; but I must say that I find no convincing evidence that McPeck commits it.

Where McPeck is right, it seems to me, is in this: The fallacies are mistakes which are functions of distinctive structural or contextual features of the arguments they infect, *different mistakes* for different structural or contextual features. So it cannot be illuminating to attribute to fallacies in *general* any proposed counterpart of what you *can* say with some utility of all argumentative mistakes in *FDL*, namely, that in every case their premisses fail to entail their conclusions. But that is not to say that we could not non-trivially characterize the various fallacies according to their general types, and I can find nothing in McPeek that undermines this point.

It is worth remembering that the reason that *FDL* has theoretical legitimacy has a good deal to do with its being a mature theory of validity. To the extent that it is also intended to be a theory of logical *error* it is not, as Massey has said, nearly as mature a theory, if theory at all. So it may be possible to construe McPeck's skepticism about fallacy theory as a variant of Massey's skepticism. If so, I would say of McPeck what a few paragraphs above I said about Massey.

5.3 McP**k

Now, McP**k, on the other hand, is quite right to scold the argumentation theorist for careless assumptions about the existence and nature of fallacies, if that is what he is guilty of. But if McP**k also wishes to go on record as saying that *before* you do fallacy theory, you must produce the fallacies and say what they are like and how they function, then he has landed himself in an extremely uncongenial situation. For suppose that we applied this advice to quantum theory. First we say what quanta are, what they are like and what they do; only then do we bother with their theory. And once erected, the theory will face and meet the usual tests of adequacy, including the elucidation of its subject matter. However, this is not at all the way that quantum physics has evolved. Although it is an empirically adequate theory and though its mathematics is deep and well-understood, quantum theory does *not* illuminate its subject matter. Quantum theory is an ontological and conceptual mess. (See [Hooker, 1985] and [Maudlin, 1994, ch 1]). By the logic of his attack on fallacy theory, McP**k should also be ready to dismiss quantum theory, concerning which I have two concluding observations. If he did dismiss quantum theory, (a) I would feel better about the prospects of fallacy theory, even under the sting of his criticisms, and (b) I would buy him a new hat, if I knew who he was!

Of course, some people might not like the analogy with quantum theory. Perhaps they are right. It doesn't matter. The better, and more basic answer to McP**k is to propose that he read chapter 1 of this book.

5.4 Scriven

Scriven's enjoinder to forego the physics of argumentation studies in favour of their engineering is rhetorically fine, but it is in other ways an unfortunate metaphor. For modern engineering is thoroughly and utterly bound up with modern physics. And modern physics is surfeited by engineering models of some of its central results. It is very difficult to do the one without doing rather a lot of the other. If, however, Scriven's message is that, like physics and engineering, argumentation theory must be understood as inseparable from real-life models, and that any theory of argument must answer to very strong conditions of empiricial adequacy, then I agree entirely.

6. Pedagogy

Part of the task of the pedagogy of argumentation theory is to divine the nature of the subject sufficiently authoritatively to enable us to state with reasonable confidence (a) in what a competent understanding of argumentation consists (and what part of that competence is behavioural), and, more prosaically, (b) by what instructional devices and in how many semesters of study is the path through competency to mastery walked? In particular, it falls to the pedagogy of argumentation theory to design good doctoral programmes in the subject, and to set forth criteria for determining its important research projects.

At the heart of this pedagogy is an authoritative understanding of the relationships between (1) the *research programme* of argumentation studies, and (2) its *pedagogy*, and (3) the design and delivery of the *classroom programme*. I have space to give only the slightest hint of how complex these relationships are, and for brevity's sake I refer the reader to the following chart, to which I attach without proof the following 'Kantian Theorem': Minimal fulfillment of the Kantian injunction, concepts without intuitions are empty and intuitions without concepts are blind, requires that *some* of (1), mediated by (2), *must* reach (3). That is, classroom practice cannot be entirely innocent of theoretical developments in the research programme. (In the chart, by the way, entries in *bold-face* refer to application models of the *argumentum ad baculum*).

Finally, it goes without saying that argumentation theory must have empirical models with which to illuminate argumentative practice. It also seems obvious, despite what one finds in so many introductory textbooks on the subject, that sound pedagogy will give little encouragement to dig for one's empirical models in *either* advertising copy or Hansard or the Congressional Record *or* newspaper editorials. As for advertising, though methodologically and strategically multivarious, its principal object is to make a product attractive to consumers. Very little of this process relies upon or resorts to *argument*, although sometimes it does employ *for effect* an argumentative idiom (e.g., "studies *prove* that Bongo clears up coughs more quickly"). Similarly, while it can be said that advertising is inherently persuasive in its design, there is seldom much in the way of rhetorical devices discernible in most copy. The single most dominant cognitive undertaking of such copy is the authoritative presentation, orally or visually, of a *deeply salient fact.* It can be a prosaic thing to do, as in the case of the flyer from the local supermarket which catalogues such facts — tomatoes in n amount cost m dollars, and similarly for tuna, chicken and bath salts. Or it can be more indirect and subtle, as in the case of the slick luxury automobile ad in an 'upscale' glossy monthly, where the most salient fact may be visually presented — e.g., that this is the most gorgeous looking car in the cosmos. True, a suitably determined argumentation theorist may hold out for an argumentational reconstruction of such cases, pleading suppressed premisses, but no advertising professional — even those who know what they are doing — sees it this way.

In his non-philosophical moments, the present writer has enjoyed an interest in advertising both strategically and operationally, and he has for some years served on the Board of a mid-sized communications company. From such a perspective, it might be helpful to make brief mention of an Anacin campaign that had been on Canadian television for two or three years in the 1980s. The design of the thing was straightforward: a pretty young woman (actually there have been two; one went on to be a news anchor in Toronto!) looks into the camera from the streets of a pleasant looking place and says, "We are in the lovely town of Orillia (or some other place), to try the Anacin test!" Cut to a front door-step scene, showing pleasant Orillians. Next: back to the first scene and the pleasant young woman who avers that over 78% of the people of Orillia say that they will switch to Anacin. The campaign varied the city to give both regional and bicultural balance (Anacin is attractive to *everybody*), and also to emphasize its *multiple* attractiveness. So its being liked in Orillia suggests that it has met the test of decent, small-town cautiousness; and its being like in Quebec City suggests that in having won the affection of an exotic, more cosmopolitan people, it is a product for modern tastes, having almost international appeal. A prominent feature of these ads is that for each of the test cities, a statistic is proudly offered in the form of "n% of these citizens will switch"; and this might encourage the thought that the ad turns on a statistical argument which, given the suppression of sample details, is fallacious. But I think not. There is *no* effort in these ads to disguise the fact that if they were intended to be statistically argumentative, they would be fallacious, blatantly so. That is not their intent. Instead, in a language that suggests scientific sobriety [this is theatre], they just *assert* their *knowledge* of the propositions that Orillia or Quebec City have fallen for Anacin. And that is not a fallacy, though it could be a lie.

By the same token, much more Parliamentary or Congressional debate is *explanatory* than argumentative. An M.P. or Congressman explains his support for a bill much more than he will construct arguments to prove it, so to speak. His support is unblushingly autobiographical, which had he been making arguments would have utterly wrecked them, for his support is registered in terms of what '*seems* evident' to him, or of what he '*believes* to be good for' Alberta or Louisiana, or of what his 'good people back in the constituency *expect* of him'. Likewise, the main job of an editorial is to state an opinion and sometimes briefly to clarify and illuminate it. Here, too, if the typical editorial were construed as argument, it would be bad argument; and even if reconstructed by a fanatical missing-premissist, the missing premiss would, in general, be improbably long and doubtful.

So it seems to me important that there can be good editorials that aren't good arguments, good speeches in the House that don't even begin to prove their points and good ads that resolutely resist argumentative recasting. This shows, I think, that editorials, congressional debates, and advertising copy are not in general *dominantly argumentative* forms of cognitive effort. If this is so, it is of immediate pedagogical consequence. For if empirical models are wanted, obviously they are best sought out in those domains in which *knowledge is sought by way of proof*, in such areas as paleontology, economics, physics or law, that is, those domains that *do* involve dominantly argumentative forms of cognitive effort. Thus, or example, it is a good and important question for argumentation theory whether paleontology is circular; and similarly whether *FDL* argument commit *a petitio*; and whether quantum field theory commits the *post hoc*.

Of course the trouble with *this* suggestion (reflected in the questions it gives rise to) is that most beginning students have insufficient grasp of these bodies of knowledge to make accurate assessments of their principal arguments especially in so little time as a semester. That being so, it seems quite clear, as I suggested in chapter 1, that there must be pedagogical room for the abbreviated, artificial, schematized and just plain 'made-up' examples that mimic instructive characteristics of the real thing, current textbook fashions notwithstanding.

7. Conclusion

Modern argumentation theory is a young subject of vexing difficulty and complexity. It is natural and to be expected that its results are often incomplete and immature, and it is also natural and to be expected that it would be up to its ears in methodological testiness. I believe that ours is not the time for methodological manifestoes. I hope that we might for a time put our ideological squabbles to one side and that we might embrace a tolerant, responsible and interdisciplinary methodological pluralism. I for one recognize that I need all the help I can get, and I will happily appropriate any idea from any source if it genuinely helps in getting the job done.

8. Appendix A

What answers are possible to the anti-*FDL* complaint?

1 Formal logic is too narrow to model anything as rich an complex as actual argumentation.
Answer: Quite right. But even in its headiest moments *FDL* never presumed to model anything but fragments of argumentational structures. See the discussion of Massey.

2 Formal logic is too abstract, as witness first order model theory in infinite domains.
Answer: Agreed; that satisfiability can be defined for infinite domains is mainly of metamathematical interest. It is not a harmful thing to know, even though it won't be of much direct help in analyzing, e.g., diplomatic argumentative discourse.

3 Formal logic squanders too much of its energy on such merely technical questions as the representation of truth-functionality by the Sheffer stroke, $|$.
Answer: Perhaps "squander" is not quite the word I want, but here too the interest is largely metamathematical. I am not aware of anyone who seriously proposes that the stroke mimics a single unified English truth-function-like expression. Mind you, the Functional Completeness Metatheorem for $\{\neg, \wedge\}$ gives an extremely deep insight into truth functionality and should not be thought of as a mere technicality. The point about the Sheffer stroke is this: its elimination of $\neg$, $\wedge$ in favour of $|$ doesn't still further deepen our understanding of the minimalist semantic character of truth functionality.

4 Formal logic is too metatheoretical, as witness its interest in such things as the Compactness Metatheorem.
Answer: What Compactness shows is that nothing deducible by first order means requires an infinity of premisses; that deducibility is finitary. At

the very least, this suggests that *FDL*-deducibility manoeuvres are of a *type* appropriate to human competence. (Not to say, however, are all human-performance capable; for, though finitary, many such routines are too long for human management.)

5 Formal logic is too counterintuitive, sacrificing credibility to algorithmicity, as witness its treatment implication and equivalence.
Answer: True, the system of strict implication cannot replicate all entailments, for example, those borne by determinate-determinable relationship, as in 'X is red' entails 'X is coloured'. And equivalence, taken as strict co-implication, certainly does not fully model equivalence as sameness of meaning. But what seems so striking about strict implication and strict equivalence is not that they fall somewhat short of what they are models of, but that their representation is as close and accurate as it is.

9. Appendix B

The Nomic Structure of the Abusive Ad Hominem: A First Pass

The abusive *ad hominem* occurs in the following three-part arrangement: (1) an arguer X (2) attributes to an opponent Y some putative fact about his person which, if true, is designed to (3) falsify a thesis T held by Y. Note that this differs from the circumstantial *ad hominem*, in which X's attribution to Y, if true, is designed to point up an inconsistency between Y's behaviour or circumstance and his espousal of T, an inconsistency of such a kind as to discredit Y in his support of T. (See chapter 6 above.) Thus, the chief difference between the abusive and circumstantial *ad hominem* is that only the former is intended to falsify T (it should also be said in passing that not all abusive *ad hominem* arguments are abusive in the ordinary sense of that term).

The structure of the abusive *ad hominem* can be set out as follows:

(i) There is a thesis T espoused by Y and attacked by X

(ii) There is a necessary condition K on the truth of T

(iii) There is a fact F about Y, attributed to him by X

(iv) F is a fact of a type to *contra-indicate* T because it calls K into question.

Definition: If F is a fact of a type to contra-indicate T then F gives a (not necessarily conclusive) reason to disbelieve T.

(v) Therefore, F gives a reason to disbelieve T.

It is clear, then, that not every abusive *ad hominem* should be rejected, e.g., where T="Jones would make a good county treasurer", K="Candidates for county treasurer should be honest", and F="Jones was twice convicted of embezzlement", or where T="Smith is fit to play (football) on Sunday", K="Players should be free of major injury if they are to play a game like football" and F="Smith has a serious groin injury". The analysis also is sternly disapproving of all the usual horror cases, as when *T*="Smith would be a good county treasurer" and *F*="Smith is a Catholic".

It emerges, then, that if the abusive *ad hominem* is to be classified as a fallacy at all, it is at most a generic fallacy. But on reflection, it seems to me that the present analysis-sketch shows or suggests that the abusive *ad hominem* is not even that, that is, not a form of argument "that has an acceptably high tendency to generate false or unfounded beliefs..." [Fogelin and Duggan, 1987]. To be sure, it is a form of argument that is open to error and thus one that does not possess structural guarantees of goodness. But no forms of argument have such guarantees, not even the valid forms of *FDL*; for they are also open to unsoundness. The abusive *ad hominem* is a generic fallacy only if it is peculiarly prone to error, only if there is some unusually high likelihood that a arguer's attribution of *F* to his *vis-à-vis* will miss the mark and not call into question condition *K* which is necessary to the truth of his opponent's thesis *T*. If I am right about this, the interest one has in the abusive *ad hominem* is not so much structural as contextual, and the means we have of avoiding the commission of erroneous cases of it is more judgemental or intuitive in character than algorithmic. And this is something that informal logicians have been saying for a long time about fallacies in general.

10. Appendix C

Here I briefly set out an example of the use of technical information which would seem to violate our two rules.

I refer to Arnold Günther's interesting paper "Some Ways of Representing Dialogues" [Gunther, 1985]. I wish to make it clear that I am not registering a complaint against his analysis of dialogues, but only against what would seem to be an inessential resort to technical formulation. Here is an example of what I mean. Günther's analysis furnishes us with six rules of a possible dialogue, each of which is set forth in clear relatively untechnical language, and each of which is followed by an interpretation, also in language as free of technical material as the rule in question can permit. Finally, each interpretation is summed up in a highly technical manner. Thus we have rule five, *R5*:

> *R5* A possible dialogue is a finite sequence of triples each of which consists of a name, a well-formed expression, and a set of triples such that each of them consists of a name, a verb of propositional attitude and a designatory a phrase [Gunther, 1985, p. 246].

Next comes the interpretation of *R5*.

Interpretation:

(a) The names are names of the dialogue partners.

(b) The well-formed expressions represent the expressions used by the dialogue partners.

(c) A triple of name, verb of propositional attitude and designatory phrase represents an occurrence of a propositional attitude (see the description above).

(d) A set of triples representing (occurrences of) propositional attitudes represents those propositional attitudes of the dialogue-partners which are relevant for the dialogue.

(e) A triple of name, well-formed expression, and set of triples represents both the speech act the person named performs and the propositional attitude connected with this speech act.

(f) The sequence represents the temporal order of the speech acts and of the emergence, charge and disappearance of propositional attitudes connected with the dialogue" [Gunther, 1985, p. 246].

And now comes the summing up.

"More formally: D is a possible dialogue according to *R5* if there is a language, L, a two-element set N of names, a set V of verbs of propositional attitude, a set P of designatory phrases and a natural number n such that for each i, $1 \leq i \leq n$, there is a name $N \epsilon N$, and expression ϕ_i of L, and a set E_i of triples of the form $((N, v,p), N \epsilon N, v \epsilon V, p \epsilon P$, such that $D=(N_i, \phi_i, E_i), \ldots, (N_\eta, \phi_\eta, E_\eta))$" [Gunther, 1985, 246–247].

I find that I am unable to discern in the formal paraphrase anything that couldn't more easily be got form the preceding interpretation. In a footnote, by the way, the author indicates that he too might share this reservation: "'*P*' and '*p*' now from 'propositions', (they are usually designated): 'V' and '*v*' from 'Verb (*very helpful, isn't it?*)" [Gunther, 1985, p. 250 n. 11], emphasis added.

Chapter 3

THE INFORMAL CORE OF FORMAL LOGIC

"As for the business of translating from ordinary speech into precise logical forms... that is a matter of applied logic if you will..."

[Peirce, 1992, p. 145].

In this chapter I want to carry on with issues raised in its predecessor. The position that I take here is that if informal logic can lay claim to a coherent research programme it is incumbent on its practitioners to display the programme concretely. In the preceding chapter I argued that a degree of formality is needed for informal logic. In the present chapter I return the compliment, claiming that informality is an irreducible part of formal logic.

1. Artificial Languages

"Logic is an old subject and since 1879 it has been a great one". So said W.V. Quine in yet another of his elegant dismissals. I will not take the time to expatiate fully on Quine's remark. It suffices for my purposes in the present chapter to concentrate on just one facet of the change in logic initiated with the publication in 1879 of Frege's *Begriffsschrift* ("Concept Writing"). Frege insisted that natural languages were not the proper subject of a logician's interest. This was also Peirce's view, a fitting concurrence given that Peirce, along with Frege, is the independent co-founder of modern quantification theory. The knock against natural languages is that they lack the precision and the unambiguousness required for logical theory. What is more, natural languages harbour contradictions. The task of logical theory had (in Frege's view) become the task of finding a home for arithmetic, including transfinite arithmetic, i.e., the arithmetic of actual infinities. The home that logic would provide for arithmetic is that of a reductive theory, according to which every truth of arithmetic would be expressible without relevant loss in the language of the new logic. The claim that such a reduction is possible was called the *thesis of logicism.* Frege's rejection of natural language turned on his conviction that natural languages were too crude to support the reductive niceties of logicism. Peirce, on the other hand was not a logicist, but he shared Frege's distrust of natural language. It is perhaps worth remarking in passing that Frege's conception of the purpose of logic bears no resemblance to that of the founder of logic. For it was Aristotle's view that logic was to be the theoretical core of a wholly general theory of argument and reasoning — of argument and reasoning as transacted in everyday Greek.

The *Begriffsschrift* was carefully designed to avert the purported defects of natural languages. Frege writes that

> to prevent anything intuitive from penetrating here unnoticed, I had to bend every effort to keep the chain of inferences free of gaps. In attempting to comply with this requirement in the strictest possible way I found the inadequacy of language to be an obstacle; no mater how unwieldy the expression I was ready to accept, I was less and less able, as the relations became more and more complex, to attain the precision that my purpose required. The deficiency led me to the idea of the present ideography [=technical notation=artificial language]. Its first purpose, therefore, is to provide us with the most reliable test of the validity of a chain of inferences and to point out every presupposition that tried to sneak in unnoticed so that its origin can be investigated [Frege, 1879, pp. 5–6].

Here, too, is Peirce, to much the same effect.

> It is true that the propositions must be expressed *somehow*; and for this reason formal logic, in order to disentangle itself completely from linguistic, or psychical, considerations, invents an artificial language of its own, of perfectly regular formation, and declines to consider any proposition under any other form of statement than in that artificial language [Peirce, 1992, pp. 144–145].

Frege and Peirce see logic in the following way. Logic sets itself the task of describing certain target properties, such as implication and consistency. These properties are without exception properties of *linguistic structures.* But since logic does not transact its business within natural languages, it follows that the linguistic properties that the logician seeks to describe are properties of *non-natural* or *artificial* languages.

Not any made-up language (Esperanto, for example) counts as an artificial language in the logician's sense. In fact, upon reflection, the logician's artificial language strains to the uttermost the very idea of a language, as we shall see. This becomes evident by examining the most elementary part of logic, known as the *propositional calculus*, or the *logic of propositions* (**PC** for short).[1] The logic that I am in process of describing is certainly not all there was to logic from Frege onwards (see, for example, [Sidgwick, 1910; Dewey, 1981]). An adjective is needed to sort it out from other developments in the one hundred and thirty-some years since the publication of *Begriffsschrift*. So I will speak of these particular developments as *mathematical* logic. There are three reasons that make this a suitable choice. One is that the propositional calculus can be interpreted in a wholly algebraic way, as Boole showed [Boole, 1847]. (Thus that part of logic is subsumed by that part of mathematics.) Another is the twist that Frege gave to the first in attempting to make good on logicism. (Thus arithmetic is now subsumed by logic!) The third reason is that the logic in question developed into largely autonomous sectors (model theory, proof theory, set theory and recursion theory) that are dominated by mathematical methods and procedures. So, then, to return to the point at hand, the artificial languages of mathematical logic are surprisingly odd structures. We illustrate this point by specifying the grammar of the propositional calculus **PC**.

2. The Language of PC

The target properties of **PC** are properties of sentences or of sets of sentences. The properties of tautologousness, contingency and contradictoriness are properties of sentences. The relational properties of implication and equivalence are properties of sentence-pairs. Consistency and inconsistency are properties of sets of sentences. Proof is a property of sequences derived from axioms; and axiomhood in turn is a property of sentences. Fundamental to the tracking of these target properties is the idea of the *sentence*. What, then, is a sentence of **PC**?

The sentences of **PC** are defined as follows.

1 First, we say that the (possibly infinite) set of symbols
$p, q, r, s, t, p_1, p_2, p_3, \ldots, p_n$
is the set of atomic sentences of **PC**.

2 Second, we say that the molecular sentences of **PC** are all possible combinations of occurrences of atomic sentences and one or more connectives, $\neg$, $\wedge$, $\vee$, $\supset$, and $\equiv$, as constrained by the following Formation Rules.

FR_1 If A is an atomic sentence of **PC** it is a sentence of **PC**.

[1] Given its elementary nature, **PC** is not even a candidate for the host language of Frege's logicism. Since logicism is not at issue in this chapter, beyond what it tells of Frege's motivation to reform logic, we shall not here pursue the nice question of how a contentful language such as the language of arithmetic translates into a contentless language of pure logic. Technically-minded readers will be aware that the answer to this question depends largely on the logicist's choice of logical particles.

FR_2 If A is a sentence of **PC** so is $(\neg A)$.

FR_3 If A and B are sentences of **PC** so too are $(A \wedge B)$, $(A \vee B)$, $(A \supset B)$, and $(A \equiv B)$.

FR_4 Nothing else is a sentence of **PC**.

The connectives $\neg$, $\wedge$, $\vee$, $\subset$, and $\equiv$ are *pronounced* as follows. 'Not' or 'it is not the case that', 'and', 'or', 'if... then' and 'if and only if'. The connectives appear to bear some similarity to the English connectives 'not', 'and', 'or', 'if... then' and 'if and only if'. For this reason is is widely but erroneously believed that the connectives of **PC** capture the meaning or anyhow one of the meanings of their English counterparts. Properly understood, the connectives of **PC** have nothing to do with the meanings of English connectives. (We shall return to this point. But we should also note in passing that a good deal of the hostility of informal logicians to formal logic turn on the perfectly obvious fact that the **PC**-connectives don't preserve the meanings of their English counterparts. This hostility is entirely misplaced.)

If the connectives of **PC** are indifferent to whatever meanings the English connectives may or may not have, how are they to be *interpreted*? And, relatedly how are the atomic letters p, q, r, s, t, p_1, p_2, etc. also to be *understood*? We deal with these questions in reverse order.

How are we to understand the p_i? Intuitively, they are wholly uninterpreted simple sentences of our artificial language. Even though they are completely devoid of semantic content (or meaning), it is stipulated that each atomic sentence has a truth value, i.e., is either true (T) or false (F), and never both. This should strike the reader as exceedingly strange. How can a symbol that doesn't say anything be true? The logician answers as follows: It is stipulated that the atomic sentences of **PC** satisfy the *minimal condition* on truth, *viz.*, a principle known as,

Convention T: 'A' is true if and only if A.

The connectives of **PC** are also defined in terms of truth and falsity. The negation $(\neg A)$ of A is T when A is F and F when A is T. The conjunction $(A \wedge B)$ of A and B is T just when A is T and B is T and otherwise is F. The disjunction $(A \vee B)$ of A and B is F just when A is F and B is F and otherwise is T. The material conditional $(A \supset B)$ of A and B is F just when A is T and B is F and otherwise is T. The material equivalence $(A \equiv B)$ and A and B is T just when A and B have the same truth value and otherwise is F.

We see in this an example of a *truth conditional semantics* for the uninterpreted sentences of **PC**. Because they lack semantic content, the closest that the **PC**-sentences can come to having a meaning is the conditions under which they are true (in the minimal sense of 'true' captured by Convention T). We can illustrate this by examining the standard truth table definition for one of the **PC**-connectives, say for '$\supset$':

	A	B	$A \supset B$
row 1	T	T	T
row 2	T	F	F
row 3	F	T	T
row 4	F	F	T

Each row in the truth table gives what logicians call a *valuation* for $(A \supset B)$ which depends on the truth values of A and B. Consider now each row in which $(A \supset B)$ comes out T. These are rows 1, 3 and 4. Each row tells us the truth values of A and B for that row. When $(A \supset B)$ is true, the row or rows in which it is true tell us the truth values of the components of $(A \supset B)$ which *make* it true. Row 1 tells us that $(A \supset B)$ is true when A and B both are true. Row 3 tells us that $(A \supset B)$ is true when $(\neg A)$ is true (because A is false), and B is true. Row 3 tells us that $(A \supset B)$ is true when $(\neg A)$ and $(\neg B)$ are both true (since A and B are both false). So, then, $(A \supset B)$ is true under three different possibilities. Each possibility is a *conjunction* ('and') of true sentences, and the totality of those possibilities is the *disjunction* ('or') of them all. Hence $(A \supset B)$ is true just in case

$$(A \wedge B) \vee (\neg A \wedge B) \vee (\neg A \wedge \neg B)$$

is also true. This latter sentence, then, exhaustively sets out the conditions under which $(A \supset B)$ is true.[2]

It is important to emphasize that if the logician is not able to contrive a way for contentless sentences to be T or F, there is no prospect of his succeeding in tracking down target properties such as implication (for pairs of sentences), and validity (for arguments constructed out of **PC**-sentences). Another way of saying this is that one of the most remarkable *technical* feats of logics such as **PC** is the way in which it is made possible for contentless syntactic structures to *have* properties such as these.

[2]Readers familiar with the technical details of **PC** will know that the disjunction of conjunctions presently under discussion is the *characteristic formula* of the truth table for '$\supset$'. It is easy to see that the truth-table for any arbitrarily selected connective also has a characteristic formula, which will always be a disjunction of the conjunctions of literals that make the sentence of the table true. This fact is fundamental to the proof of the functional completeness of **PC**. (**PC** is functionally complete in the sense that every truth functional connective whatever is definable in terms of $\neg$ and one of the pair $\{\wedge, \vee\}$.)

3. The Utility of PC

Not even the most ardent or doctrinaire mathematical logician believes that human reasoners should or could abandon their mother tongues. Frege and Peirce would grant that for everyday concerns — concerns that lie beyond the reach of the exact sciences — one's mother tongue is the unavoidable medium in which they are to be addressed. What Frege and Peirce would also have insisted on is that human reasoning in a human language is something that lies beyond the ambit of logic.

This is one of the things that disappoints (actually, infuriates), informal logicians about mathematical logicians. It is their insistence that 'ordinary reasoning' has no logic. However some formal logicians are heedful of such disappointment and are minded to respond to it in the following (and somewhat conciliatory), way.

The basic idea is this. Take an episode of human reasoning or argument-making which has been transacted in a piece of (say), English. Subject to certain constraints, that piece of English can be tightly paired with a counterpart bit of the language of **PC**. The tightness of the fit is consequential. It allows us to say that if the **PC**-structure has one of our target logical properties, so too does the English structure have it. The process of finding an English structure's counterpart in **PC** is called *formalization.* When a formalization works as it should, it maps an English structure to its *logical form* in **PC**. And, again when the formalization has been properly contrived, the property possessed, by an English structure's logical form is *reflected backwards* onto the English structure itself.

To better see how this works, here are the standard formalization rules for English and **PC**.

1 Simple sentences of English, and only they, are formalized by arbitrarily selected atomic sentences of **PC**.

2 The English construction 'not' is formalized by '$\neg$'.

3 The English constructions 'and', 'or', 'if... then' and 'if and only if' are formalized respectively by '$\wedge$', '$\vee$', '$\supset$' and '$\equiv$'.

In this way, the formal logician thinks that he has the means to reassure his critic. In its most basic sense, the assurance is this: If we can formalize some reasoning (or arguing), in English in some or other logical system, then, since formalization has the backwards reflection property with respect to target concepts, if the logical form instantiates that concept, so too does the English construction.

Essential to the case for the backwards reflection of target properties is that the connectives of English stand in the following truth conditional relationship. Let C be an English connective which is a candidate for formalization; and let K be its formalization in **PC** if it has one. Then C is properly formalized as K, provided that for any K-sentence that is false in **PC**, its corresponding C-sentence in English is also false. In other words, the falsity of a K-sentence is sufficient for the falsity of the corresponding C-sentence. But not conversely; the falsity of the C-sentence is not sufficient for the falsity of the K-sentence; nor does the truth of the respective C-sentence and K-sentence suffice for the truth of the other. So K can be absolutely the right formalization of C in **PC** without there being any need for C and K to be logically equivalent to one another, to say nothing of synonymous. We see, then, that the fact that our **PC** connectives do not mean the same as their English counterparts, or do not capture at least one of their meanings (in case they happen to be ambiguous), has nothing intrinsic to do with whether the formalization rules have the backwards reflection property with regard to our target concepts.

But what is to be done when an English connective C fails this test? Then the rule is that the connective in question does not *have* a formalization in **PC**, nor does any sentence of English in which C occurs. The rule can now be explicitly stated.

> *The Connective Rule* — Where C is an English connective and K its counterpart in **PC**, K is the formalization of C in **PC** if the falsity of any K-sentence implies the falsity of its corresponding C-sentence; otherwise C has no formalization in **PC**.

4. The Irreducible Informality

Backwards reflection is really quite remarkable. It shows that formalizing a content*ful* argument in English as a content*less* argument in **PC** allows us to determine with certainty whether the *English* argument is valid.

Some people are of the view that it is too good to be true. To see what their reservation comes down to, it is necessary to emphasize that there are constraints on what can be inputs to our formalization rules. For example, we are not allowed to apply these rules to interrogative sentences of English, nor are we allowed to formalize molecular sentences of English as atomic sentences of **PC**. And we are not permitted to formalize any connective of English other than 'not', 'or', 'and', 'if... then', and 'if and only if', and those that can be defined in terms of these connectives.

Consider, in particular, the rule that only simple sentences of English can be mapped to atomic sentences of **PC**. Are there any other constraints on this atomic rule? Consider the argument:

1 If Sarah has been awarded the first university degree, then Sarah is a bachelor.

2 If Sarah is a bachelor then Sarah is an unmarried man.

3 Therefore, if Sarah has been awarded the first university degree, then she is an unmarried man.

If we now apply the formalization rules to this argument, we see that its logical form in **PC** is

1.* If p then q

2.* If q then r

3.* Therefore if p then r.

This matters. Here is an English argument with a valid logical form in **PC**. But the English argument is invalid. We wanted validity to have the backwards reflection property, but the present example shows that it doesn't.

We can solve this problem by noticing that our English argument equivocates on the ambiguous term 'bachelor'. In premiss (1) it means one thing, and in premiss (2) it means something quite different. This suggests a way out of our difficulty. We can impose upon the formalization of English sentences the *Disambiguation Rule.* The rule says that for any expression of English which has more than one meaning, its different meanings require a mapping for each sentence in which it occurs to different expressions of **PC**. Applying this rule to our present example, we see that the correct logical form of our English argument is

(a) If p then q

(b) If s then r

(c) Therefore, if p then r.

This gives us the desired result. The logical form is invalid. So it would appear that when we add the Disambiguation Rule to our rules of formalization, validity does indeed have the backwards reflection property; for we no longer have an argument which is valid in **PC** but invalid in English.

What about *in*validity? Does it too have the backwards reflection property? Consider the following argument.

1 The shirt is red.

2 Therefore, the shirt is coloured.

The premiss is a simple sentence of English which entails the conclusion, also a simple sentence of English. Its logical form in **PC** is

1.* p

2.* Therefore, q

which is invalid.

Here is a second case to consider.

(a) The figure is a triangle

(b) The figure is a circle

This is an inconsistent set of sentences in English. But the logical form of this set in **PC** is $\{p, q\}$, and $\{p, q\}$ is a consistent set.

In the first case, invalidity fails to have the backwards reflection property. In the second case, consistency fails to have the backwards reflection property. But we want *all* our target properties to satisfy the backwards reflection condition. We want this because we want **PC** to be useful in the appraisal of real-life reasoning and real-life argument.

As it happens, we can recover the backwards reflection property with regard to invalidity and consistency if we agree to impose a further condition on our formalization rules. I call the

> *Logical Inertia Rule*: Simple sentences of English to which the formalization rules apply may not either imply one another or be inconsistent with one another. In other words, the simple sentences that are inputs to the formalization mechanism of **PC** must be *logically inert.*

It seems, then, that we have recovered backwards reflection of validity by imposing the Disambiguation Rule, and likewise that we have recovered the backwards reflection of invalidity and consistency by imposing the Logical Inertia Rule. Even so, there is a cost to these recoveries.

Let us deal first with the cost of imposing the Logical Inertia Rule. This requires us to be able to recognize implications and inconsistencies between simple English sentences in a principled way. This is equivalent to saying that we must have a *theory* of implication and inconsistency for English. But this is what our formalization rules were supposed to provide. **PC** would analyze the properties of implication and consistency, and our formalization rules would reflect them back into English. So **PC** together with the formalization rules would be a theory of implication and consistency for English. But, as we now see, we can't run the formalization rules in a principled way until we have a theory of implication and consistency for English. And we don't have a theory of implication and consistency for English until we've executed the formalization rules. So we have a *bootstrapping problem.*

Faced with this kind of difficulty, most logicians have in effect withdrawn the Logical Inertia Rule and, in so doing, have abandoned the hope that invalidity and consistency would have the backwards reflection property. This, as Bertrand Russell said about another matter, has all the virtues of theft over honest toil.

The Disambiguation Rule is also a serious matter. If it goes, then we lose the backwards reflection of *validity*. So the question now is whether we are able to apply this rule in a principled way. The answer is no. We do not yet have a theory that permits us to recognize ambiguity in the general case. (In fact, a good many philosophical problems have turned out to have resulted from *undetected* ambiguities.) This leaves us with Hobson's Choice. Either we can give up on the backwards reflection of validity or we can try to apply the Disambiguation Rule in other than a principled way, that is to say, *intuitively*. Most logicians opt, in effect, for the second option. This also matters. No one doubts that a native speaker of English is adept at recognizing large numbers of violations of the Logical Inertia Rule. This is part of what fluency in a language consists in — the intuitive ability to notice elementary logical connections. The same can be said for the fluent speaker's untutored capacity for making and recognizing logical deductions in his own language. These are hugely important capacities, part of what in chapter 1 I called our "rational survival kits". The question is whether we have theories of such skills, and if so what they would look like. Gerald Massey has suggested that one place to look for theories of such capacities is in the Natural Logic of writers such as Lakoff [Massey, 1975a; Massey, 1975b; Massey, 1981]. I do not know whether Massey has retained his enthusiasm for Natural Logic over the years; but it is now clear, if it wasn't before, that Natural Logic has not attracted anything like a large and settled theoretical consensus.

Natural Logic aside, two other points should be made. One is that nowhere in the capacious writings of *informal logicians* do we find any attempt to construct theories of (logical) implication and (logical) consistency for natural languages. The other is that monitoring the Disambiguation Rule and the Logical Inertia Rule is not the business of (or within the capacity of), formal logic. If formal logic is to offer any principled guidance to natural language reasoners and arguers, it will be able to do this only on the basis of an irreducibly informal deployment of the requisite constraints on formalization. Informality is prior to formality, and not displaceable by it. After the attempts of the previous chapter to sort out the informal-formal matrix, it now strikes me that the reasoning of these present pages brings us closer to a satisfactory answer. I also recant my assertion in Woods that "the formal logician has a coherently formulable programme. The informal logician, on the other hand is simply out of business" [Woods, 1995, p. 193]. I make this recantation unabjectly. My claim, I now say, was mistaken, but it was not massively mistaken. I think that it is nothing if not regrettable that informal logicians do not seem much interested in tackling the issue of how best to analyse implication and consistency for natural languages. But this is not to say, as I now believe, that this indifference has put informal logic out of business. A case in point is the problem of quantum logic.

5. Quantum Logic

Here is the nub of the problem. There are quantum states that are complexes of simpler structures. Some of these states are describable by sentences in the form "$X \oplus (Y \otimes Z)$", in which '$\oplus$' denotes an operation which set theorists call *join* and '$\otimes$' denotes an operation which they call *meet*. Now join is a kind of disjunction and meet a kind of conjunction. So, *informally speaking*, it would not be wrong to read our sentence as "X or (Y and Z)".

Classical logic (i.e., the standard first order predicate calculus including our friend **PC**) sanctions the Law of Distributivity. Expressed in purely natural language terms, this Law provides that if "X or (Y and Z)" is true, so necessarily is "(X or Y) and (X or Z)". But in quantum physics, if "X or (Y and Z)" is true it does *not* follow that "(X or Y) and (X or Z)" is true. This leads some theorists to the view that the validity of the classical distributivity principle is over-turned by certain established facts about the quantum domain. According to these theorists, since the distributivity principle fails in the quantum domain it is not universally valid. It may be valid in the macro-and meso-domains, but not in the interior of the atom. So the micro-domain will require us to specify a different law that invariably regulates mix of meet and join in that domain. In other words, the quantum world requires its own purpose-built logic — it requires *quantum logic.*

It is easy to see that the case for quantum logic turns on an elementary mistake. It is an error that involves a fundamental misconception about formalization.

The claim in question is that there are true quantum physical sentences of the form "$X \oplus (Y \otimes Z)$", in which '$\oplus$' denotes disjunction and '$\otimes$' denotes conjunction, and 'X', 'Y' and 'Z' are elementary quantum sentences. It is experimentally demonstrable that there are cases in which a sentence of the form "$X \oplus (Y \otimes Z)$" is true and yet "$(X \oplus Y) \otimes (X \oplus Z)$" is not true. It follows from this that for sentences 'X', 'Y' and 'Z' and for operators '$\oplus$' and '$\otimes$', the distributivity rule fails in the micro-domain. The question is whether this shows that it also fails in classical logic. If so, then there will be valuations which make

$$A \vee (B \wedge C)$$

true, and which make

$$(A \vee B) \wedge (A \vee C)$$

false, where 'A', 'B' and 'C' are **PC**-sentences and '$\wedge$' and '$\vee$' are the connectives for truth functional conjunction and disjunction. It takes no more than a simple review of the truth table definition of '$\wedge$' and '$\vee$' to see that no such valuation exists. There *is* no valuation which makes "$A \vee (B \wedge C)$" true and "$(A \vee B) \wedge (A \vee C)$" false. The distributivity law is valid in classical logic.

Here is an objection that the promoter of quantum logic may now find it tempting to raise.

> *Objection:* Since classical formalization has the backwards reflection property with regard to validity and since the validity of distributivity is *not* reflected backwards onto quantum sentences, it follows that the distributivity principle cannot be classically valid after all.

It is an interesting move; but it is wrong. It arises, as I said, from a misconception about classical formalization. To see that this is so, it suffices to revisit the Connective Rule. We may take it that the symbol for join ('$\oplus$') denotes a kind of disjunction, and that the symbol for meet ('$\otimes$') denotes a kind of conjunction. This being so, they are *prima facie* candidates for formalization by '$\vee$' and '$\wedge$' respectively. Let us suppose that the formalization is correct. Then, by the Connective Rule, there is no false sentence in **PC** with connectives '$\vee$' and '$\wedge$' whose counterpart sentence in English is not also false. But now consider the **PC** sentence which is false for every valuation

$$(A \wedge (B \vee C)) \wedge \neg((A \wedge B) \vee (A \wedge C))$$

and its quantum counterpart

$$(X \otimes (Y \oplus Z)) \otimes \neg((X \otimes Y) \oplus (X \otimes Z))$$

which, for some interpretations of 'X', 'Y' and 'Z' is *not* false. Hence, by the Connective Rule, '$\oplus$' and '$\otimes$' have no formalization in classical logic. Since these quantum sentences can't be formalized classically they can't do classical logic any damage.[3]

[3]Some logicians also think that intuitionistic logic shows certain classical laws to be incorrect (double negation, for example). But consider the classically false sentence, "$\neg(A \supset \neg\neg A)$" and its counterpart sentence "neg($A \supset$neg neg$A$)", in intuitionistic logic. But whereas there

The case against quantum logic is an exercise in informal logic. As Peirce rightly says, the formalization theory for a formal logic is not itself part of that logic. There is no law of **PC**, no logical truth and no theorem, that tells you whether this, that, or the other piece of English (including quantum English) is formalizable in **PC**. That task, which is central to the question of what **PC** is good for, is an intrinsically informal matter, and fully deserves the name of informal logic. Thus the examination we have just made of the claim that the truths of quantum physics falsify some of the laws of classical logic turns out to have been an exercise in informal logic. Given the importance of that issue, the last thing to be true is that informal logic is out of business.

6. Further Work for Informal Logic

Even the most vigorous critics of formal logic, including those who decry the logocentrism of argumentation theory (see, e.g., [van Eemeren et al., 1996]), will concede that at least the fundamental principles have an appropriate, albeit qualified, place in the analysis and appraisal of arguments. It therefore comes as a shock to all concerned when counterexamples are proposed which purport to overturn these basic laws. In this section I shall consider the extent to which these alleged counterexamples are genuinely damaging. I wish to emphasize that the work of each of the cases we shall examine is very much an exercise in informal logic, as we shall see. In every case, the counterexamples are natural language arguments which fail to have the inferred from this validity that their respective logical forms are thought to possess. It is that the validity of those logical forms is only apparent. Thus the argument that is central to these counterexamples is this:

1 English argument E is clearly not valid

2 F is the logical form of E (in **PC**, for example).

3 Therefore, F is not valid, never mind that it may seem to be valid.

By the lights of this argument, the classical forms that it indicts are *fallacies*. They are argument forms (in **PC**), that appear to be valid but are not valid in fact.

---

is no classical valuation which makes the first sentence true, there are intuitionistic valuations which make the second true. So, by the Connective Rule, intuitionistic logic is not formalizable classically. This might have been anticipated, given that the informal meaning of intuitionistic negation is not "it is not the case that" but rather "there is no proof that". We also note that the informal logician has reason to think that the sentences of quantum physics satisfy the Disambiguation Rule. Quantum languages have been carefully refined to eliminate lexical ambiguity. That is reason to think that quantum sentences will not contain certain terms which require disambiguation by the formalizer. Even so, quantum sentences fail the Connective Rule, and that is decisive.

I make no pretense of having collected all the counterexamples of this kind, but what follows is a representative list.[4]

Example 1. (An apparent counterexample to *modus tollens* [Adams, 1988].)

English Argument	*Logical Form* in **PC**
1. If it rained, it didn't rain hard.	1. $p \supset \neg q$
2. It rained hard.	2. q
3. So, it didn't rain.	3. $\neg p$

Example 2. (An alleged counterexample to *modus ponens* [McGee, 1985].)

English Argument	*Logical Form* in **PC**
1. If a Republican will win the election, then if it is not Reagen who wins, it will be Anderson.	1. $p \supset (\neg q \supset r)$
2. A Republican will win the election.	2. p
3. Therefore, if it is not Reagan who wins, it will be Anderson.	3. $(\neg q \supset r)$

(*Historical Note:* in the U.S. Election of 1980, Ronald Reagan ran for the Republican party and the incumbent Jimmy Carter ran for the Democrats. John Anderson was a third candidate who ran as an *independent* Republican. Anderson lagged behind both other candidates in the polls, and was considered as having no chance. As it happened, he finished a distant third.)

Example 3. (An apparent counterexample to the transposition rule [Stove, 1986].)

English Argument	*Logical Form* in **PC**
1. If Baby cries then we beat him.	1. $p \supset q$
2. So, if we don't beat him, Baby doesn't cry!	2. $\neg q \supset \neg p$

Example 4. (An apparent exception to the repetition law [Jacquette, 1999].)

English Argument	*Logical Form* in **PC**
1. This argument is unsound.	1. p
2. So, this argument is unsound.	2. p

(*Exposition:* If the premiss is true, then by the definition of unsoundness, the argument's premiss is false. But the conclusion of the argument is

[4] I am indebted to Andrew Irvine and Dov Gabbay for discussion of these would-be counterexamples.

the very same proposition as the premiss. So it is false, if the premiss is true. Hence the argument is invalid.)

It is important to note that in each case, whether the apparent counterexample is genuine pivots on whether in fact the English argument does in fact have a logical form in **PC**. Even if it does, it does *not* follow that those logical forms are invalid. What actually follows, if these counterexamples are genuine, is that *either* those logical forms are invalid, *or* they are valid, but validity does not have the backwards reflection property under the formalization procedures of **PC**.

Are the counterexamples genuine? No. In every case but one the English argument fails to be admissible input to the formalization procedures of **PC**. In the other case, the English argument is not invalid.

In example 1, the English argument is:

1 If it rained, it didn't rain hard.

2 It rained hard.

3 So, it didn't rain.

The argument can't be inputted. It contains occurrences of an adverb ('hard'). Adverbs cannot be formalized in **PC** and cannot be part of any sentence formalizable there. Whatever we think of the English argument, it does no damage to the validity of *modus tollens* in **PC**, nor to the claim that validity has the backwards reflection property.

The argument of example 2, is:

1 If a Republican wins the election, then if it is not Reagan who wins, it will be Anderson.

2 A Republican will win the election.

3 Therefore, if it is not Reagan who wins, it will be Anderson.

The argument presents a counterexample to *modus ponens* only if it is invalid and yet has in **PC** the logical form of *modus ponens.* It satisfies the second condition but not the first. The English argument is not invalid. Here is why. Suppose that it were invalid. Then the conclusion would be false and the premisses true. If the conclusion is false then its antecedent (the if-sentence), must be true and its consequent (the then-sentence), must be false. So the only way in which the conclusion can be false and premiss 1 concurrently true is to make premiss 2 false. So the English argument is not invalid.

The argument of example 3. is:

1 If Baby cries, then we beat him.

2 So, if we do not beat him, then Baby does not cry.

Here the implausibility (and apparent invalidity), of the English argument turns on implied temporal indicators. To get an interpretation under which the argument comes out badly, we need to read it as follows.

1 If Baby cries *now*, then we beat him *afterwards*.

2 So, if we don't beat him (tenselessly), then Baby does not cry (tenselessly).

The argument is another input-casualty. It offends against the prohibition of adverbs. It also violates the *Ambiguation Rule*, since the verbs of the premisses are temporally indexed, and the same verbs occur tenselessly in the conclusion. So even if we waived the objection about adverbs, the English argument *would* have had a logical form in **PC**, but it would not have been transposition. It would instead have been,

1 $p \supset q$

2 $\neg r \supset \neg s$

which, while invalid, is not the law targeted by the putative counterexample.

Finally, the argument for our fourth example is:

1 This argument is unsound.

2 So, this argument is unsound.

If we flip back to review the exposition I gave of what makes the argument invalid, we see at once that it involves a massive failure of the *Principle of Logical Inertia.* The argument fails only on the basis of logical connections involving the semantic content of the two occurrences of this simple English sentence. This is precisely what the *Logical Inertia Principle* precludes.

Consider now a further example for which I am indebted to Rami Lill Anjum.[5]

1 If God does not exist then it is not the case that if I pray, then my prayers will be answered.

2 I do not pray.

3 Therefore, God exists.

This argument has a valid form in *PC*

a. $\neg G \supset \neg(P \supset A)$

b. $\neg P$

---

c. G

There is no assignment of truth values to the atomic letters of this schema for which the conclusion is false and the premisses both true.

---

[5] *Viva voce* at the conference on Mistakes of Reason, University of Lethbridge, April 2002.

On the other hand, we would seem to have it that, provided we assume that it is *possible* that God doesn't exist and yet that the premisses of the English argument are true. If this is right, then since the English argument is invalid notwithstanding that it has a valid form in *PC*, validity does not, after all, have the backwards reflection property.

How solid is this would-be counterexample? Does it actually stand up to scrutiny? It does not; although, to be fair, nothing we have expressly said so far about backwards reflection directly rules it out. The nub of our question is this. Is there a reasonable interpretation of the sentences of the English argument under which the premisses are true and the conclusion false? If so, what is it that makes premiss (1) true? Premiss (1) says that if God does not exist, then its not the case that if I pray, my prayers will be answered. Certainly it could be true that God doesn't exist and also true that if I pray (to God), my prayers *will* be answered (by my attentive parents). In that case, (1) would be false. But the interpretation under which it is clearly true is one in which "I pray" means "I pray *to God*", and "My prayers are answered" means "My prayers are answered *by God*". Clearly, if God doesn't exist then, even though I pray to God, it cannot be that my prayers are actually answered by God.

What the present interpretation reveals is not only that there is topical overlap among the atomic sentences of premiss (1), but that this overlap is semantically significant. What shows this very clearly is that premiss (1), as presently interpreted, cannot but be true, which means that when its formal counterpart in *PC* is evaluated as *false*, premiss (1) is evaluated as *true*. But since this violates the Connective Rule, premiss (1) is not an admissible input to the formalization mechanisms of *PC*, and the counterexample fails.

It also bears mentioning that since the atomic sentences of premiss (1) exhibit semantically significant topical overlap, they violate the spirit of the Logical Inertia Rule. This in turn shows a not wholly surprising link between this Rule and the Connective Rule. It also suggests the advisability of strengthening the Logical Inertia Rule. To that end, we might stipulate that *meaning connections* exist between atomic sentences when

1 one entails another;

2 one is inconsistent with another; *or*

3 they share semantically significant topical overlap.

The strengthened rule can now be re-issued as the

> *Logical Inertia Rule**: Atomic inputs to *PC*'s formalization devices may not include those that bear meaning relations to one another.

In our discussion of quantum logic, it fell to informal logic to come to the defence of (classical) formal logic. Here, too, informal logic has saved the bacon of formal laws.

Ironic, isn't it?

7. Informal Theories of Implication and Consistency

It may strike some readers as over-hasty that I should have claimed that informal logic has made no effort to produce informal theories of implication and consistency. Perhaps some people will think that such theories can't be produced; that the informal concepts of implication and consistency are theoretically intractable, as Michael Scriven has conjectured about the concept of relevance, and Frege (and before him, Kant) held about the concept of truth. In fact, this was precisely Aristotle's approach to implication (necessitation as he called it). Implication is a theoretical primitive in Aristotle's logic [Woods, 2001, ch. 1].

It lies in the spirit of this question that the argumentation theorist may find my demand for theories of implication and consistency as unavailing as a demand for a theory of truth directed to Frege or Kant. After all, we have very rich intuitions about how the concepts of implication and consistency work in natural language contexts, and the same is true for truth. So it follows from this, does it not, that anyone at home in his native tongue has command enough of the requisite concepts to enable him to make defeasibly accurate input-decisions regarding the formalization mechanisms of **PC**, or whatever other logical system might be in question. This is tantamount to the highly significant claim that people have a defeasibly reliable command of an apparatus that reflects *validity* backwards onto large sets of natural language arguments. Although validity is a *decidable* property in **PC**, it is not that in English; but it is a defeasibly *recognizable* property of English, which is certainly nothing to sneeze at.

There is something to be said for this line of reasoning. It takes note of Grice's point that it is one thing to have a theory of the thing denoted by a word, and another thing to have an understanding of the word. Even so, the present proposal is a bit of a letdown. Bearing in mind that it is the informal logician who takes the decisions that map subsets of English to formal structures in **PC**, then, on the present suggestion, we are invited to see the informal logician giving the following advice:

> Look, if you want to determine whether an English argument is valid, one way to do so is to determine whether it has a valid form in **PC**. To determine that, you must also determine whether *Logical Inertia* is satisfied by any pair of simple sentences of the English argument in question. To do *that* you must

> decide for each such pair whether they imply or implied by each other and whether they are inconsistent, i.e., such that either implies the negation of the other. This final task is discharged in the following way. Whatever you *think* is so, let that determine your decision.

Impressions may differ as to how much a letdown this is. For my own part, I am minded to think that those who devote their scholarly and analytic energies to a better understanding of argument-adequacy, should, on the issues at hand, be expected to offer something rather more than a 'logical' version of the old maxim, "Let your conscience be your guide". In any case, no one will deny that there is one rock-firm respect in which this imaginary advice from the informal logician is wholly unavailing. It offers no guidance to native speakers who in particular cases disagree about whether sentence-pairs conform to the *Logical Inertia Principle.* For we may assume that both parties are guided by their logical consciences. That, of course, is part of the trouble with the deployment of intuitions in contexts of disagreement (see here [Woods, 2002b]).

It may have appeared that in attributing to informal logicians a failure to produce or attempt to produce informal theories of implication and consistency, I was scolding informal logicians, and hectoring them for their neglect of important matters. If so, then I was scolding and hectoring myself. The last time I seriously sought to make some progress with these issues was in a series of papers in the late 1960s and early 1970s, the results of which I tried to bring to a head in [Woods, 1974b]. So who am I to talk? I admit to having ideas about how now to proceed. But they are half-baked ideas. This is work for another time.

PART II

THREATS AND INTIMIDATION

The next two chapters investigate a general type of argument that is often given short shrift by the theoretical community. Arguments of this type are at bottom instances of what we might call *non-cooperation arguments.* These we understand not in the classic sense pioneered by John Nash, i.e., as arguments in which participants don't share information and don't collude with one another [Nash, 1997]. What is intended here are arguments that are actually hostile to the addressee's interests. The fundamental question for the argumentation theorist is whether, and the extent to which, such arguments may make a rational claim upon compliance. In a rather general way, the chapters to follow ask whether, in departing the Goody Two-Shoes model of argument, the course of rational argument is compromised. More concretely, we may say that our question requires that we determine how "bad" are arguments that turn on threat and intimidation.

Chapter 4

AD BACULUM AND PASCAL'S WAGER

In the present chapter and the one that follows I attempt to work out an account of *ad baculum* argument which evolves from what we might call the traditional conception, then to a mid-position, and lastly to a final position. ('*Baculum*' is Latin for 'stick'.) On the standard conception, an *ad baculum* argument is always a fallacy, since it substitutes a threat for good reasons. The mid-position is that, in certain perfectly legitimate contexts, *ad baculum* argumentation is *not* fallacious. According to the final position, two chapters hence, *ad baculum* reasoning is not fallacious at all (although it can be inept, stupid or unsuccessful in still other ways). I do not regard the standard conception as tenable; so we begin here with the mid-position.

In the Prologue we spoke briefly of Goody Two-Shoes arguments. We said that such arguments are exercises in "open, objective enquiry in which parties behave (or are expected to behave) reasonably, honestly, helpfully and equably". We may speak of *non-cooperation arguments* as arguments that deviate from the Goody Two-Shoes model, the greater the deviation the greater the degree of non-cooperation. In the present chapter, we also examine *ad baculum* arguments as examples of (rather vivid) non-cooperation arguments. We also re-engage the distinction between alethic (truth-orientated) and strategic rationality.

1. The Standard Treatment

I shall take it for granted that the *ad baculum* is a special case of prudential argument; and in so saying I venture the modest hope that before long the Standard Treatment of this fallacy will have lost its allure. True, there is recent evidence of the dogged persistence of the

Standard Treatment. In [Copi, 1986, p. 91] Professor Copi has it that the "*argumentum ad baculum* is the fallacy committed when one appeals to force or the threat of force to cause acceptance of a conclusion". Carney and Scheer hold the same view and express it in almost the same words in [Carney and Scheer, 1980, p. 39]. Professor Kahane echoes the standard view but he also makes a significant departure from the tradition when he suggests that the *ad baculum* is not a fallacy, not at least according to the standard definition of the term [Kahane, 1969, pp. 251–252]. Woods and Walton suggest that to "meet the requirements for an instance of *ad baculum*, a sequence would have to be (i) an argument, (ii) a fallacious or incorrect argument, and (iii) a threat or appeal to force... [Woods and Walton, 1976, p. 140]. We will not scruple to say that it would take the rationality of a Tertullian commingled with the black power of Svengali to actually to pull the thing off.... Until the thing is produced, we remain unconvinced".

2. Prudential Argument

Now, sixteen years later (which is too much later for my taste) I shall try again to "produce" the thing. To do so, we will need the idea of a prudential argument. A prudential argument is an n-person dialogical exchange, having some such features as the following, which for ease of exposition will be illustrated in a 2-person setting. There is an arguer and, if you will allow me, an arguee. The arguer seeks to make upon the arguee a claim which, in accepting it, leads the arguee to comply with a deontic or hortatory conclusion of the arguer, something in the form "Do X" or "Would that your good sense would prevail and prompt you to the doing of Y", or some such thing. It is clear at once that prudential argumentation need not be bullying, lunatic, irrational or wicked. The skillful framer of a prudential argument will try to make his claim upon compliance a rational one, one that cohere's with his best estimate of his *vis-a-vis's* interests, and so one that is fashioned with due regard to the "the standard mechanisms of probabilistic decision theory" [Rescher, 1985, p. 16].

At the heart of prudential argument is the interpretation of a person's (or a group's) interests so as to make a rational claim upon him (or it) to act in fulfillment of the deontic or hortatory thrust of the argument. Plainly, such arguments can succeed or fail. A prudential argument may fail if it misinterprets the arguee's interests; or if it proposes an action which in fact does not cohere with those interests.

Consider now prudential arguments in which there is a conscious, deliberate and essential factor of threat or the prospect of dire consequences. In the Standard Treatment, these arguments would commit the *ad baculum* and would be fallacious, that is, *bad arguments*, just on that account. Let us see. I wish to discuss three cases, which for ease of exposition, I shall present rather informally.

3. Case One: The Heist

A says to *B*, pointing a pistol at him, "Your money or your life!"

Now if *A* has any of the technical proficiency of the hold-up man, he will have bet that *B*'s interest in survival outweighs his interest in retaining his wallet, and, further, that there are no special features of the situation that would annul or seriously weaken the dominance of the one interest over the other. Does *A* make a rational claim upon *B*; and does *B*'s compliance make for anything but good sense? *A*'s criminality, though morally and legally regrettable, is an essential condition of the rationality of the claim upon *B* and of the sensibleness of *B*'s compliance. There is no fallacy in this case; it is a good prudential argument.

What would a bad prudential argument of the "stick-em-up" type be like? Two cases come to mind, (a) one in which the interests targeted for violation by the threat of the arguer are not in fact dominantly held by the arguee ("Your money or I shall be moderately displeased with you for the next four or five minutes"), and (b) another in which the interests threatened by the arguer are expected to dominate over the interest of the arguee in not being robbed, but the arguee does not, in fact, hold those interests dominantly ("Your money or your life!", "To hell with it, fire away!").

If there is a fallacy in the first case, it would seem to be a fallacy of relevance; for how is the arguer's modest, temporary displeasure a relevant inducement to acquiesce in a hold-up? And, though the second argument fails, it fails because the arguer has made a factual error; he mistakenly supposes that survival will dominate over the disinclination to be robbed, and oddly or tragically, it does not.

4. Case Two: The Anti-smoking Commercial

I leave it to the reader to reconstruct any standard version of the anti-smoking argument. It suffices here to notice that although the argument contains and essentially turns on reference to the probability of dire consequences, the reference is not a threat, except metaphorically. In Case One, the dire consequences, on which the arguer's position turns, are within the arguer's power to bring, about; in Case Two they are not: the Canadian Cancer Society does not cause smokers to experience or to risk serious illness. Here, too, there seems to be ample room for such arguments to go wrong. They can fail on a misreading of interests-dominance in the target group; and they can fail for citing irrelevant factors, as in "Smokers are disapproved of by official decree of the American Nazi Party". There may, I suppose, be some grounds for finding such reasoning susceptible to fallacies of relevance, but it is important to notice that the "fallacious" cases are those in which the appeal to force or to dire consequences is made ineffectively, carelessly, ignorantly or stupidly, *not* that it was made at all.

5. Case Three: Pascal's Wager

Pascal's infamous argument has been seriously misunderstood and has been wrongly dismissed as an aberration of an otherwise gifted thinker, the discoverer of the Law that bears his name, who strayed into hostile theological territory.

No, Pascal's Wager is a substantial contribution, not to theology, but to probabilistic-decision theory. Pascal's intention is not to establish or prove the truth of any theological claim, least of all of the existence of God; nor does he propose the absurdity of overcoming theological skepticism by an act of the will, by exercising the will to believe. Pascal knows well that belief lies beyond volition and constraint. Neither should the Wager be thought of as an isolated and eccentric quirk of Europe's intellectual tradition. It derives from the skepticism of the Middle Academy and is part of the tradition of Kant's practical reasoning and James' pragmatism.

Whatever the complete analysis of Pascal's Wager, it is clear that it is and is intended to be a prudential argument — an argument concerned with the rational justification of action by an appeal to self-interest; and just as clearly it contains an essential reference to dire consequences which, in turn, are the purport of the threat of a third party. So Pascal's wager is an *argumentum ad baculum.* There is, I think, more to Pascal's Wager than an *ad baculum*, as Professor Rescher's admirable *Pascal's Wager: A Study of Practical Reasoning in Philosophical Theology* has made me believe [Rescher, 1985]. But for present purposes I shall concentrate on its *ad baculum* features.

It will be illuminating, I think, to construe the Wager as follows. Pascal directs his argument to a certain kind of skeptic, to him who though he does not believe the doctrines of seventeenth century Western European Catholicism, *does* believe that they are, in James' words, a *live option.* Pascal seems to think that such a skeptic will also find them to be a *momentous* and *forced* option. In any event, Pascal seeks to attract the skeptics attention with a cost-benefit argument, central to which is a catastrophic threat and concerning which there is "a measured gamble, a probabilistically *calculated* risk" [Rescher, 1985, p. 33]. If Christianity is true, yet not believed, then there is the dire consequence, threatened by God Himself, of infinite suffering; if Christianity is false, yet believed, there is the non-trivial cost of one's life arranged inconveniently and exposed to various importunities and anxieties. However, the risk of unbelief of the true vastly over-reaches the risk of belief of the false. Consequently, the skeptic is enjoined, not to believe (which is absurd), but rather to put himself in the ways of faithfulness, that is, "to follow those devices that will induce the belief that has so far not implanted itself.... Pascal's point is that someone without faith would do well, considering matters prudentially, to acquire it. But what this says he should acquire is the genuine article, not some simulation of it. The rituals are to be adopted in the first instance not because following them is faith, but because following them may *induce* faith" [Penelhum, 1971, p. 156].

6. Doxastic Surrender

'As long as there is a finite chance of God's existence — no matter how small — the expectation of the 'bet-and-believe' alternative outweighs that of its rival [Rescher, 1985, p. 15]. Pascal tries to motivate belief rather than validate it. His argument occupies a central place in practical reasoning and it adumbrates the "leading idea of modern decision theory (of which, after all, Pascal was the founder), [which] is to treat the acceptance of theses as a *decision* that can quite appropriately be made in the light of practical considerations" [Rescher, 1985, p. 53].

How should we assess the Wager? Insofar as it is a menacing *ad baculum*, does it make a rational and definitive claim upon the skeptic? It depends on what, if any, countervailing considerations dislodge the *ad baculum* from decisive dominance.

On the face of it, we are met with a conflict between rational self-interest and the requirements of truthfulness. We are met with a competition between and among the virtues, between and among self-protection, truthfulness and intellectual courage. We also seem to be faced with the issue of self-deception or something rather like it; for Pascal encourages the skeptic to try and come to believe propositions that he (now) believes he has no reason to believe. This is doxastic surrender, in which the attitude of truthfulness is abandoneded to the blandishments of self-interest, in which doxastic manipulation is counselled, even in the face of insufficiency of reason to believe. At issue is the question whether the requirements of truthfulness can ever rationally and morally surrender to the (legitimate) claims of self-interest. I have no doubt that much of what we believe, and are rewarded for believing, we do not know, and perhaps do not know how to come to know. For present purposes I shall concede that in certain of these doxastic circumstances no offence is given to rationality or morality. I think that we might also allow that the way of decision (via doxastic surrender) might tolerably dominate over the way of discovery, as in an attempt to repair a shattered marriage. Equally, however, doxastic surrender must often be resisted even at high cost to self-interest. Which way is it, then, with Pascal's Wager?

7. Inaccessibility to Reason

Central to Pascal's argument is the idea that the truths of Christian teaching are inaccessible to reason, that neither they nor their negations can be known or evidentially warranted.

Whether Pascal is correct in so thinking is, I believe, vitally important for a sound appraisal of the Wager. In fact, we have two versions of the Wager, one more purely Pascalian than the other; and it may be that the two versions assay *differently.* The purely Pascalian version of the Wager is one to which is adjoined the premiss that Christian truth is inaccessible to evidential reason. The less Pascalian version is one to which is adjoined the negation of this premiss.

The purely Pascalian Wager sometimes seems rather irresistible. For consider:

(i) There is no point in the skeptic's maintaining his skepticism, since there is no possibility of epistemic pay-off. Skepticism is not doxastic indifference, but rather the seriously considered withholding of assent and dissent until the impasse is resolved by the weight of evidence. But inaccessibility to reason implies that there can *be* no weight of evidence; the impasse is evidentially irresolvable.

(ii) Consequently, there can in this case be no rightful place or the attitude of truthfulness to dominant over doxastic surrender; the attitude of truthfulness is a pose.

(iii) Finally, there is likewise no toe-hold for the exercise of intellectual courage for the subordination of self-interest to the imperatives of dispassionate discovery.

The skepticism of Pascal's *vis-a-vis* is thus pole-axed, and the latter does not even have recourse to doxastic indifference, since Christianity is for him a live and momentous option. It is a forced option as well, owing to God's omniscience and His steadfastness in punishing unfaithfulness.

I do not say (yet) that the purely Pascalian Wager is knock-down. It gives rise to many questions and possibilities which a fuller treatment than this one would need to attend to. These I shall leave for another occasion, as also any further examination of the less Pascalian version of the Wager, in which it is allowed that the truths of Christianity are knowable by evidential warrant.

I wish instead to develop just one line of argument which suggests that the purely Pascalian Wager should *not* be regarded as knock-down.

8. The State, the Party and the Wager

I am thinking of programmes of "correct thinking" and of party recruitment in technologically efficient, modern totalitarian regimes. It is not necessary to produce concrete examples here; they are numerous and known, sometimes in punishing detail, by us all. The totalitarian state is ruled by and is in thrall to the Party, the calculated influence of which is pervasive, intrusive, comprehensive, unyielding and effective. Information technology, social engineering, behaviour modification, all driven by "correct thinking", provide the state with a stunning capacity to know and control its people. Dissent is discouraged or suppressed in many ways, some subtle and bloodless, others blatant and brutal. The objectives, purposes and, indeed, the very *raison d'être* of the state is set forth in a body of doctrine that brims with lofty principles and noble ideals in the service of the human good. True, such ideologies provide harsh remedies for enemies of the state, but such measures are necessary and justified by the larger obligation to serve the human good.

Totalitarian regimes have their skeptics. Their skepticism is not unlike that of Pascal's fence-sitter. In both cases, the skeptic is part of a cultural *mileu*, which reinforces the very doctrines that he does not accept, but which he nevertheless takes to be a "a real possibility", a live option.

Imagine, then, that the Party is recruiting students from the universities, and that *X* is one of these. Most of *X*'s family are Party loyalists, as are close friends and many of the senior professors at the university. There is ample evidence, known to *X*, that the Party brings to its rule of the State, high ideals, noble principles and intelligent, skillful and industrious social and political practice. There is for the faithful citizen promise of an abundant, comradely, safe, and civilized life. Admittedly, the State is also capable of, and entirely confident in, dispensing cruel discouragement of dissent, and this it does with terrifying effect.

The young student, *X*, is a skeptic about Party doctrine and practice. His reservations need not be chronicled here; they are the usual ones and will be known to the reader. However, *X* also takes Party doctrine to be a live and momentous option; and, as we shall see, forced. The Party's recruiter approaches *X* with the following argument:

> *X*, he contends, cannot afford this skepticism of his; it is tantamount to repudiation of the Party and treachery to the State. One appreciates that his doubts are honest ones; they are also very dangerous. They expose him to certain harassment, poverty, disgrace; and eventual torture and death. That

is a monstrously stupid cost for his wholly unnecessary coyness toward something he freely regards as a *real possibility*. Therefore, he should renounce his skepticism, put it out of his mind, and put himself in the ways of faithfulness. He might start by signing this Party membership card.

It is not necessary to spin out the recruiter's argument in any further detail. But let us notice that:

(i) For X *simulated* faithfulness will not adequately serve his interests, for such simulations are *dangerously* detectable and, in fact, detected with terrifying frequency and, so far as one can tell about these things, astonishing accuracy.

(ii) The recruiter needs to point out that the truths of the Party are the gift of belief, the fruit of correct thinking; but that they are inaccessible to reason and do not admit of evidential warrant or disconfirmation. They are normative propositions. Therefore, there is no point in X maintaining his skepticism, since there is no possibility of evidential pay-off.

(iii) Similarly, for X there can be no room for the attitude of truthfulness to dominate over doxastic surrender.

(iv) And, finally, there is no toe-hold for intellectual courage to serve the risky course of the attitude of truthfulness.

A final word about the kind of totalitarian set-up in which our recruitment-argument is made. It is a state entirely free of the more or less routine lunacies that have dotted the actual human record of totalitarian criminality and idiocy. In the state that I am imagining, extreme and brutal measures are reserved for domestic enemies and for *no* other class or category of citizen; and domestic enemyhood is defined and measured exclusively by reference to deviation from or resistance to civic obligations that fall consistently (though with unavoidable hierarchical adjustment) upon all citizens.

However, I admit that the present case is not one in which the candidate for recruitment faces an option that is altogether forced, for the totaliarian regime, though adept at invading the privacy of its citizens, is not omniscient, as God is. A clever and brave actor might well attempt to fool that State, as he could not fool God, with simulations of fidelity. Still, it is easy to imagine an improvement in psychotechnology — one which may actually occur, and soon — in which the polygraph test is made absurdly cheap, very quick to administer and virtually 100% accurate. If the State routinely tested everyone, simulations of fidelity would be made hyper-risky, and the recruit's option would, I think, be forced.

I am also assuming, for simplicity's sake, that our recruit has no interests that would redeem his torture and death.

If I am right in thinking that a purely Pascalian Wager works equally well in the Christian and the totalitarian cases, then it becomes clear that good prudential arguments can commit, perhaps innocently, grave moral error, by which their claim upon compliance is surely vitiated. This faces the purely Pascalian Wager with a huge and, some would say, undischargable onus; for before it can rationally induce the ways of Christian faithfulness, it must be shown that Christian moral theology is free from serious moral error. If the onus is not accepted, then it seems to me that the Wager is a cynical and depraved manoeuvre, in which the object, as Rescher says with surprising approval, is "to lead people to God through an appeal to their fundamentally base nature" [Rescher, 1985, p. 130].

I shall conclude with a quick word about the *ad baculum*. Back in 1976, Douglas Walton and I wondered whether the *fallacy* of the *ad baculum* could actually be "produced". I now think that it cannot; it is not a fallacy. It satisfies the condition that it be an argument for it is a prudential argument. It satisfies the condition that it incorporate a threat or reference to dire consequences, but it does not satisfy the requirement that, as such, it offends logic, never mind what else it may offend.

Chapter 5

APPEAL TO FORCE

In the previous chapter, I tried to show that not all *ad baculum* arguments are fallacious. I want to develop this claim more concretely in the present chapter. As was the case in its predecessor, the present chapter lays heavy emphasis on the distinction between strategic and alethic (truth-oriented) rationality, and is a further exploration of what we have been calling non-cooperation arguments, made so by the intrinsic factor of threat.

1. Arguments From the Stick

Ad baculum arguments are those which pivot on appeals to threat or dire consequence. Though not called such, *ad baculum* arguments seem to have been first recognized by the Port Royal logicians, as witness Arnauld and Nicole [Arnauld, 1964, p. 289]. As we saw in the preceding chapter, modern writers have not done very well with the *ad baculum*. Carney and Scheer characterize it as the error — a special case of the *ad populum* fallacy — of inferring or concluding the truth of a claim on the basis of an appeal to force, for it is the "fear of force [which] cause[s] acceptance of a conclusion" [Carney and Scheer, 1980, p. 390]. The same words are found in Copi and Cohen. "The argument *ad baculum* [... is] the appeal to force to cause the acceptance of a conclusion...." Here too the fallacy is thought to be the mistake of concluding the truth of some claim on the basis of a threat. Most authors regret *ad baculum* fallacies as "the abandonment of reason" [Copi and Cohen, 1990, p. 105].

In these writings, and elsewhere, it is plain that *ad baculum* fallacies are thought to be fallacies twice over. For the arguer who knowingly promotes his addressee's fear of force, the fallacy can be seen as a case of trickery or deception, an attempt to dupe the other party into reasoning erroneously. For the addressee, who succumbs to the arguer's trick, the fallacy is delusional in character; the addressee is tripped by his anxiety into thinking that his acceptance of the truth of the claim in question is well-justified. In each case the fallacy is thought to be dialectical.

Though they are right to notice dialectical aspects, such accounts are troublesome. They seem not to be believed by their own sponsors. How can "so obvious a fallacy" trick anybody [Copi and Cohen, 1990, p. 105]? It is also doubtful that anything answering to the present notion would qualify as an argument [Woods and Walton, 1976].

Solider analyses of the *ad baculum* are possible, and they require that six types of case be attended to.

2. Case One

1 (Proposition) P or I'll punch your nose.

2 ∴ P is true.

Case one puts considerable pressure on the idea that the *ad baculum* fallacy is a deceptive argument since, for one thing, (1) is transparently irrelevant to the truth of (2). Also strained is the suggestion that case one presents an argument at all. Suffice it to say that if we did allow that case one did present an argument, the *ad baculum* fallacy would be a fallacy of relevance. A more straightforward judgement is that the idea of the *argumentum ad baculum* is undefined for instances of case one.

3. Case Two

1 Believe P or I'll punch your nose.

2 ∴ Believe P.

Here we meet with a significant departure. The *ad baculum* is now seen as a prudential argument, an argument not to the effect that a certain fact obtains or that a certain state of affairs is so, but to the effect rather that a certain course of action would be prudent to pursue. The essentially dialectical character of the *ad baculum* is also discernible in case two, for we imagine that the maker of the argument directs it to an interlocuter who is counselled to do something the prudence of which consists in averting the damage of the arguer's threat. Still, case two involves a serious error. Since belief is not a creature of the will, it cannot be summoned up voluntarily. The arguer of case two makes the mistake of bidding the addressee to do the impossible.

4. Case Three

1 Accept P or I'll punch your nose.

2 ∴ Accept *P*.

Actually, a better formulation of case three is one which makes more explicit its dialectical character. Imagine, then, that *X* and *Y* are arguer and addressee; then case three can be represented as the following dialectical exchange.

X: Accept *P* or I'll punch your nose.

Y: OK, I accept *P*.

Case three is less dismissible than the previous two. Acceptance, unlike belief, is a creature of the will; so *X* is not counselling *Y* to do the impossible. Even so, there is something wrong with *X*'s argument. It bids *Y* to violate a sincerity condition. The condition in question is that acceptance should be sincere and, in particular, that a rational agent ought not accept claims for which he has contrary evidence or insufficient positive evidence. So *X*'s error is that of suborning dialectical insincerity in *Y*.

5. Case Four: Risk Aversion Strategies

As before, *X* and *Y* are arguer and addressee. *X* puts to *Y* the following case.

1 You (*Y*) agree that *P* is a real possibility.

2 If *P* is true, and yet you do not believe that *P*, then by *P*'s very truth you will suffer momentous disadvantage.

3 On the other hand, if *P* is false and yet you believe that *P*, you will be inconvenienced, but well below the level of momentous disadvantage.

4 Given these risks, it would be better if you believed *P* rather than not.

5 Therefore you should try to acquire the belief that *P*.

Case four is essentially Pascal's Wager to the Christian agnostic, which we discussed in the previous chapter. *P* here is Christian doctrine. The momentous disadvantage of disbelieving *P*, in case *P* should be true, is eternal damnation, for that is what Christian teaching teaches. The inconvenience of believing *P*, should *P* be false, is a life constrained by Christian strictures. Given that *Y* acknowledges that *P* is a real possibility (never mind that he does not now believe *P*) the prudent course is a risk aversion or a minimax strategy designed to minimize the worst that could happen to *Y* in case *Y*'s belief in regard to *P* turns out to be mistaken.

Case three throws up a number of interesting issues. It is worth noting, for one thing, that the threat on which *X*'s argument pivots is not made by *X* himself. Rather it inheres in the truth of Christian dogma should Christian dogma be true. It is God's threat. Moreover, *X* does not make the mistake of advising that *Y* do what he cannot do, namely, to summon up a belief in Christianity by an act of will. Instead he counsels *Y* to try to get himself to believe, that is, to employ techniques of attitude-modification , such as returning to a life of Christian practice and hoping for the best. A third point is that there is nothing at all fallacious or untoward as such about X's argument. This is not to say that such an argument cannot go wrong. It would go wrong if, for example, *X* was mistaken in attributing to *Y* the belief that *P* is a real possibility. The concept of a real possibility is central to the present analysis and should be explained.

Intuitively Y's situation with regard to *P* might be thought of as follows. *Y* used to be an informed, practising Christian, but now his faith has lapsed. It has not lapsed into total atheistic certitude, however. So his rejection of Christianity could be contrasted with his refusal of, say, Druidism. Though *Y* could acknowledge the possibility that either is true, the possibility of Druidism is for him wholly empty, and of Christianity less so. Borrowing William James' metaphor, Druidism is a dead option for *Y*, whereas Christianity is, to some extent at least, a live option. So real possibilities can be equated with live options. Or, more formally, let us define *not* being a real possibility for Y, as follows.

Definition: *P* is *not a real possibility* for *Y* if and only if

(i) *Y* disbelieves *P*

(ii) *Y* believes that he knows not-*P*

(iii) *Y* acknowledges the possibility that *P* is true

(iv) *Y* disbelieves that the truth of (iii) provides any reason, however slight, for believing P.

And so,

Definition: *P* is a real possibility for *Y* if and only if it is not the case that it isn't a real possibility for Y.

6. Case Five: The Stick-Up

The mugger approaches the visitor to New York, points his Magnum 44 and says, "Your money or your life." Although the threat is the mugger's own, the present case bears a striking resemblance to the previous one. The mugger invites his victim to do some quick cost-benefit calculations to facilitate the appropriate risk aversion behaviour. What the gunman proposes is a certain minimax strategy. There can be little doubt that in the general case the mugger's intervention, though a crime, makes a perfectly rational claim on his victim's compliance. The gunman presses his case competently to the extent that he specifies alternative options for his addressee, to each of which he assigns the costs and benefits attaching to the exercise or not of those options, and this he does in the light of what he takes his addressee's interests to be. The costs and benefits turn directly on the credibility and performability of the arguer's threat. The addressee's options are to hand over his wallet or to refuse. The cost of refusal is death, provided that the victim finds the threat credible and performable; and the benefit of refusal is perhaps his hanging on to his money for a bit longer than had he handed it over in a timely way. The cost of compliance is some degree of impoverishment, and the benefit is that the victim, though poorer, stays alive.

There is plenty of room in these cost-benefit prudential arguments for mistake and for fallacy too. But no fallacy is committed just because the argument pivots on a threat. Besides, it is not by any means the case that just any mistake made in the forwarding of an *ad baculum* argument qualifies as a fallacy. It is certainly a mistake to try to mug a visitor to New York with a banana rather than a Magnum 44, and likewise a blunder to threaten the would-be victim not with death but with the gunman's mild disapproval for the next two minutes. Fanatical traffickers after fallacy might feel that the gunman's mild disapproval is not a relevant consideration in the context of a stick-up. Thus they might plump for attribution of a relevance fallacy. There is perhaps no great harm in speaking this way, but deeper descriptions are available. If I attempt to underwrite a stick-up with a banana, I fail to produce a credible threat. If I offer him my brief and modest scorn, I fail the credible-cost condition. If we are to yield to the suggestion of a relevance fallacy, it is necessary to keep in mind that irrelevance here is the failure of that particular condition. It is also prudent to note that, in other contexts, irrelevance is very much something else. In any event, Carney and Scheer say of gunman-like examples that they "are not or do not involve fallacies" [Carney and Scheer, 1980, p. 39]. Talleyrand, in one of history's more celebrated wisecracks, said of the execution of the Duc d'Enghien, "It's worse than a crime, it's a mistake." This touches our analysis at two places. The gunman case shows that a crime need not be a fallacy and also that a mistake need not be a fallacy. One has to fight hard to convict the *ad baculum* of fallaciousness, and it is worth repeating that whenever an *ad baculum* argument is fallacious it is never so simply on account of its being *ad baculum.*

Prudential arguments abound in everyday life. Who can think of the upbringing of children without resort to threats or bribes? Bribes are close kin of *ad baculum* arguments, except that they substitute rewards for threats. In all other respects they are the same: cost-benefit prudential arguments in promotion of risk aversion — for the risk with promised rewards is failing to qualify for the thing promised. If *ad baculum* arguments are arguments from the "stick", we might well speak of bribes as *ad carotam* — from the "carrot".[1] Standard arguments against smoking are *ad baculum* and standard arguments for the purchase of this washing powder rather than that are *ad carotam.* Either way, fallacy does not inhere in their basic structures. They are fallacious, if at all, only in details of their particular formulations.

The gunman case affords the opportunity to resist a common misconception. Even among those who acknowledge that the mugger's argument is fallacy-free and that it makes a perfectly rational claim upon compliance, there is nevertheless something very seriously wrong with it, dialectically speaking. It, and all other arguments from the stick, are grave defections from cooperation, what with the routines of rational consensus-building displaced by strategies of intimidation. It is worth emphasizing that the intended contrast has not quite been found in the wording of the preceding sentence. For what better way to build a consensus than force it and what more rational way to maintain it than by attaching sanctions? Of course one knows the contrast that is wanted, but care must be taken against drawing it vacuously. It might be proposed, for example, that consensus building is cooperative just in case it is done freely, that is, in ways untainted by factors *ad baculum.* But now the claim that the *ad baculum* is a defection from cooperation is true by the prior stipulation.

If it is a misconception to think that threats are always dialectical defections from cooperation, it is perhaps best to reveal the misconception head-on, and without the need to find characterization of the contrast between cooperative and uncooperative consensus formation. This is easily enough done by considering negotiations.

7. Case Six: Negotiations

Negotiations have the following features:

(a) They are reciprocal minimax strategies in which each side attempts to secure an outcome that minimizes the worst disadvantage that it is in the power and intent of the other party to exact. With collective bargaining in industrial settings, for example, it is routinely the case that the maximum disadvantage within the (lawful) power of management to exact is the lockout, and of labour, the strike.

[1] *Carotas* has not had a flourishing lexical career. It seems to have first cropped up in the Roman cookbook called *Apicius* after the gourmet Marcus Gabrius Apicius and published in the century following his death in the first century. *Carotas* has occurred since, but with notable infrequency. Perhaps this is explained by the fact — or what appears to be a fact — that the carrot of contemporary celebrity seems not to have won its place in Rome's descendent culture until the second decade of the century just past. It may have had something to do with the introduction of night-flying.

(b) They are discussions held to conventions of rational case-making. Although the threat of disadvantage is at the heart of this form of discussion (and intended to be decisive), it is nevertheless expected that each side will give reasons for its position apart from the prudential factors rooted in their reciprocated threats. That is, each side is required to 'explain itself'. This we might call the Explanation Condition on negotiation.

(c) There is also involved a further requirement of cooperation enshrined in the fundamental law on collective bargaining. It provides that settlement should carry a cost for both parties, these costs being in turn reciprocal benefits. It follows, then, that a negotiator's opening position cannot be his closing position. In many jurisdictions this is a requirement of civil law and is called the Good Faith Condition.

It is interesting that the fact that negotiators are rational minimaxers, when taken in conjunction with the Explanation Condition and the Good Faith Condition, suffices to commit the cooperative negotiator to ritual lying or at least to the systematic misrepresentation of his real position, as reflected in diminishingly inaccurate representations of it in a series of on-the-table offers.

Negotiations may nevertheless seem to stand out as paradigms of irrationality and uncooperation in two significant ways. They are *ad baculum* by virtue of the fact that the costs which both parties seek to avoid are threatened by each and allegedly within the power of each to exact. And conformity with the procedural rules entails the systematic use of misrepresentation. In fact, however, it seems extremely implausible that negotiators stand convicted of either irrationality or uncooperation. There is a reason for this. Negotiations have their conventions and conventions are solutions of coordination problems [Lewis, 1969, pp. 5–51]. Whether expressly acknowledged or not, negotiators are pledged to the fulfilment of those conventions, the whole point of which is to coordinate opposing costs and benefits in such a way that agreement can be a rational outcome, never mind that doing so involves ritual uncooperation, an admittedly paradoxical-seeming situation. The air of paradox vanishes, however, to the extent that negotiations are held to the fundamental law on collective bargaining. For if it is an essential requirement that negotiators split their differences, they must have differences to split. Where such differences do not inhere in their respective closing positions, that is to say, at the point of settlement, they will have to be fabricated in order to avoid violating the fundamental law.

It can now be seen that, though a good prudential argument, a mugging is at best a limiting case of a negotiation. An efficiently transacted mugging always violates the Good Faith Condition, the legal requirement that collective bargaining be transacted by way of *genuine* negotiation.

When this condition is met, parties are said to be bargaining in good faith.

At this juncture, the analysis of the *ad baculum* intersects in an interesting way with the pragma-dialectical account of fallacies ([van Eemeren and Grootendorst, 1984, ch. 8] and [van Eemeren et al., 1987]). The primary insight of the pragma-dialectician is that any violation of the rules of rational conflict resolution counts as a fallacy. So, for example, one way in which a discussant can commit the *ad baculum* fallacy is to break the dialectical rule which bids discussants not to prevent the advancing or criticizing of a thesis. This carries the consequence that if a discussant sought to discourage an opponent's pressing his case by singing "I've got you, Babe" at the top of his lungs, this would count as a fallacious *ad baculum* under pragma-dialectical construal.

Whether readers will be convinced that such is a particularly rich analysis of this fallacy, it does seem quite clear that "fallacies of negotiation" could be specified pragma-dialectically [Woods, 1991]. In particular, one could plead that violations of either or both the Explanation Condition and the Good Faith Condition deserve the name of (cooperation) fallacies. The pragma-dialectician gives our discussions an interesting additional turn. Collective bargaining, for example, will turn out to be infested by what he calls cooperation fallacies, and in so saying he will be seen as calling down a non-traditional concept of fallacy. However, it will also be apparent that the pragma-dialectical account restores a feature of the traditional understanding of the *ad baculum* which our own analysis contradicts. For, in the case of a mugging, the role of the threat is precisely to discourage the victim's making a case for the mugger to desist, and in the case of collective bargaining, reciprocal threats make the respective opening positions entirely useless to persist with. Thus although the concept of fallacy is very different from the traditional one, the pragma-dialectical account re-imports into the *analysis* of the *ad baculum* a traditional feature, the idea, namely, that threats in rational argumentation are always the occasion of fallacy. An interesting and ironical twist, to be sure.

8. Case Seven: Veiled Threats

The *ad baculum* is perhaps at its most insidious in the context of veiled intimidation. A case in point is of Hitler's use of intimidation, as described in [Zbynek, 1973]. It will aid exposition to press a distinction between two different types of argument. Roughly speaking an *alethic* argument is an argument about what is the case; a *prudential* argument is, as we have seen, an argument about what should be done. If, however, truth is disquotational, conclusions stating what should be done can always be represented in the form "It is true that such and such should be done" and, by our rough distinction, prudential arguments collapse into alethic ones. Such collapse is averted by introduction of the notion of 'deontic formulation'. A conclusion about what is the case is given a deontic formulation just in case it asserts or enjoins that a given action should, must, or ought be done or not. An argument whose conclusion is, for example, that it is the case that 2 is a prime number will count as alethic, whereas an argument whose conclusion is that it is the case that you ought to give me your wallet will count as prudential or, as we can now say, as deontic.

Equipped with this distinction we can consider an example of case seven from Carney and Scheer.

> In 1961 the Western Powers held that they had certain rights in West Berlin, while the Soviet Union denied that they had such rights. Other nations were placed by both powers in the position of having to take sides. At that time the Soviet Union announced and tested a nuclear bomb that was about a hundred times more powerful than any United States bomb [Carney and Scheer, 1980, p. 39].

On the stated facts, it is possible to reconstruct two arguments, one alethic and one prudential (or deontic). In the alethic argument, the conclusion is that the Soviet Union is objectively correct in saying that the Western powers lack rights in West Berlin. In the deontic version, the conclusion is that the Western Powers should abandon any claim to rights in West Berlin. It is clear that the alethic argument is not much good, whereas the deontic argument fares much better. That said, it is now possible to characterize the veiled-threat *ad baculum* as an argument forwarded as an alethic argument but whose unexpressed premisses or content also present a prudential argument. The veiled-threat *ad baculum* can therefore be regretted on three counts.

(i) In representing itself as a purely alethic argument, it suppresses significant portions of the actual argument by leaving the deontic portion unexpressed. This violates Grice's Quantity Maxim which bids arguers not to say less than is appropriate for the adjudication of the issue at hand.

(ii) It encourages the suggestion that the alethic argument is correct by virtue of the suppressed truth that the unexpressed deontic argument is or may be correct. Thus it equivocates on the notion of correct argument.

(iii) It misrepresents the significance of Soviet nuclear superiority. Offered as objectively extinguishing the West's claim to rights in Berlin, in fact it makes of such superiority a disguised threat in the form "If such rights are not abandoned, we the Soviets will or may resort to such superiority". We see, then, that in each three cases, the veiled-threat *ad baculum* can be characterized as a fallacy of misrepresentation.

Having argued that *ad baculum* arguments are not always fallacies, we are now at a point at which it is reasonable to ask whether they are *ever* fallacies.

9. Practical Arguments

Our point of departure in this section is the practical syllogism. The invention is Aristotle's and the interpretation we give it is Anscombe's [Anscombe, 1957]. As is well-known, the standard syllogism is a discursive entity, a sequence of declarative sentences, of which the terminal member is the conclusion and the rest are premisses. In contrast, a practical syllogism is a mixed structure, part discursive and part non-discursive. The difference shows up in the conclusions of the two structures. In a standard syllogism, the conclusion is a sentence; in a practical syllogism the conclusion is an action. It is useful to compare practical syllogisms with deontic or prudential arguments. (Deontic arguments are arguments about obligation; prudential arguments are arguments about self-interest.) A simple example of such is

1 If you are late getting home from the movies, you'll irritate and worry your mother.

2 So, you shouldn't be late.

It is easy to construct what we could call the *practical syllogisation* of this argument. It is the ordered pair in which the first member is the premiss of the deontic argument

1.* If you are late getting home from the movies, you'll irritate and worry your mother.

and in which the second member, the conclusion, is not the sentence which *bids* the addressee not to be late, but is simply the addressee's *not being late*. Thus the conclusion of our practical syllogism is the action advocated by the conclusion of the preceding deontic argument.

The distinction between deontic-prudential arguments and practical syllogisms calls to mind the old maxim that talk is cheap. 'Cheap' in turn suggests 'suboptimal', and, in some respects, this is precisely what can be claimed by deontic-prudential arguments in contrast with their practical syllogisation. It is one thing to get an addressee to concede that he should do such-and-such; it is another, and often better thing, that he actually do it. Better the cheque in the mail than 'The cheque is in the mail'. We may say in a quite general way that a practical syllogism is the *consummation* of a deontic arguer's intent.

In this chapter we propose to expand on the concept of practical syllogism in a slight but natural way. I shall attempt to show that, modest though the extension might be, it produces results of genuine consequence for the theory of argument. In this proposal, a second way of being a practical syllogism is one in which one or more of the *premisses* is an action rather than a sentence. It is a point worth emphasising that the conclusions and, as we now may say, the premisses that make for practical syllogisms are *role-specific.* Any action by any agent at any time make for a true proposition, namely the proposition ascribing that action to that agent at that time. Any of these truths is available in principle as the conclusion or as a premiss of some or other bit of argument that may chance at a time to bubble out of the dialectical soup of the human community. Such arguments are not made into practical syllogisms in consequence of this fact; for it is the actions themselves, not the sentences they make true, that are the irreducible components of practical syllogisms. In Aristotle's conception of it, the action that is the conclusion must be the action of the party to whom the argument is addressed. In our extension of it, the action that is the premiss of a practical syllogism must be the action of the maker of the argument, not his addressee. So there is an agent-specific asymmetry between, as I shall now say, *conclusionally practical syllogisms* and *premissorily practical syllogisms.*

There is a further asymmetry. Let <'*P*', *A*> be a conclusionally practical syllogism. Let '*A*' be the sentence in which action *A* is attributed to its agent. Then, in general, a *standard* syllogism <'*P*', '*A*'> is not preservable from the practical syllogism <'*P*', *A*> under replacement of *A* by '*A*'.[2] On the other hand, consider a simple case of a premissorily practical syllogism. Suppose that Joe will be elected Treasurer if and only if Henry, Sarah, Frank and John vote yes. Imagine that John is attempting to construct an argument whose conclusion is that Joe is elected. Joe adds the following true premisses.

1 'Henry has voted yes'

2 'Frank has voted yes'

3 'Sarah has voted yes'

The desired conclusion that Joe will be elected Treasurer requires a further premiss. So:

---

[2]That this is so is indicated by the fact that the more natural candidate for the standard syllogistic counterpart of the present practical syllogism would be the deontic argument whose conclusion is, 'You *ought* to do *A*'.

4. [John simply votes yes.]

In this, two asymmetries are evident. For one thing, the action which serves as the clinching premiss must be John's, the speaker of the argument, rather than Sarah's or Henry's. But, secondly, if our previously practical syllogism is correct, there is a correct standard syllogism got by replacement of the action that constitutes premiss (4) of the former with the sentence 'John votes yes', which correctly attributes that action to John.

Essential to both types of practical syllogism, and corresponding to the parameter of role-specificity, is the element of *participant control.* If my conclusionally practical syllogism that you *do* so-and-so is good, then that it is so lies essentially in your power, not mine. All that rests with me is to show that you *should* do so-and-so. But if my intent is to produce a practical syllogism rather than a deontic-prudential argument, the premisses are up to me to select and present; but the final conclusion is up to you. This other-party dependency is missing in the case of premissorily practical syllogisms. To recur to our example, the argument cannot succeed without premiss (4), i.e., 'John voted yes'. But *that* premiss is true if and only if John votes yes; which, again, is entirely up to John.

10. *Ad Baculum* Reasoning

It is perhaps not surprising that fallacy theorists and argument-analysts should have been preoccupied with the idea that there is something inherently defective about *ad baculum* arguments. As will be evident by now, my own view is that those comparatively few writers are correct who, like Walton, see the *ad baculum* as a form of prudential argument which, when bad, cannot have been made bad simply because it pivoted on the factor of threat [Walton, 992a]. We have discussed the prudential aspect of *ad baculum* reasoning in the past two chapters. (But here see [Wreen, 1995].) I want now to take the Woods-Walton Approach a step further. I shall attempt to show that

1 *ad baculum* arguments are systematically connected to premissorily practical syllogisms;

2 they are in a sense to be explained always in a more benign and welcome form of argument than their counterpart practical syllogisms;

3 in vindication of something theorists such as Walton have been saying — perhaps with insufficient explicit motivation — arguments from negative

consequences are not as such *ad baculum* arguments (that is, *ad baculum* arguments are a proper subset of negative consequence arguments);[3] and

4 (recurring to point (2)), although some theorists have been aware of the importance of utility functions in the analysis of *ad baculum* arguments, there are always utility-functional considerations which favour recourse to *ad baculum* argument over their counterpart practical syllogisms.

Let us now see how it is that our analysis of *ad baculum* arguments give rise to these four consequences. We again consider in turn three arguments of a type well-known in the recent literature. They are: (1) collective bargaining arguments, (2) the mugger's argument, and (3) anti-smoking arguments.

10.1 Collective Bargaining

For expository convenience we consider a simplified case. We assume that in the present example both parties, workers, and management alike, are satisfied that a threat to strike is sincere and that a strike would encumber management with higher costs than would a settlement in the near vicinity of the union's most recent offer. Even so, consider the following *action-matrix*, an ordered pair.

1. The workers strike (S)
2. The management yields (Y)

Schematically our action-matrix is

1.* S

2.* Y

S and Y are abbreviations of (1) and (2), which in turn report certain action-facts. In the circumstances of the case, the episode characterized by $<S, Y>$ is costlier to each party than an available alternative. In real-life situations, this is not always the case, of course, and in any event, calculating the actual cost-benefit spread over actual option spaces can be a fairly complex matter. Even so, we know that one of the alternatives is the one we now describe; and we also know that in general it appears to yield a better cost-benefit payoff for both parties. We represent this option as a dialogue between the workers' representative W and management's spokesman M. As before, S denotes the strike-action

[3] I note in passing Douglas Walton's view that arguments that turn on dire consequences which are not in the arguer's power to enact are better thought of as negative consequence arguments, and that name of *ad baculum* should be reserved for arguments which pivot on the arguer's own threats. This is certainly a nontrivial distinction, and readers who are sufficiently moved by it are free to adopt the narrow usage of "*ad baculum*" and adjust the discussion here, which is easily done.

and *Y* management's action of yielding to the present demands of the workers.

W 'If $\neg Y$, then *S*' (where '$\neg$' denotes negation)

M *Y*

As we see, *W* makes an explicit threat. It is a conditionalization of *S* on the negation of *Y* in our action-matrix <*S*, *Y*>. *M*'s response to *W* is an action, a capitulation to *W*'s demand. If we could think of the sequence <'If $\neg Y$, then *S*, *Y*> as an argument, then not only is it a cross-agent argument, it is a conclusionally practical syllogism. Its most distinctive feature, however, is that it is a substitute for a premissorily practical syllogism, which is what our action-matrix <*S*, *Y*> in effect is. In the *W*-*M* dialogue (or quasi-dialogue), an action which is in the control of *W* to perform and which, if performed, would serve as a premiss in the practical syllogism <*S*, *Y*>, is only threatened. In our simplified example, the threat is justified on simple cost-benefit grounds. It is less costly to threaten to strike than to strike, and it is no more costly to yield to the threat of a strike than to a strike.

We propose that dialogues or quasi-dialogues of the *W*-*M* type are prototypes of *ad baculum* argumentation. If so, it is easy to see the systematic link between *ad baculum* arguments and practical syllogisms.

The threat that constitutes the dialogue as an *ad baculum* threatens an action which is within the threatener's power to effect, and which if effected would produce a premiss in the practical syllogism <*S*,. . . >, where '. . .' holds a place for, but does not guarantee, the appearance of the intended *M*-action *Y*.

It is also apparent that *ad baculum* arguments have clear advantages over the premissorily practical syllogisms, to which they are systematically linked. Here is a case in which 'Talk is cheap' is a virtue. The threat to strike possesses at least the following advantages over striking. Even an efficacious threat to strike is in general, as we have seen, a less costly inducement to yield than yielding to an actual strike. Moreover, talking about striking, rather than striking, provides the contesting partners with a larger deliveration-space than simply striking. Thus *ad baculum* contentions are dialectically more efficient (to say the least) than the premissorily practical syllogisms to which they are linked. We take it, then, that the characteristic features (1) and (2), cited above, may now be claimed for *ad baculum* exchanges. These same features will be apparent in our next example, the mugger. Before turning to that case, it is well to observe that the term "non-cooperation argument" is something of a technical term. It is easy to see that, as in the case we have been discussing, that sometimes a non-cooperation argument is benign in relation to its real-life alternatives. It is also true that collective bargaining arguments could not possibly occur without a good deal of procedural cooperation. So non-cooperation arguments are not arguments *wholly* lacking in convention. But they are not, as in the present case, open enquiries. They are exercises in power (or the prospect of it).

10.2 The Mugger

Here too, there are two parties. M, the mugger, and V, the victim, and an action-matrix $<K, T>$ in which K is the killing of V by M and T is M's getting V's money. As before $<K, T>$ can be likened to a premissorily practical syllogism, and as before it is a less good thing than its counterpart *ad baculum*, in which the action-premiss K is replaced by a discursive premiss which threatens K. It is bad enough to be threatened with death, but for most people in most circumstances it is a better thing than death itself. The mugger's *ad baculum* achieves two things at once. It identifies a situation in the joint option space which itself is constituted by the premissorily practical syllogism $<K, T>$. And it gives the addressee the option of replacing the muggers' practical syllogism with his own cross-agent *conclusionally* practical syllogism

(M) 'If $\neg T$, then K'

(V) T

As before, the conclusionally practical syllogism confers on M all the benefits conferred by the premissorily practical syllogism, yet sparing V the extreme cost of that option. Either way, V loses his money. But in only one of these ways does he lose his life.

10.3 Anti-smoking Arguments

Again we simplify. We shall take it that in some non-trivial sense, habitual cigarette-smoking shortens a smoker's lifespan. If so, then we could expect to find instances of the matrix

Sm
D

Sm for 'The subject was an habitual smoker' and D for (The subject died earlier than would have been the case otherwise'). Is there an *ad baculum* counterpart of this sequence $<Sm, D>$? If so, it would be something like:

P1: 'If $\neg D$ then Sm'
P2: D

(where P1 and P2 are respectively the *ad baculum*-maker and his addressee). We see the absurdity of this reconstruction as self-announcing.

This is tantamount to a proof that negative-consequence arguments are not just as they stand *ad baculum* arguments. If this is right, the rejection of the present example by the *ad baculum* model will show up in structural features of the model.

If the anti-smoking argument were an *ad baculum*, then the sequence

Sm
D

would be construable as a premissorily practical syllogism. For this to be so, two conditions require fulfillment. One is that *Sm* be an action-premiss, and the other is that *Sm* be the action of the *argument-maker*. But as the example shows, this is not the case. Similar difficulties, and then some, apply to the interpretation of the would-be *ad baculum*

P1: 'If $\neg D$ then *Sm*'
P2: *D*

The reader will note that we have conformed the present example to the *ad baculum* structure recognised in our model. It is significant that it gives rise to such nonsense. For one thing, it is hard to conceive of 'If you don't die, then you are an habitual smoker' as any kind of threat. (In fact, it may be wondered what are the truth conditions of this fabulous conditional!) For another, *D* can hardly be represented as P2's action-conclusion, since in no direct way is his death in his own control. Thus the sequence <'If $\neg D$, then *Sm*', *D*> is not representable as a conclusionally practical syllogism. We see, then, that if we opt for an analysis of the *ad baculum* in which premissorily and conclusionally practical syllogisms play a load-bearing role, the anti-smoking argument cannot be made out to be *ad baculum.* Central to this result is the fact that even if Nature herself threatens a certain fate for the smoker and even if I know this, I cannot threaten the same thing on Nature's behalf, so to speak. I am not Nature's agent, nor is Nature my principal (although she is indeed my mistress).

The same holds of Pascal's Wager, in which the Wagerer cites God's threat to the Christian sceptic. But citing a threat is not making a threat. Once cannot issue God's threats, except that one is God, anymore than one can catch Yogi Berra's catches except where one *is* Yogi Berra. I conclude, therefore, that contrary to recent speculation in chapters 2 and 3, Pascals' Wager is not an *ad baculum* for Pascal, though it would be for God. This being so, Walton is right to say that negative consequence arguments are not in general *ad baculum* arguments.

10.4 Utilities

I said that utility functions play an important role in the analysis of practical argument. At one level, there is a constant utility-functional component. The arguer seeks to give the addressee the option of conceding, hence of avoiding the cost of looking stupid (not to put too fine a point on it). In other respect, utility-functions bite more differentially. When the mugger makes an intervention *ad baculum*, he is predicting his victim's deployment of utility functions in a context of menace imposed by the mugger himself. But, as we have already suggested, there is a further respect in which *ad baculum* arguments are the result of utility functions of both arguer and addressee alike. For the arguer to forward an *ad baculum* is a reflection of a cost-benefit analysis which induces the arguer to favour the *ad baculum* over its counter part premissorily practical syllogism. On the other hand, the *ad baculum*-maker also anticipates a favourable cost-benefit determination by his addressee in which it is obviously preferable to yield a benefit under threat of death than to suffer the loss of the benefit as a result of one's death.

The above discussion shows that utility, actions and time play a central part in our understanding of the *ad baculum*. There is an underlying logical model, which is propositional logic enriched by temporal flow and action symbols. The next section develops such a model in some generality. Models of this kind can be applied to other areas such as the analysis of natural language conditionals, but we shall leave that for another time. Because space is limited I shall not dwell in detail on the way the analysis of Sections 1 and 2 is represented in the model. It will, in any case, be obvious to the interested reader. Those not much interested in formal details can get the gist from a light reading of (especially) subsection 11.1 onwards.

11. Description of a Basic Model

We imagine that we are moving through a flow of time. Time is discrete and moves by the performance of actions. So if we are at time t we can move to time $t{+}1$ by performing some action a.

Let us assume we have two players A and B. A is capable of actions $a_1, a_2, a_3, \ldots$ and B can do $b_1, b_2, b_3, \ldots$. Since time moves by actions, assume we are initially at state (time) t. The database (description of the world) is a theory Δt, in the language of a logic L, which can be *FDL* (which is Ralph Johnson's acronym for standard systems of *formal deductive logic*. It is pronounced 'fuddle'). We move to state $t+1$ by someone, e.g., A, performing actions $\{a_{t,i}, b_{t,j}\}$. If A performs $a_{t1}, a_{t2}, \ldots$ and B performs $b_{t1}, b_{t2}, \ldots$, each action has preconditions α_x

and postconditions β_x. Then the action is *allowed* at time t if $\Delta_t \vdash \alpha_x$, if the action is allowed and it is performed, then at $t+1$, β_x holds. The new theory is Δ_{t+1}.

We should note that if D_t partially describes the state of the world at t then Δ_{t+1} is an update of D_t; in symbols

$$\Delta_{t+1} = \Delta_t * \beta_x$$

where * is an update operation. Only when β_x is consistent with Δ_t do we have $\Delta_{t+1} = \Delta_t U\{*\beta_x\}$.

The operation * can be assumed to satisfy some rationality postulates, e.g., AGM, or to be given some specific algorithms [Alchourrón et al., 1985]. We also assume that we have cost functions $U_A(\Delta_t)$, $U_B(\Delta_t)$, giving a $ (positive or negative) number which is a utility value for each player at time t. So, for example, player A may be unhappy with Δ_t. He may perform action a=(α_a, β_a) such that $U_A(\Delta_t)¡U_A(\Delta_t\beta_a)$.

11.1 Example

A and B are arguing about something.

The database contains

1 $p \wedge q \to r$

2 p

B is desperate to deduce r. He controls an action whose postcondition is q. By performing this action he is practically inserting q into the database and thus enabling the deduction of r. In the language of Section 1, B is in the process of constructing a premissorily practical syllogism.

11.2 Example

Jobless John (*JJ*) has an old car which he insured with the Universal Insurance Company. According to the terms of the insurance his coverage expires December 31st, 2000. As is common practice with many insurance companies, if *JJ* pays his premium by 31.1.01, his insurance coverage is renewed from 1.1.01 to 31.12.01. So for example, if *JJ* forgets to pay on 1.1.01 and has an accident on 15.1.01 he can still pay his premium on 20.1.01 and be covered for the 15.1.01 accident.

In our story, *JJ* has no money and does not pay his premium. On 15.1.01 he bumps into Richy Rich's (*RR*) Rolls Royce, causing extensive damage. *RR* now has a problem. It is clear that *JJ* cannot pay his premium. If he doesn't, then he is not covered, and *RR* cannot collect from *JJ*'s company. *RR* cannot of course collect from *JJ*. On the other hand, *RR* collects from his own insurance company, he will lose his 56% no-claims bonus.

Let us give some utility values.

JJ's premium is $500
RR's loss of the no-claims bonus is worth $3,000
RR's damage is assessed at $8,000.

It is clearly worthwhile for *RR* to pay *JJ*'s premium provided *JJ* is co-operative. We assume the factual circumstances of the accident, $\Delta_{accident}$, strongly support $q =$ '*JJ* is at fault'. This means that the database can probably prove that *JJ* is at fault even if *JJ* denies fault. However, it is much simpler if $q_1 =$ '*JJ* admits fault' is available.

We now construct the story somewhat more formally.

11.3 Propositions

p	$=$	*JJ* damages *RR*'s car
q	$=$	it is *JJ*'s fault
q_1	$=$	*JJ* formally admits fault
c	$=$	*JJ* is covered
r	$=$	*JJ*'s insurance pays
$\Delta_{accident}$	$=$	facts about accident.

11.4 Actions

b	$=$	*RR* gives *JJ* $750.
α_b	$=$	q_1 holds and furthermore *JJ* commits to perform action a.
β_b	$=$	*JJ* performs action a.
a	$=$	*JJ* pays his premium before 31.1.01.
β_a	$=$	c

Let $t_0 =$ accident time and let Δ_{t0} be the following database:

1 $p \wedge q \wedge c \rightarrow r$

2 $\Delta_{accident}$

3 p

4 $q_1 \rightarrow q$

RR wants r to follow. For that he needs to generate q and c. The simplest course of action is for *JJ* to make q_1 true at *t+1* and give a commitment to perform a, then get $750 from *RR*, then go ahead and perform a; whereupon *RR* can claim from the insurance company.

JJ can threaten *RR* that he will not renew his insurance unless he receives $750 from *RR*. This is a legitimate threat. Although *JJ* is at fault and may face a damages claim from *RR*, it is clear that simply by not renewing his insurance, he creates a premissorily practical syllogism of particular consequence for *RR*. This is why, in effect, *RR* is trying to persuade *JJ* to produce a different premissorily practical syllogism, in which the practical premiss of not renewing is replaced by its action-negation.

We can now summarise what an *ad baculum* fallacy is. It is *not* a fallacy. At worst it is an incompetent threatening move, which is either illegal (since preconditions do not hold) or ineffective (since it has a low utility threat for the postcondition).

PART III

ARGUMENTS INVOLVING REFERENCE TO PERSONS

In as much as real-life arguments are interpersonal exchanges, its is wholly understandable that in dealing with one's interlocuter reference to his or her personal circumstances is a factor whose commonness borders on invariability. Notwithstanding the traditional hostility toward *ad hominem* arguments it is interesting that neither Aristotle, who coined the term, nor Locke, who borrowed it, shared in this opprobrium. In the chapter to follow an effort is made to find something for the opprobrious to be hostile about. In the chapter that follows it, a case is made for the argumental integrity (and indispensability) of certain sorts of *ad hominem* arguments. In the chapter after that we expand our reach to a special case of *ad hominem* argument. These are arguments that appeal to the authority or sayso of others. The linking thread is the notion of *bias*, widely regarded as an intrinsic blighter of argumental integrity.

Chapter 6

DIALECTICAL BLINDSPOTS

The modern circumstantial *ad hominem* owes nothing to Aristotle's treatment of it or to Locke's generalization of it. In the latter two cases, an *ad hominem* argument is one in which the arguer derives from his opponent's concessions a logical consequence which contradicts the opponent's thesis or (in the Lockean case) counts in some possibly non-deductive way against it. In the chapter to follow I take up the Aristotelian-Lockean conception of the *ad hominem*. In the present chapter the modern circumstaintial conception is our target. As we shall see in due course, the Aristotelian-Lockean *ad hominem* pivots on a notion of *inconsistency*. So does the modern conception, but on a quite different conception of inconsistency.

It is widely held that the *ad hominem* is a dialectical fallacy, that is, it is a dialectical fallacy when it *is* a fallacy. This is true, but it doesn't take us very far. The *ad hominem* has been stingy about disclosing its structure. It has been falsely modest and something of a tease.

I shall here try and subdue this reluctance. I will suggest that *ad hominem* pronouncements possess highly complex structures. Interesting in their own right, these structures, once exposed, reveal how *deeply* dialectical the *ad hominem* is.

1. Not Practising What You Preach

I want in the first instance to consider a range of cases loosely collectible under the rubric "defections from one's own policy".[1] So I am

[1] I intend the word "policy" rather slackly. If defections are cases of "not practising what you preach", then policies are anything "preached".

not here interested in cases in which the *ad hominem* fallacy is directly definable via a notion of negative relevance. If I ask you for your support in the upcoming election for County Treasurer and, suppressing my natural modesty, assert that I would make a good County Treasurer, you might rejoin with "But you've been convicted of embezzlement three times in the last eleven years". If true, your attribution is negatively relevant in the sense that it damages the likelihood that my claim is true.[2] Had you replied instead, "But you wear size eight-and-a-half shoes", the *ad hominem* would have misfired significantly. Though true, my shoe-size possesses no negative relevance for the claim in question. Whereupon a definition drops out:

> *Definition One:* In accurately attributing fact F to person Y, the attributor X commits a fallacious *ad hominem* with regard to Y's thesis T to the extent that F is not negatively relevant to T.

Neither shall I be concerned with a class of cases in which the *ad hominem* is offered as the refutation of a dialectical adversary's position "out of his own mouth", that is, by way of his own concessions, elicited in the course of critical enquiry. Such are subcases of Definition One, for which fallaciousness is definable as follows:

> *Definition Two:* In accurately attributing F to Y, X commits a fallacious refutation-*ad hominem* if and only if F is in the form "But you, Y, concede that Q", and X claims of Q that it contradicts a prior concession K of Y's, yet Q does not in fact contradict K.

Here, too, fallaciousness reduces to, and is a mismanagement of, negative relevance. X holds falsely that Y's concession Q is negatively relevant to his prior concession K in the extreme sense that it contradicts it. X claims that either Q is refuted or K is, but he is mistaken either way. (For more on this kind of case, see the chapter to follow.)

Neither shall I be dealing with situations in which the *ad hominem*-er accuses his interlocuter of bias. (Bias is discussed in detail in chapter 8.)

As I say, it is *not* cases of this sort that I shall be concerned with here. In so saying, I should note with some emphasis that there is not always a well-marked divide between the cases I discuss and those I don't discuss.

Imagine now that a father and his teenaged son are arguing about smoking. Father is strongly opposed, the son is not so sure. As the father is putting his case, his son intervenes with, "But you smoke, Dad". It is at once noticeable that, contrary to traditional accounts of

---

[2]We might say that the conditional probability of my claim given the truth of yours is less than the probability of my claim taken alone. But this is problematic. We lack good theories of relevance. See [Johnson and Blair, 1994; Gabbay and Woods, 2003].

it, there is nothing intuitively speaking to reproach the boy for in the case at hand. Clearly he is "onto something". What might it be? I shall suggest that what is of central importance here is not the father's inconsistency (though it too is important, obviously), but rather what his son — his dialectical adversary — *makes* of his father's inconsistency, and, in turn, what the father himself makes of it. What is made of the inconsistency is problematic; for the son leaves this part of the exchange unexpressed. Whereupon

> **Thesis One:** The problem of analyzing *ad hominem* remarks alleging defections from policy is significantly one of specifying an interlocuter's unexpressed content.

There is a twofold dynamic here in play. Our task is to try to figure out what the son is up to, and what the son is up to is trying to figure out what the father is up to. So twice-over we are looking at inference to the best explanation. It seems clear that the son is entertaining one or both of two possible explanations, one of which calls down a sincerity assumption and the other of which invokes a rationality assumption. Before considering these further, we would do well to take note of three particularly interesting features of the present kind of case. One is that in defecting from one's own policy it appears that one is somehow inconsistent with it. I shall take it as a condition of adequacy on the present account that this notion of inconsistency be elucidated. Another factor to take note of is that sometimes a retort in the form, "But you yourself do so-and-so", is in some sense a relevant remark. I shall take it as a further condition on the present account that it specify truth conditions appropriate to this conception of relevance. A third feature of the cases which presently occupy us is that sometimes an *ad hominem retort* intensifies or enlarges the burden of proof on the person to whom the remark is addressed. A third adequacy condition is to specify conditions under which this happens.

2. Insincerity and Irrationality

Defection from policy suggests the insincere espousal of it. Suggests but does not prove. The boy has two options to consider: (i) The espousal is insincere and his father doesn't believe what he says; and (ii) the espousal is sincere after all, and the father (somehow) does believe what he says.

Here it would be useful to review what it is to be a Grice-cooperator. A Grice-cooperator is a discussant who complies with Paul Grice's maxims of conversation. These include the "supermaxim" which bids the discussant to be truthful. Another obliges the discussant to be relevant. A third requires him to be perspicuous (i.e., to be clear, unambiguous, brief and orderly). Yet another obliges the discussant to make his utterance as informative (no more and no less) as it needs to be, given the kind of discourse that is underway [Grice, 1989, pp. 26–28].

We are now in a position to characterize what the son is up to. He is trying to be a good Grice-cooperator; and so that he stints on attributions of adversarial insincerity. If so, his task is to take the measure of his father's inconsistency. A conjecture awaits him. It is that his father does not believe his own thesis *in the manner in which he has formulated it*, but that he does believe some reformulation of it. If the thesis as stated is,

(T): People should not smoke

it may be that the intended thesis is either an adverbial intensifier of T as in

(T*): People should not smoke *heavily*

or an adjectival cutdown of T as in

(T**): *Young people* shouldn't smoke.

It is significant that under either conjecture the father's inconsistency is not explained away. Rather it is precluded by the conjecture at hand: there *is* no inconsistency to explain away. Thus we have:

Thesis Two: There are cases in which the sincerity factor forces the *ad hominem*-maker into Grice-uncooperation. The dynamics of such inducement is this. In an effort to Grice-cooperate with respect to his adversary's sincerity, it is necessary for the *ad hominem*-maker to be a Grice-defector with respect to his adversary's fidelity to the Quantity Maxim.

The Quantity Maxim obliges discussants to be appropriately informative in their assertions. Required to say neither too much nor too little, the father would be saying too much if he asserted T while intending that T* or T**. The son himself offends against Gricean strictures by attributing to his father offence against Gricean strictures.[3]

In the event, however, that the son did invoke the insincerity assumption and hence conjectured that his father doesn't believe what he says (and means), his assumption would reverse the field, so to speak. His assumption would now restore the inconsistency of the father's position, but it would also explain it away — for the father doesn't believe the policy from which he defects. But this explanation is made at substantial cost. That is,

[3]True, the Quantity Maxim is not usually taken as discouragement of the overstatement or exaggeration of one's case, but there is no harm in extending its provenance thus.

Thesis Three: Sincerity explanations of the type in question force the *ad hominem*-er into another kind of Gricean defection. For the benefit attaching to the explanation-away of his adversary's inconsistency and thus the fulfilment of the Gricean desideratum not to think your opponent a simpleton, the *ad hominem*-er must gravely defect from sincerity assumptions.

A third possibility is available to the son and it trails off quickly into the precincts of rationality explanations. He may think both that his father is sincere and that he believes exactly what he says, that is, that it is *T* that he believes, not T* or T**. If so, it is open to him not to explain-away his father's inconsistency but rather to explain how he *fell into it.* He might postulate a deficit of rationality. This is important in two ways. It convicts the son of a still further Grice-defection, and it *might* be taken as negatively relevant in a very particular way. The Principle of Charity[4] requires adversaries to give one another's positions appropriate weight. Each gives to the other sufficient standing to keep their dispute from lapsing into triviality, or just lapsing, for that matter (more Grice-defection still). As a first pass, Charity bids us to concede that the very fact that an opponent attributes *F* counts as some reason for thinking that *F* might be a true fact. Charity commits dialectical adversaries to implicit recognition and tentative acceptance of reciprocal (and tentatively legitimate) *ad verecundiam* manoeuvres. This sounds wrong, of course, as the following objection suggests. Stated as in the two sentences above, it is clear that the Charity Principle is given unacceptable providence. For surely it cannot be that anything anyone says is, just by virtue of his sayso, shown to be more likely to be true than if not asserted at all. What if someone asserts that the Holocaust is a hoax? Surely his saying so is entirely without prospects of plausibility-enhancement. This is quite true. To understand how the Charity Principle here operates, it is necessary to tease out some tacit assumptions about context. If Harry and Sarah are dialectically serious discussants seeking to fulfill Gricean cooperation maxims, then, among other things, they will both be semantically serious. That is, they will not forward claims if they seriously think that they are untrue and they will not offer support of the claims forwarded unless they think that the support is, to some extent at least, telling. Even where Harry and Sarah strongly disagree about some thesis *T*, the very fact that they are having a Grice-cooperative argument about *T* implies that, for the purposes of the argument, each allows that his opponent's thesis is a real possibil-

[4]For an illustration of how Charity works as a guiding principle of Grice-cooperative measures for specifying missing premisses, see van Eemeren and Grootendorst [van Eemeren and Grootendorst, 1983, especially pp. 134–135]. See also [Woods, 1992b; Woods, 1993a] reprinted here as chapters 1 and 14.

ity.[5] Real possibilityhood is a technical notion. It is loosely described as follows. In the course of their discussion, the very fact that Harry claims *F* must count for Sarah as evidence that Harry has a case to make for the claim that *F* and, given Gricean-constraints, that the case needs to be examined. It is in this way that Sarah accepts *pro tem* that Harry's case for P has weight; that is, it has weight enough to be heard and answered. And so: something *P* is a real possibility for someone, *S*, when it is not, just or even, a logically consistent proposition, and not, just or even, a proposition compatible with everything that S presently knows. Rather it is a proposition for which S is prepared to recognize a case, and a case which he concedes that a rational person of *S*'s own broadly cultural orientation could come to accept.

None of this makes procedural sense unless underpinned by a further assumption, namely, that Harry and Sarah accord to each other cognitive wherewithal sufficient to handle their debate competently. Without it, the idea that an opponent's case has weight enough to require answering collapses. And it is just this rock-bottom assumption that the son is at risk of rejecting on our present interpretation of his reaction to his father's defection from policy.

So we have

> **Thesis Four:** Irrationality explanations necessitate suspension of the Charity Principle (which is a cost) in return for a verdict of modest negative relevance with regard to the defecting behaviour (which is a modest benefit).

It is now apparent that our paradigm requires semantic description, where "semantic" is used in the *logician's* sense of "pertaining to truth".

> **Thesis Five:** The analysis of a dialectician's response to the pragmatic inconsistency of an opponent obliges the *ad hominem*-er to entertain at least two semantic possibilities. One, that his opponent's thesis is not true and not espoused as originally formulated. Two, that his opponent's defection may be negatively relevant to his opponent's own thesis.

I shall return shortly to the business of irrationality assumptions. What now needs bringing out is that the unexpressed content of an *ad hominem*-er's retort, for the large range of cases considered here, lands the retorter every time in the dilemma of having to be a Grice-defector in order to be a Grice-cooperator. By virtue of the dilemma's either horn the *ad hominem*-er is doing something wrong; and this is some encouragement of the traditional view that an *ad hominem* remark is always untoward. And if dialectical infelicity *were* the heart and soul of fallaciousness, then an *ad hominem* would always and as such stand convicted of fallaciousness. Whereupon,

[5]The idea of 'real possibility' is discussed further in chapter 12.

Thesis Six: An *ad hominem* is a dialectical infelicity just in case and to the extent that the costs of Grice-defections outweigh the benefits of Grice-cooperation.

The trouble is that neither in theory nor in practice is it possible to make the cost-benefit determinations of thesis six. The dialectician is left to blunder about unaided by his pocket-calculator. I dare say that it is an attraction of the present analysis that it nicely reflects the ambivalence of the real situation in which it is often unclear even to him who makes it whether an *ad hominem* remark is kosher and to what extent.

It is time to play a further card which removes some of this ambivalence and improves the *ad hominem*-er's prospects noticeably. It is significant that the *ad hominem*-er's various conjectures and assumptions about his opponent's inconsistency are here represented as part of the unexpressed content of his retort. This alone relaxes substantially the grip of the dilemma recently noted.

On the face of it, this cannot be true, since it is a serious violation to leave unexpressed considerations as important as these. In fact, however, if our dialectician knows what he is doing, these considerations are handled on two levels at once. At the level at which they are unexpressed, they are unasserted; they are only entertained or conjectured as possibilities. What are not expressed are considerations which if expressed would give Gricean offence, and that alone is a Gricean reason not to express them. There is also a level at which they are expressed. They are expressed by way of the subsequent attempt to verify whether the conjectures are true. This is done by *asking* ("Do you mean, perhaps, that only heavy smoking should be avoided?", and so on). The fact that one is asking rather than asserting all but removes one from the jaws of Grice-defection. That said, it is necessary that we now adjust our previous theses in ways that take much of the sting out of them. Doing so makes them less charming but more true, and it rids them of the air of phoney paradox. What is now clear is

Thesis Seven: In the preceding theses, incidents of Grice-defection were misattributed. What should have been said is that the *ad hominem*-er is forced into unexpressed consideration of possibilities which *if asserted* would have involved him in Grice-defections.

And so,

Thesis Eight: It cannot be true, as suggested above, that it is by way of forced Gricean defections in the general case that *ad hominem* retorts of the kind here considered have the air of fallacy.

The idea of the fallacious *ad hominem* is better pursued in an altogether different way. Basically, what the *ad hominem*-er wants is that his opponent explain himself. It is not his opponent's inconsistency that nags at him but rather that it is explanatorily opaque.

3. Explanatory Opacity

The notion of explanatory opacity can be elucidated with the help of *compulsive* and *akrasiacal* (i.e., weak-willed) behaviour. We need to recognize that the father could have given a further kind of response to his son. He could have said, "I can't help it". That is, he could have pleaded that he smokes either because he is addicted to smoking or because he is weak-willed. Some critics think that the distinction between compulsive and akraisiacal behaviour marks only a difference in degree. True or not, it does not matter for present purposes. It is also contended that akrasiacal actions are irrational. This seems to me absurd, but, again, it is unnecessary to settle this question for the business at hand.

Consider now the case of the notorious and very noticeably alcoholic father who boozily cautions his daughter against drinking. It is striking that her *ad hominem* rejoinder, "But you drink, Dad", is so off the mark. Resorting to it seems a cruel thing to do, or perhaps is a kind of bad faith. The reason for this is that the father's inconsistency does not require explanation; it is made readily intelligible by various background and contextual features of the case in question. His defection from policy is not for her explanatorily opaque. An *ad hominem* retort is an assertion in the form "You do action A". Its dialectical function is wholly indecipherable in the surface structure of such sentences. Postulations are needed. One is that the *ad hominem*, "You do (or don't do), action A", is always an ellipsis for

(α) Kindly (or I challenge you to) explain your action A.

The challenge is misdirected unless the facts of the case fulfill a presuppositionwhat I will call its "enabling presupposition" — namely, that

(β) Action A is a defection from policy.

We can now see that the *ad hominem* can go wrong in two quite different ways. As suggested just above, an *ad hominem*-er can make an unneccessary request for an explanation of a defection from policy. Such a request is unneccessary when the defection in question is made intelligible by factors already understood by both participants. We could characterize such infelicities as fallacies, but I am unable to discern any advantage in doing so. In any event, it is easy to see the infelicity in question is a dialectical move the enabling presupposition of which — that is, β — is confirmed and explained by the shared facts of the case.

It is exactly the other way with the *fallacious ad hominem*. An *ad hominem* is a fallacy when it is a remark whose enabling presupposition is false. Whereupon, further definitions present themselves.

> *Definition Three:* In accurately attributing F to Y, X commits a pointless *ad hominem* to the extent that the explanation tacitly called for is already present in the participants' common understanding of the facts of the case at hand.

And,

> *Definition Four:* In accurately attributing F to Y, X commits a fallacious *ad hominem* just in case F's enabling presupposition is false.

The fallacy of the *ad hominem* is thus the fallacy of false presumption. In the smoking example, we might (though not without difficulty) imagine the son's making the *ad hominem* retort, "But you wear size eight-and-a-half shoes". The potent intuition that "this has nothing to do with that" can now be given utterly precise expression: The remark's enabling presuposition is false. The son thinks, or appears to think, that his father's "shoddy" behaviour is a defection from his no smoking policy. It is no such thing.

Those with a fondness for relevance may resist the claim that the *ad hominem* fallacy is at bottom a fallacy of false presumption. They will notice that definition four provides clarification of the notion of "this having nothing to do with that". So it does. In fact, we might even propose

> *Definition Five:* In accurately attributing F to Y, X commits a relevance fallacy just in case F's enabling presupposition is false.

With definition five at hand, it would appear that we have discharged the first of our three adequacy conditions.

Although it gives the general picture, definition four handles the notion of fallacious *tu quoque* remarks a bit heavy-handedly. It convicts "But you wear size eight-and-a-half shoes, Dad", as a fallacy which the son could have committed only by thinking that his father's wearing shoes of that size counted as a defection from his no-smoking thesis. Those fond of Gricean strictures will protest. It reserves fallacious *tu quoques* for simpletons, which alone is a Gricean reason to think that they don't in general occur. So it does, *in extremis*. But the problem corrects itself under the gentle guidance of an ancient view of fallacies: a fallacy involves a mistake which seems not to be one. The earliest recognition of such a conception is Aristotle's in *On Sophistical Refutations*, where fallacies are arguments which appear to be syllogisms but are not in fact. The core idea, at any rate, can be incorporated in a revision of definition four, as follows:

Revised Definition Four: In accurately attributing F to Y, X commits a fallacious *ad hominem* just in case F's enabling presupposition is false, yet appears not to be.

This is better. It is tailor-made for the type of case briefly noted by Whately. If Harry's thesis T is now that blood sport is unconscionable, and Sarah's retort is that Harry "feed[s] on the flesh of harmless sheep and ox," two interesting points come to light [Whately, 1848, bk. III p. 15]. One is that the new definition allows for, without precluding, the commission of the *tu quoque* fallacy without the intention to deceive. The other is that the more that a false enabling presupposition seems not to be false, the narrower is the divide between fallacious and non-fallacious attributions *tu quoque.* This is interesting; the greater the likelihood that Sarah is committing a fallacy the greater the appearance of an or enlarged burden of proof on Harry. Either the attribution is a fallacy or not. If it is not, Harry is called upon to deal with his own incontinence (see below). If it is a fallacy, then Sarah is called upon to show that it is; and the more it is a fallacy, the less it will appear so, with attendant difficulties for discharging the enlarged burden of proof.

4. Blindspots

It would be a welcome efficiency if that were all there is to fallacies of the "you smoke" type. Though effortlessly concedable, saying of an *ad hominem* remark that it is fallacious just in case it falsely characterizes an adversary's apparently incontinent behaviour as defection from his own policy leaves something central out of the account. It fails to take the measure of behaviour which is a defection from policy and yet which is not and cannot be made explanatorily intelligible. This is the case in which Dad offers no explanation of his defection and rebukes his son with, "Don't do as I do, do as I say".

It is altogether striking that Dad's dismissive response, "Don't do as I do, do as I say", would be absolutely the right reply to challenges of the kind made by his son if some traditional accounts of the *ad hominem* were correct.[6] After all, Dad is saying, it is the argument that we should be interested in, not the arguer. But, as anyone knows who has been on the receiving end of such admonitions, the last thing that they are is the absolutely right reply. What are we to make of them?

I am here considering an irreducibly core situation. It is the situation in which

[6]Thus: "The circumstances of the opponent are not the issue in serious argument;... such premisses are not to the point" [Copi and Cohen, 1990, p. 90].

A: Nothing said by S^* (the proponent of T — by Dad, for example) and nothing in the shared understanding of S^* and S makes S^*'s defection explanatorily intelligible to S.

B: Moreover, S^* shows no discernible interest (or worse, denies an interest) in making his defection intelligible. Dismissiveness is the mark of such indifference. Thus, "Don't do as I do..."

In these core situations, if S^* could be brought to say something which explains his dismissiveness (not his defection), he might say something like, "What I do is irrelevant to what I say". Now, putting to one side theoretical difficulties about how to analyze irrelevance, it is clear that what S^* means by it in the present context is that

1. What I say (i.e., my thesis T) is true or false independently of what I do

and so it is.
Two important facts co-obtain:

2. S^*'s dismissive, "Don't do as I do...", is *not* perfectly all right

and

3. What S^* does is indeed irrelevant to T; that is, T is true or false independently of S^*'s behavioural defection.

It can now be ventured that no account of the *ad hominem* can pretend to adequacy so long as it leaves unexplained the co-existence of these two facts. So something further needs saying.

It lies at the heart of the matter that (2) and (3) are co-satisfiable. They pass in the dark, so to speak. The fault of the present case lies not in the defection (which might be made intelligible in any of several ways) but in the dismissal. Since fact (2) pertains to the dismissal, and fact (3) pertains to the defection, it is perhaps not surprising that they should co-obtain.

What, then, is wrong with the dismissal? Suppose that our proponent S^* suffers a confessional turn and confides that

4. I shouldn't smoke, but I do.

Things like this are commonly said. Diet-conscious people say this *sort* of thing all the time. The admission takes on a particular interest when made in the context of advocacy for T, i.e., on behalf of "No one should smoke, including you, (and me, too)".

5. Trust

It might be said that deontic advocacy is underwritten by a kind of moral authority which admissions such as (4) cancel or compromise.

They give rise to protest: "Who are you to talk about that?" In so saying, it sounds as if *S* may be complaining that the admission, (4), cancels *S**'s moral authority to advise. I don't doubt that this is sometimes so, though specifying the conditions under which it is so would prove to be uncommonly vexing. In any case, when it is so, it is a case comprehended by a more general situation. This is the situation in which *S**'s advocacy turns at least in part on claims offered and accepted by *S* on epistemic trust. If Dad's case against smoking involves citation of statistical studies on the damage that smoking does, he may offer them knowing that his son lacks independent access to them. He is asking his son to trust his word.

Similar considerations weigh on the present point. Advice-giving is, often enough, invested with presumptions of epistemic trust, but also with considerations of expertise or judgement. "Don't do that", I say, expecting you to heed my wisdom, my expertise, my special position to know or, if I am your parent, my authority to instruct.

Thus the protest, "Who are you to be telling me that?", is sometimes a rhetorical device for recording the conviction that features of the other party's situation undermine his presumption to advise. It is as if the protester were suggesting that

5. There are features of *S**'s situation that make of *S**'s invitation to heed his word an invitation to commit an *ad verecundiam* fallacy, the fallacy of the illicit appeal to authority or expertise (concerning which see [Woods and Walton, 1989, ch. 2 and 14], and [Woods et al., 2000, ch. 11]).

This will be so to the extent that *S**'s advocacy in support of *T* turns non-trivially on the authority of his sayso and yet *S**'s circumstances (e.g., Dad's smoking) undermine that authority.

These remarks apply, of course, to factors of defection. Yet, as I keep saying, the greater interest lies in dismissal or in *S**'s indifference to his own defection. In bidding S to ignore his defections from *T*, *S** can be taken as saying

6. *T*, but *T* has no weight for me.

And (6), it is clear, announces its kinship with Moore's paradox:

7. *P*, but I don't believe that *P*.

Moore's paradox has attracted a huge diagnostic literature. I shall not attempt to deal with it here, except in the following way (but see [Sorensen, 1988, ch. 1]). Virtually all commentators are of the view that Moore's paradox is no contradiction, that the proposition expressed by "*P*" is consistent with the proposition expressed by "*S** doesn't believe that *P*". I agree with this and am prepared to accept it as a fact. Let us call this fact 'fact three'. There is another thing that the commentators agree about — and I, too; it is that the consistency of those propositions, that is, the content of fact three, is insufficient to make it perfectly all right for someone to assert the Moorean sentence. Let us call this 'fact four'.

It is instructive that facts four and three parallel in a certain way, indeed they mimic, facts one and two. This alone inclines me to the view that in its dismissiveness Dad's "Don't do as I do..." remark approximates to Moore's paradox. 'Approximates' is the right qualification. In (7), P is any sentence whatever appropriate to the speech act of assertion. But in (6), T must be construed as a deontic claim of sufficient generality as to include the speaker in the range of its (implicit) quantifier. If this is so, it would be helpful to know the offence that fact four gives, notwithstanding the benignness of fact three. Granted that it is entirely co-possible that P and yet that S^* doesn't believe that P, there is still something wrong with S^* himself saying so. "P, but I don't believe that P", as said by S^* to S, presents S with an intractable interpretation problem. Without supplementary clarification, S doesn't know *what* S^* believes. This is a claim that constitutes a blindspot for S, to adopt Sorensen's attractive metaphor.

This is a diagnosis we may wish to extend to our present case. What we are after is the significance of dismissal. S^* has said that no one should smoke and yet that his own smoking doesn't matter. There is no inconsistency between the proposition that no one should smoke and the proposition that S^* does smoke. If S^*'s dismissal is understood to be making just this point, then it is fine as far as it goes; fine and uninteresting. Beyond this point, things are far from fine. For then S^*'s dismissal is construable, again, as

6. T, and T has no weight for me.

Like our Moorean sentence, (6) afflicts anyone to whom it is asserted with a blindspot. The addressee is not able to see what S^* is *advocating*. As long as S^* persists with his dismissal of his own defection from policy, he can be taken as holding something in the form of (6). This is important. We said at the beginning how deeply dialectical the *ad hominem* is. That was not quite right. It is the dismissal of the defection that is deeply dialectical, not the complaint of it. And its depth consists precisely in this: Blindspots are inimical to the very condition of advocacy itself. For as long as S^* persists with (6), S cannot slip the blindspot it creates for him. Advocacy has lost its object.

Advocacy's loss of its object is a matter of losing one's dialectical purchase. It afflicts both dialectical adversaries but in different ways. For S^*, dialectical purchase is lost just in virtue of his having cancelled the conditions for his own advocacy. For S, his addressee, dialectical purchase is lost just in virtue of his having been deprived of something to answer. The pathology of dialectic is now deranged, and S is left with only weak protests to make: "I don't know what you're saying". "Your position regarding T is unintelligible to me". In the free fall from dialectical purchase, something is lost. Retained is the thesis T, lost is an identifiable position with respect to T that S is expected to engage. Loss of advocacy's object, then, is not loss of T but loss of an *answerable position* with regard to T.

We now have resources at hand to fulfill our two remaining adequacy conditions

> *Definition Six:* S^*'s doing A is pragmatically inconsistent with his espousal of thesis T iff S's utterance "But you, S^*, do A" has a true enabling presupposition and S^*'s utterance "T, but that I do A carries no weight for me" constitutes a blindspot for S.

And,

> *Definition Seven:* S's utterance "You, S^*, do A" *intensifies or enlarges the burden of proof* on S^* iff there is a thesis T which S^* espouses, S's utterance is true, and S^*'s doing A is or appears to be pragmatically inconsistent with his espousal of T.

We might wonder, by the way, about the "or appears to be"-qualification. If S^*'s behaviour, A, appears to be, but isn't, pragmatically inconsistent with his espousal of thesis T, then S's charge "But you do A" has an enabling presupposition which, though false, appears not to be. the more it appears not to be, the more it is a fallacy. And yet the more it is a fallacy, the harder it will be for S^* to show it so.

The great blunder of certain traditional accounts of the *ad hominem* is committed not by the *ad hominem* but by the tradition itself, in comparison with which *ad hominem* retorts are the acme of dialectical innocence. The tradition's blunder is to have answered the *ad hominem* complaint with the dismissal of it. In so doing, the tradition asserts (6) on S^*'s behalf. Doing that afflicts the *ad hominem*-er with a blindspot and advocacy is sent into oblivion.

Chapter 7

AD HOMINEM

In the previous chapter, I mentioned a class of *ad hominem* arguments which I said that I would deal with later. I called them "refutation-*ad hominems*". They are the business of the present chapter.

1. Johnstone to Locke

It has been said that philosophy is made distinctive by two of its features, one broadly epistemological and the other rather more dialectical. The epistemological feature is non-cognitivism: philosophical statements "do not convey genuine knowledge" [Johnstone, 1978, p. 13]. The dialectical feature is that philosophical arguments are inherently *ad hominem*, and that "no genuine *argumentum ad rem* is available for philosophical controversy" [Johnstone, 1978, p. 8 *et passim*]. These are views for which Henry Johnstone was notorious. On the principle that, concerning one's published view, it is better to be noticed and scorned than not to be noticed at all, perhaps notoriety is no bad thing. But in Johnstone's case it is misplaced. It is not his position that philosophical arguments, as such, commit the *ad hominem* fallacy; so Johnstone's is not the more or less received understanding. Johnstone was a Lockean about the *ad hominem*, just as Locke in turn was an Aristotelian about it; and so too Johnstone if the requisite transitivity holds.

In the celebrated passage, entitled "Of reason", of chapter seventeen, book four of the *Essay*, a passage which everyone seems to cite and hardly anyone seems to read, Locke distinguishes *ad hominem* arguments from arguments *ad judicium*, which are "proofs drawn from the foundations of knowledge or probability". Unlike *ad hominem* arguments, these alone bring "true instruction with [them] and advance [...] us in our way of knowledge" [Locke, 1975]. In contrast, arguers *ad hominem* merely "press a man with consequences drawn from his own principles or concessions" [Locke, 1975, p. 686]. Far from fallacious, the pressings do not, however, bring with them true instruction; they do not advance us in our way to knowledge. Henry Johnstone's metaphilosophical claim is that philosophical argument is always Lockean in this present sense. There is in this a positive thesis, as well as a negative one. The positive thesis is that philosophical argument is always argument in which a claim is pressed with consequences of its espouser's own principles and concessions. The negative thesis is that this is not the way to knowledge. Thus the non-cognitivism of philosophy.

There is not much doubt that Johnstone held that the negative thesis follows from the positive thesis. What may not be clear is why it does. An answer can be found, or so I believe, in Aristotle's conception of refutation, which is nothing if not a pure case of Locke's *ad hominem*.

2. Aristotelian Refutations

Aristotle embeds his account of refutation in contexts of dialectical argument. There are two participants, Q, a questioner, and A, an answerer. A proposes a thesis T.[1] Q's role is to question (*erotan*) A, putting questions to him, answers to which are formatted as simple (non-compound) declarative sentences, or propositions (*protaseis*) in Aristotle's technical sense of the term. A's answers are thus available to Q as premisses of a syllogism, (let's call it *Ref*), which it is Q's role to construct. If *Ref* is constructed and if its conclusion is the contradictory, $\ulcorner\neg T\urcorner$, of A's original thesis, then Q's argument is a refutation of T. In this we see the pure form of Locke's *ad hominem*, for *Ref*s premisses are A's own principles and concessions (and nothing else); and *Ref*s conclusion, the contradictory of A's thesis, is got by pressing those concessions with their consequences.

It is necessary again to say a brief word about how Aristotle marshalls the concept of syllogism. Syllogisms, here, are taken in a broad or generic sense, discussed by Aristotle in the *Topics* and *Sophistical Refutations.* They include but are not restricted to ordered triples of categorical propositions, which are the structural focus of a kind of completeness proof worked up in the *Prior Analytics.* Syllogisms are always valid arguments; that is, their premisses always necessitate their conclusions. Strictly speaking, there can be no such thing as an invalid

[1] A word of caution. In Aristotle's usage a thesis is a paradoxical claim. This is not here its intended sense. Aristotle's word is *problema.*

syllogism. To qualify as a syllogism, a valid argument must meet three main conditions (and others that derive from them). First, a syllogism must have multiple premisses and non-multiple conclusions. Second, a syllogism can contain no superfluous premisses; that is, they are valid arguments whose validity is lost under premiss-deletion. Third, if an argument is a syllogism then its conclusion may not repeat a premiss; hence, in one contemporary sense of the term, syllogisms are required to be non-circular.

For a correct understanding of Aristotle's logic it is essential that syllogisms not be confused with (merely) valid arguments. Aristotle's theoretical interest is concentrated on syllogisms, not validities. Even so, it is interesting to speculate on two subsidiary questions.

1. How did Aristotle understand the concept of validity?
2. Notwithstanding the preoccupation with syllogisity, does validity have any loadbearing role to play in Aristotle's logic other than its being a condition on syllogisms?

I will assume without argument that Aristotle's validity is classical in the modern sense (see [Woods, 2001]). That is, an argument is valid if and only if the set of premisses and the negation of its conclusion is inconsistent. The second question, I defer to another occasion (see [Woods, 2001]). For concreteness, we now imagine a simple case of refutation. *A*'s thesis is *T*. *Q*'s refutation is

C
B
—
$\neg T$

in which C and B are *A*'s concessions and $\ulcorner \neg T \urcorner$ is syllogistically derived from the premisses. It is a noteworthy feature of our case — a feature which generalizes to all refutations — that *A*'s thesis *T* cannot be a premiss of *Ref*. We cannot suppose that *T* is either a missing premiss of *Ref* or that the result of adding *T* to *Ref*'s premiss-set is itself a syllogism. To see that this is so, it suffices to recall that Aristotle imposes on syllogisms the requirement that they contain no superfluous premisses. Syllogisity thus is a nonmonotonic property for Aristotle.[2]

If *Ref* is, as it stands, a syllogism, then its extension, E^{Ref}

T
C

[2]A reminder: A property *P* of an argument-type is *monotonic* if and only if any argument of that type which has *P* continues to have it under arbitrary supplementation of its premisses in arbitrary but finitely many applications.

—

$\neg T$

is not a syllogism (though it is a valid argument if validity is monotonic). On the other hand, if this new argument E^{Ref} were a syllogism then its subargument, i.e., *Ref*

C

B

—

$\neg T$

would be valid. This being so, $\{C, B, T\}$ would be a consistent set, whereas, by the validity of E^{Ref}, $\{T, C, B, T\}$ is an inconsistent set. But this is a contradiction: $\{T, C, B, T\} = \{C, B, T\}$. The contradiction is pointful: No argument is a syllogism if a premiss is the contradictory of its conclusion. That is, no argument of the form

A_1

$\vdots$

—

$\neg A_i$

is a syllogism (or a refutation either).

Let us revisit our original refutation *Ref:*

C

B

—

$\neg T.$

Since *Ref* is a syllogism, $\{C, B, T\}$ is an inconsistent set. Hence at least one sentence of the three is false. Because *Ref* is a refutation *of* T, it is attractive to suppose that *Ref* establishes that it is T that is false (for what else would *Ref*s refutation of T consist in but showing that T is false?) It won't do. Saying so is fallacious. It is the fallacy of distributing negation through conjunction, said by some to be Aristotle's fallacy of Noncause as Cause [Woods and Hansen, 2001]. (It isn't, but let that pass.)

What we require is some principled reason to pick out T, *rather than* C or B, as the unique proposition refuted by *Ref.* How are we to do this?[3] There are two possible answers to consider. The first will prove attractive to people who favour a broadly dialectical conception of fallacies. The second will commend itself to those who think of fallacies as having a rather more logical character. We examine these possibilities in turn.

[3]The same problem visits the idea of counterevidence to a scientific theory construed *holistically.*

First Answer (Dialectical): The first answer proposes that question-and-answer games of the sort played by Q and A are subject to the following pair of linked rules. They are dialectical rules.

> *Premiss-Selection Rule:* In any dispute between Q and A in which Q constructs a refutation of A's thesis T, Q may use as a premiss of his refutation any *affirmation* of A provided that it is subject to the *no-retraction rule.*

> *No-Retraction Rule:* In any such dispute as above, no answer given by A to *a question of* Q may be given up by A.

It is entirely straightforward that among A's affirmations, germane to his defence of T, T itself is uniquely placed in *not* being subject to the No-Retraction rule. If it were, then once affirmed it could not be retracted. But if it couldn't be retracted, it couldn't be refuted (or, more soberly, couldn't be given up for having been refuted). This would leave refutations oddly positioned. The rules of the game being what they are, a refuted proposition would be precisely what the refutation couldn't make (or allow) A to abandon. Thus it may be said that, according to answer one, if our refutation game is a procedurally coherent enterprise, then, of all of A's relevant affirmations, T alone stands out. It is the only affirmation that A *can* retract and it thus qualifies for the proposition which Q's refutation *Ref* refutes.

It may be thought, even so, that it is unrealistically harsh to hold all of A's other affirmations to satisfaction of the No-Retraction rule. A little reflection will discourage the complaint. In the give-and-take of real life argument some latitude is given to A. He is allowed to retract some affirmations some of the time. There are limits on such sufferance. If there were not, then any thesis would be made immune to any refutation of it. If A always had the option of cancelling one of Q's premisses, rather than giving up his own T, then T would be made strictly irrefutable by any Q whose opponent were prepared to exercise the option. Thus the No-Retraction rule can be considered an idealization of this limit.

Second Answer (Logical): According to our second answer, that T is the uniquely positioned proposition that Q's *Ref* refutes can be explained with greater economy as follows. Looking again at *Ref*,

$$\begin{array}{l} C \\ B \\ \hline \neg T. \end{array}$$

we see that T is distinguished as the proposition refuted by *Ref* precisely because it satisfies a certain condition. Before stating the condition, it is necessary to introduce a further fact about syllogisms. Aristotle requires that syllogisms always have consistent premiss-sets. For suppose that they did not, and let

A
$\neg B$
—
B.

be a syllogism. Then since syllogisms obey an operation called argument-conversion, and since conversion preserves syllogisity, our imaginary syllogism converts to

A
$\neg B$
—
A.

in which the conclusion repeats a premiss. This is explicitly forbidden by Aristotle's definition:

> A *syllogism* rests on a certain statement such that they involve necessarily the assertion of something other than what has been stated, through what has been stated (*On Sophistical Refutations*, 1, $165_a 1 - 3$; [Barnes, 1984]).

Whereupon, since the second argument is not a syllogism and yet is an argument converse of the first, neither is the first a syllogism. This ends the aside.

The condition proposed by the second answer to our question is:

> *C1:* Since *Ref* is a syllogism, its premiss-set is consistent. Let *Aff* be the set of *A*'s affirmations with respect to *Ref*. Thus *Aff* is the set $\{T, C, B\}$. Since *Ref* is valid *Aff* is inconsistent. It is easy to see that *Aff* possesses exactly three maximal consistent subsets: $\{C, B\}$, $\{C, T\}$, $\{B, T\}$. We will say that a maximal consistent subset of *Aff* is *excluded* by *Ref* if and only if it does not *syllogistically* imply *Ref*'s conclusion, $\ulcorner \neg T \urcorner$. Thus $\{C, T\}$ and $\{B, T\}$ are excluded by *Ref*, and $\{T\}$ is their intersection. Thus T is *the* proposition refuted by *Ref* precisely because it is the sole member of the intersection of all maximal consistent subsets of *Aff* excluded by *Ref*.

It is worth noting that the requirement that syllogisms be consistently-premissed bears directly on our present question. T is dignified as the proposition which *Ref* refutes by virtue also of the requirement that *Ref* have consistent premisses. Since *Ref* is a syllogism then $\{C, B, T\}$ is inconsistent and $\{C, B\}$ is consistent. So too are $\{C, T\}$ and $\{B, T\}$ since if they weren't, C would entail $\ulcorner \neg T \urcorner$ and B would entail $\ulcorner \neg T \urcorner$, a happenstance precluded, each time, by *Ref*'s syllogisity. Thus three ideas cohere: (1) the idea that T is unique in *Aff* precisely because it is T that cannot be a premiss of *Ref*; (2) the idea that syllogisms must be consistently-premissed; and (3) the idea that $\{T\}$ is the intersection of all maximal consistent subsets of *Aff* excluded by *Ref*.

Just how narrowly 'logical' is this characterization? Or, more carefully, how non-dialectical is it? Suppose that we said that the idea of the proposition that a refutation refutes could be analyzed without any reference to dialectical procedure. Then any syllogism whatever would count as the refutation of the contradictory of its own conclusion. Every syllogism would be a refutation, and a great many refutations would be refutations of propositions no one has ever proposed or will. There is no great harm (in fact, there is considerable economy) in speaking this way. But it is not Aristotle's way. Taken his way, refutations can arise only in question-and-answer games of the sort that we have been considering. We may say, then, that refutations have a *dialectically minimal* characterization. That is, (a) their premisses must be answers given by A to Q, and (b) what they refute must be theses advanced by A. Beyond this dialectical minimum there is no need to go. In particular, there is no need to invoke the dialectical rules, Premiss-Selection and No-Retraction. Premiss-Selection is unneeded in as much as there is an entirely non dialectical reason for excluding T as a premiss of any refutation of T. As we have seen, no argument of the form

$$\begin{array}{l} A_1 \\ \vdots \\ A_n \\ \hline \neg A_i \end{array}$$

is a syllogism.

Nor, as we have also seen, is the No-Retraction Rule needed to enable the specification of T as the unique proposition refuted by a refutation. Thus the idea of refutations as refutations of some unique T can be specified without exceeding what we have been calling the dialectical minimum. It is in precisely this sense that our target notion is a "broadly logical" matter. And it is, in any case, well without the boundaries of the conception of logic that I briefly set out in the Prologue.

Whether we find ourselves drifting toward a dialectical explication of that which a refutation refutes or to a more narrowly syllogistic specification of it, it is fundamentally important that on neither construal is a refutation of T *definitively probative* for T. Refutations do not in the general case falsify what they refute. (However, see below.) In contrast to this, *reductio ad impossible* arguments are probative, but they are not refutations in the present sense. They lack the requisite forms.[4] It

[4]For one thing, *reductio* arguments admit of a special class of non-premisses as lines in the proof, i.e., *assumptions*. For another, the conclusion of a *reductio* is a pair $\{\Phi, \neg\Phi\}$, which though a contradic*tion*, is not the contradic*tory* of the proof's assumption.

is not that refutations don't falsify something. They always falsify the conjunction of the members of *Aff*. And it is not that refutations are indeterminate. They always refute some unique single proposition. What refutations do not manage to do in the general case is to bring what they falsify into *alignment* with what they refute. What they falsify and what they refute are different propositions. If Locke were with us today, he would chuckle. "I told you so all along", he might say: "refutations are *ad hominem* arguments and no *ad hominem* argument leads those who would have it into truth". If Henry Johnstone were with us today, he would not chuckle; he is too much the gentleman for that. But I daresay that he would notice with modest satisfaction that there exists a proof in the ancient doctrine of refutation that his negative thesis is true if his positive thesis also is. That is, if philosophical argument is inherently *ad hominem*, then no satisfactory philosophical argument can disturb his conviction that philosophy is noncognitive.

3. Falsifying Refutations

I have said that refutations do not falsify what they refute, in the general case. We must ask whether there might be exceptions to this. There are. Consider the case in which $\ulcorner \neg T \urcorner$, the conclusion of a refutation Ref^t, is a logical truth. Would not the refutation have established the falsehood of T, which is its conclusion's contradictory? The answer would appear to be that Aristotle's theory of the syllogism precludes the syllogistic derivation of logical truths. Let L be such a truth. Then L follows from the empty set of premisses, which in turn is a proper subset of the premisses of Ref^t, whatever they may be. Hence the premisses of Ref^t is not a refutation. On the other hand, where $\emptyset$ denotes the empty set, even if

$$\frac{\emptyset}{L}$$

were a syllogism, it could not be a refutation, since it contains none of A's answers as premisses.

All the same, it can't be ruled out that sometimes the premisses of a refutation, Ref^t, are (known to be) true. If so Ref^t constitutes a proof of the falsity of T, when T is the proposition that Ref^t refutes. Does it not follow from this that although refutations do not as such demonstrate the falsity of what they refute, sometimes they do?

Suppose that Ref^t is

X
Y

$$\frac{}{\neg T}$$

in which X and Y are (known to be) true. As before, $\{X, Y, T\}$ is an inconsistent set. But there is a difference. Ref^t falsifies T.

This makes for a curious dialectical symmetry between A and Q. A's obligation to answer honestly is the obligation to offer to Q premisses which he (i.e., A) believes to be true. Thus for any rule-compliant answerer against whom there is a successful refutation, Ref^t, given that A is required to believe, he must also believe that Ref^t falsifies his own thesis T. Though this is what, in all consistency, A is required to believe, it does not follow that it is true, nor need A himself believe that its truth follows from what he is required to believe.

Q, on the other hand, is differently positioned. Q has no role to play in the semantic characterization of A's answers, hence in judging the truth values of his (i.e., Q's), premisses. Of course, Q will often have his own opinions about the truth or falsity of A's replies, hence about the truth or falsity of his (A's), own premisses. Whatever such opinions are, they have no role to play in the construction of Q's refutation. There is in this a strategic point. Since the success or failure of Q's refutation of A's thesis T is independent of the truth-values of the refutation's premisses, it might be thought that the most appropriate stance for Q to take toward those premisses is the one that shows A to best advantage. In fact, there is *no* semantic stance which Q can take towards A's answers, because there is none which shows A to best advantage. Consider the options.

Option 1: Q may suppose that all of A's answers are true. But this commits Q to holding that his own refutation is maximally damaging, i.e., is a falsification of A's thesis T.

Option 2: Q may suppose that at least one of A's answers is false. This commits Q to holding that although his own refutation of T is not a falsification of T, it fails to be maximally damaging because A has given a false defence of it.

Option 1 and Option 2 constitute a dilemma for Q. If he exercises Option 1, he attributes to A the semantic virtue of acquiescing in a sound argument that utterly demolishes A's thesis. If Q exercises Option 2, he makes it possible for A to evade the heavy weather of an argument that demolishes his thesis but only at the cost of attributing to A a failed defence of it. Q's dilemma is that with respect to the truth values of A's answers there is no wholly benign assumption that one can make on A's behalf. Hence Q must make no assumptions about the truth values of his own premisses. That being so, the most that Q can suppose about his own refutation is that it refutes *T* but does not falsify it.

The asymmetry between A and Q is now sharply specifiable. For A, given what he must in any case believe, he must also believe that Q's refutation falsifies *T*. For Q, given what he must *not* believe, he must not believe that his refutation falsifies *T*. (Even so, there is nothing in the dialectical canon that precludes Q from knowing that A must believe that A must believe that Q's refutation falsifies A's *T*.)

This asymmetry between A and Q holds as a general rule, but it is interesting to note that Aristotle expressly allows for exceptions to it, concerning which a brief word may prove interesting.

4. Peirastic Refutations

As we have seen, it is a condition on transaction of question-and-answer games that A gives an answer to Q's propositions which he sincerely believes. Answers must be 'actual beliefs' (*ta dokounta*), of the answerer (*Sophistical Refutations* 2, 165^b 4-5, cr. *Gorgias* 495, A-B, *Crito* 49D and *Protagoras* 331C). Aristotle is sometimes at pains to show that a refutation of A must establish not only that A is wrong about some particular claim (*viz.* *T*), but also why he is wrong (*Sophistical Refutations* 8, 169^b, 20-29, 5, 167^b 21 ff., and 6, 168^b 22-26). So sometimes there is a further condition on premiss-eligibility. Not only must Q's premisses be A's actual beliefs, they must also be "things which are necessarily known by *the one who pretends to have expert knowledge*" (*Sophistical Refutations* 2, 165^b 5-6). It is fairly clear that Aristotle does not intend this as a general condition on refutations, but rather on a special case of these, known to Plato as *peirastic* refutations, which examine "not someone with knowledge but someone who is important and pretends to know (*Sophistical Refutations* 11, 171^b 5-6. Cf. *Apology* 23D, *Gorgias* 46A-D and *Euthyphro* 13C). *Peirastic* refutations are ably examined by R. Bolton, [i.e., to have expert knowledge]" [Bolton, 1993, pp. 121–152].[5]

When a refutation is *peirastic* the answerer's answers not only must be believed by him, but also known by him (and by his questioner), to be true. We see, then, that a *peirastic* refutation is also a falsifying refutation. But saying so leaves a hard question. If A's answers are all true and known to be so, and if A, in giving these answers, is purporting

[5]Suffice it to say that if this were a general condition on refutations, the preceding account would need to be refashioned so as to reflect its provenance. I myself think that there are theoretical advantages in treating *peirastic* refutations as a special case. For one thing, we derive a more generally applicable notion of refutation than would otherwise be possible.

to be an expert about their subject matter, how can it be the case that the refutation establishes that A's expertise is mere pretense? True, A may be wrong *about* T, but if he is right about everything else, it is not unconvincing to insist that the refutation reveals his expertise to have been mere pretense? If we do suppose so, we lose the distinction between

(α) a genuine expert being mistaken about some particular claim in his are of expertise

and

(β) an allegedly genuine expert being shown up as no expert at all, even as a kind of faker or fraud.

The key to Aristotle's intent is the "necessarily" in "things which are necessarily known by the one who pretends to have expert knowledge". Consider a case. Suppose that someone is posing as a doctor, as an expert in the field of surgery, S. For ease of exposition, imagine that S is the set of all surgical truths. What is it for someone to be a surgical poseur? Is he a poseur if for *all* propositions $\Phi \epsilon S$ he holds $\ulcorner \neg \Phi \urcorner$. No; such a person is not only innocent of what it takes to be an expert surgeon, he is a total ignoramus about surgery, so epistemically disabled as to make it doubtful that he has so much as the concept of surgery. Aristotle is pointing out that to be a fake-expert with respect to a discipline *Dis* a faker must, as a condition of his own fakery, know and proclaim a goodly number of propositions of *Dis*. These propositions are the enabling prerequisities of his fakery. A Faker is someone who gets the *details* wrong, in sufficient numbers and patterned in such ways as to lose his license, in a manner of speaking. Thus we have

The Fakery Condition: To be a fake-expert or a pretend expert in a discipline *Dis* it is necessary that there be some truths of *Dis* that the pretender actually knows.

Call these propositions that it is necessary for a faker to know N. Then Aristotle's point is that in refutations of the *peirastic* kind A's answers must be drawn exclusively from N. If A's thesis T (i.e., some particular claim about *Dis*), is to be shown false, its contradictory must derive from propositions which he and Q (and any fake-expert), must know to be true.

I have suggested we should not take *peirasis* as canonical for all refutations. Doing so would hobble an otherwise agreeably general concept of argument. It is satisfying to discover some textual confirmation of such restraint. If *peirasis* were all there were to refutations, then one way for a refutation to be sophistical would be for it to appear to be *peirastic* but not be in fact; that is, to appear to be a refutation of a would-be expert but which isn't. Aristotle recognizes such a case, but not every type of sophistical refutation that Aristotle recognizes fits it.

5. Proof *Ad Hominem*

Aristotle discusses refutations (ελεγξσσ) at some length in the *Topics* and *Sophistical Refutations.* In several passages he seems to think that refutations are proofs, but in a looser sense of "proof" then *reductio* arguments (*Sophistical Refutations* 167^b 8-9ff. Cf. *Prior Analytics* B27, 70^a 6-7 and *Rhetoric* TI3, $I414^a$ 31-37). In other places, refutations appear to be proofs in no sense of the word (see *On Interpretations*, 11, 21^a5 ff. Cf. Δ5, 1015^b8, and *Posterior Analytics* A9, 76^a 13-15). For

> I mean, 'proving by way of refutation' to differ from 'proving' in that, in proving, one might seem to beg the question, but where someone else is responsible for this, there will be a refutation, not a proof (*Metaphysics*, 1006^a 15-18).

This is the Ross translation. Barnes has it this way:

> Now negative demonstration I distinguish from demonstration proper, because in a demonstration one might be thought to be assuming what is at issue, but if another person is responsible for the assumption we shall have negative-proof, not demonstration [Ross, 1984]. Cf. *Sophistical Refutations* 6, 167^b 21-27; 1, 165^a 2-3; 15, 174^b 19-23; and *Rhetoric* B22, $I396^b$ 22-27, 23, 1400^b 26-29, T17, 1418^b 1-4; and *Prior Analytics* B20, 66^b 11).

Thus

> In such matters there is no proof simply, but against a particular person, there is (*Metaphysics*, K5, 1062^a 2-3. Cf. ^{a}30-31 and T4, 1006^a 25-26).

This is Ross. In Barnes' version we have it that

> About such matter there is no proof in the full sense, though there is proof *ad hominem.*
> (περι τών τοιούτων άπλώζ μέν ούκ έστιν άποδειξιζ, πρόζ τόνδε δέ έστιν.)
> (Cf. *Sophistical Refutations*, 22, 178^b 17, 89, 170^a 13, 17, 17-18, 20, 177^b 33-34; 33, 183^a 22, 24; and *Topics*, Φ11, 161^a 21.)

It is hardly imaginable that there should be any contention about the origins of the phrase '*ad hominem*' to characterize a particular class of arguments. We owe the concept not to Locke (as Locke himself expressly said), not to Galileo, not to Boethius, but to Aristotle himself. It is clear that Aristotle is of two minds about the *ad hominem.* He is tempted to think of *ad hominem* arguments both as proofs of no kind and as proofs of some kind. Arguments 'against the man' Aristotle expressly contrasts with arguments against a man's position, and the former are considered substandard in *some* way, as witness *Sophistical Refutations* 20, 177^b 31-34, and 22, 178^b 16-23.

At times (see e.g., *Sophistical Refutations*, 8, 170^a, 12-19), Aristotle tries to draw a distinction within the class of refutations between those that turn on *ad hominem* moves and those that don't. The former he condemns outright as sophistical refutations (cf. *Sophistical Refutations*, 1).

Thus we have two contrasts to keep track of, and whose confusion is ruinous for a correct understanding of Aristotle's position. There is the contrast (*Metaphysics* K5), first, between a proof 'simply' and a refutation or argument *ad hominem*. In relation to this contrast, the following things can be said:

1 *ad hominem* arguments are not sophistical or fallacious

2 *ad hominem* arguments are refutations, hence not proofs 'simply'.

What is it to fail to be a proof simply? There are two possibilities, and Aristotle anticipates them both. One way of not being a proof simply is being what we have called a *non-falsifying* refutation. Non-falsifying refutations are in no sense proofs against the propositions they refute. Another way of not being a proof simply is being what we have called a *falsifying* refutation. Falsifying refutations are proofs in some sense, but they are not proofs in every sense. For example, they are not demonstrations in which there is a strict epistemic priority rating on the premisses of the refuting syllogism.

That is the story of the first contrast. The second contrast is another matter. It is a contrast between *ad hominem* in the first sense and those cases of them that are *ad hominems* in the a *second* sense. In the first sense, an argument is *ad hominem* just in case it qualifies as a refutation. An argument is *ad hominem* in the second sense just in case it purports to be, and appears to be, an *ad hominem* in the first sense, but is not in fact. Thus arguments that are *ad hominem* in the second sense are defective would-be *ad hominems* in the first sense and, as such, can reasonably be regarded as fallacious. Aristotle does not list the *ad hominem* in his catalogue of thirteen fallacies, as set out in the *Sophistical Refutations*; at least he does not give any of the thirteen the name '*ad hominem*'. Even so, the account that he does give, such as it is, leaves plenty of room to accommodate *ad hominems* in the second sense in the category of *ignoratio elenchi*. Also possible in principle is accommodation of *ad hominem* fallacies known to a much later tradition as 'circumstantial', 'abusive' and '*tu quoque*'.

This does not change the fact that, for Aristotle, the dominant notion of *ad hominem* argument is Lockean (if the anachronism may be forgiven). It is a concept of *ad hominem* that is nicely captured by the structure of refutation. With refutations as such there is no question of fallaciousness. The problem rather is how closely refutations resemble proofs. Aristotle has two answers: refutations resemble proofs not at all; and refutations resemble proofs loosely. As we see, this is a distinction nicely preserved by the distinction between non-falsifying and falsifying refutations.

Chapter 8

AND SO INDEED ARE PERFECT CHEAT

> [A]ll the artificial and figurative application of Words Eloquence hath invented are for nothing else but to insinuate wrong Ideas, move the Passions, and thereby mislead the Judgment; and so indeed are perfect cheat; And therefore, however laudable or allowable Oratory may render them in Harangues and popular Addresses, they are certainly, in all Discourses that pretend to inform or instruct, wholly to be avoided...
>
> John Locke, *Essay on Human Understanding*, Book III.

In this present chapter, we shall see how the concept of arguments that make reference to personal factors extends to a class of arguments more usually associated with the notions of *ad populum*. In their most elementary and *ad verecundiam* sense. These are arguments that rely on a structure of relations having to do with how other people behave. Part of this story involves an examination of bias.

The received view is that there is no fit place for bias in our cognitive and argumental lives. The complaint against bias is that it is an emotional factor, and that emotions have a natural tendency to cloud judgement and destabilize arguments. There isn't the slightest doubt that some emotions sometimes have such effects. But they do not have them *intrinsically*, as I shall attempt to show in this chapter.

1. "Fallacies" that aren't Fallacies

Among the gang of eighteen there are fallacies which appear to involve the factor of *bias* or *prejudice*. The emotive fallacies, for example, turn on matters of human feeling which are thought to intrude bias into what otherwise would be impartial reflection and deliberation. In some forms, *ad populum* arguments are direct appeals to bias, and there are instances of hasty generalization which pivot on the factor of pre-conceived opinion or prejudice.

My task in this chapter is to achieve some understanding of these notions of bias and prejudice. We saw in the previous chapter something of the human reasoner's disposition to act in ways that disconform to theories he holds true, and we said that such behaviour could not always be condemned as unreasonable or unjustified. In chapters 4 and 5 we saw that there are cases galore in which prudence trumps the alethic virtues, and that it is far from clear that the trump is unreasonable or unjustified. A further task of the present chapter is to show that neither bias nor prejudice is an intrinsically unjustified position for a human reasoner to be in. This is very much a matter to be considered in a book on fallacies, especially a book in which it is repeatedly asserted that reasoning that instantiates one or other of the gang of eighteen is *not* intrinsically fallacious. We meet with fallacy theory in its most difficult challenge when we attempt to show what the factors of mitigation are, when something that has the *name* of a fallacy isn't actually a fallacy. So it is with our present targets of bias and prejudice.

2. Bias

English displays a lexicographical ambivalence toward bias. "bias", the noun, denotes any preconceived opinion, and preconceived opinions, inevitable in any case, that are innocent until convicted. The verb tells a different story. To bias is to impair the validity of a claim or statement, and although some claims deserve to have their validity impaired (namely the invalid ones), a certain caution is called for. In this use, "to bias a claim" indicates a dubious strategy, suggesting that the impairment of validity by the means at hand may not be appropriate even if the claim isn't valid. This caution is nicely caught by the allied technical notion of *biased statistics*. A claim may be challenged on statistical grounds, the claim (say) that universities have practised persistent and systematic discrimination against women. If the challenge is underwritten by biased statistics, there is plenty of room for complaint. But the complaint cannot succeed if it is merely that the original claim is impaired. If the original claim is untrue, it deserves to be impaired. Rather the objection is, or should be, that it ought not to have been impaired in that way — by those statistics.

The cautionariness of the idioms of bias is also reflected in the kindred expressions of "prejudice". Parties to a disagreement will sometimes forward concessions "without prejudice" to the main issue. Such concessions can be said to carry conversational implicatures (in the sense of

[Grice, 1967]), which, if left standing and without appropriate justification, bias the main issue, that is, count against it in ways at odds with the dialogical purposes of the person whose conversational implicature it is. The force of the intervention "without prejudice" is to cancel the conversational implicature, and so to remove the bias and spare the main point the damage that would otherwise be done it.

The English lexicon gives us desultory guidance at best. It cannot be said that it leaves no distinction between impairing a claim's validity and illicitly impairing it. So something further wants saying about biased argument. We might tentatively propose the following as a core notion.

Definition One: P is a *biased belief* for S if and only if

(i) S believes P, and

(ii) not-P is not a real possibility for S.

Definition One pivots on the notion of something's not being a real possibility for a cognitive agent. This notion is also central to the analysis of a large subclass of negative consequence arguments — that is, arguments turning on a threat or reference to dire consequences — which are known as *risk-aversion* arguments, of which the historically most celebrated is surely Pascal's Wager (See, for example, chapter 5.) For present purposes we could propose in a slack and tentative way:

Definition 2: P is *not a real possibility* for S if and only if

(i) S does not believe P, and

(ii) S is not prepared either to hear the case for P or to make the case against P.[1]

By the lights of definition 2, claims that are not real possibilities for S are those he has "no use for", to speak loosely; they are those that don't engage him.

It is well to note that definition 1 obscures certain intuitive distinctions. In particular, it counts as biases alike both an Albanian's belief that Albanians are the smartest people in the world, and everyone's belief that a bullet in the head is disarranging. But surely, it will be said, our common sense notion of bias is best reserved for beliefs whose contraries *should* engage us but don't. I want to resist the suggestion, at least for now.

On this view, a bias is a person's belief whose contrary he or she has not and is not prepared to dismiss by argument. This still makes for

[1]Thus, for Pascal's "skeptic", Christianity is a real possibility, never mind that he disbelieves it, whereas Zoroastrianism (say) is not a real possibility.

a rather inflexible notion of bias, since it makes no provisions for temporal or contextual factors which make of real possibilities contextually sensitive phenomena.

A more realistic interpretation is got by relativizing the bias of definition 2. Thus

Definition 3: P is a *biased belief for S at time t in context C* if and only if

(i) *S* believes *P* at *t* in *C*, and

(ii) not-*P* is not a real possibility for *S* at *t* in *C*.

Thus relaxed, the notion of bias collects a large range of actual cases, including the usual ones. Of special note is the fact that definition 3, while it allows that some propositions might be biases for everybody and at all times and in all circumstances (e.g., "There is something rather than nothing", or "A bullet in the head is disarranging", etc.),[2] also provides that for all but the most pathological of across-the-board dogmatists, a bias will turn out to be a function of what is believed, by whom, when and under which circumstances. More particularly still, something will qualify as a bias or not depending on the context of argument in which the proposition is forwarded.

At this juncture it will be convenient to have at hand a principle of reasoning which Harman, dubs the *Clutter Avoidance Principle* [Harman, 1986, p. 12]. Here is how it works. Anything that a cognitive agent might believe at a given time possess a great many implications, in fact, an infinity of them. If we now call into play the assumption rules that a cognitive agent *S* is committed to all the logical consequences of what he believes, we can see that, on account of just his belief that *P*, *S* is committed to the infinity of *P*'s implications. Several things may strike us as obvious. One is that no human being is capable of representing anything like an infinite number of consequences and is certainly incapable of cognitively processing such a multitude. At any given time, a rational agent must content himself with at most a small proper subset of his own commitments. Any serious attempt to deviate from this capacity-limit clutters the agent's mind and tends to paralyze thought and action. So it would seem to be a rule of rationality that clutter should be avoided. And relevance, in one good sense of the term, will be that in virtue of which clutter is separated from non-clutter. The dual of the Cluster Principle is what we will call the *Principle of Informedness.* It bids the reasoner not to have too *little* in his mind for the purpose at hand.

[2]For a contrary view see Fain and Griffiths: "There is no proposition so compactly true that no one could miss it" [Fain and Griffiths, 1972, p. 10].

Jointly the Clutter Avoidance Principle and the Principle of Informedness require the rational agent to limit his consideration of commitments to those that suffice for the successful prosecution of the cognitive or strategic tasks that he has underway. A merit of the joint deployment is that it makes it plain that the basic idea can be violated in two different ways — by taking account of too much for the task at hand, or too little. It is a cost-benefit principle to the effect that an agent is rational to the degree that he is efficient, and it enjoins him to make the least cognitive or strategic effort compatible with the successful performance of the task at hand. There are no known algorithms for this. The prescription, "Confine yourself to relevant commitments", is a good prescription *pro forma*, but it gives no instruction for the effective recognition of relevant commitments. Still, there can be no doubt that rational agents are capable of such recognition and that rational complainers are able to recognize the absence of relevance when they see it, so to speak.

Further features of bias now come into play. We might propose as a rule

> **Bias Rule 1:** The holding of a bias, *P*, at time t is inadmissible for *S* if there is at least one *Q*, incompatible with *P*, such that *S*'s refusal to give probative consideration to *Q* is a violation of either the Clutter Avoidance Principle or the Principle of Informedness.

Rule 1 comprehends two different but related situations. Both can be characterized dialogically, although in one case the attribution is somewhat metaphorical.

> **Situation 1:** *S*'s holding the bias *P* under the challenge of Nature.
> **Situation 2:** *S*'s holding the bias *P* under the challenge of a dialogical adversary.

Situation 1 covers a range of cases in which a *P*-believer's refusal to consider or even acknowledge a contrary position is rendered unreasonable by how the world is and by what *S* knows or should know of this. (Thus "under the challenge of Nature".) Situation 2 covers a range of cases in which a P-believer's refusal to consider a contrary position reflected in an opponent's challenge is unreasonable because of its relevance to *P*, where recognition of the contrary position is guaranteed by the challenger's challenge when it is intelligible to the person against whom it is pressed. In each case, to persist with *P* is to persist with a bias, and in each case it is a Clutter Avoidance fallacy (as we must say), whether against a dialogical adversary, as in Situation 2, or against Nature herself, as in Situation 1.

Even so, given that bias is sometimes the occasion of cognitive and dialogical mismanagement, would it not be the shrewder strategy to rule against it totally and as such? (One thinks of the Christian admonition against putting oneself in the occasion of sin.) The injunction not to have or hold biased beliefs is the injunction to have or to hold no belief without considering, or being disposed to consider and refute, any contrary of it. But, as lately suggested, any such disposition would be a disposition to violate the Clutter Avoidance Principle. Of course, some people can handle more clutter than others, such are their talents and their circumstances. So clutter avoidance will weigh differentially across the grain of such differences. Still, there can be no question of our being set up for limitless clutter, whence a

> **Limitation Theorem:** bias is wholly avoidable only at the cost of ransacking the Clutter Avoidance Principle.

Perhaps it will be said that here, too, the Limitation Theorem is true, as far as it goes, but that it doesn't go far enough to be genuinely interesting. The Limitation Theorem expresses hardly more than the platitude that not every possibility can or should be considered. A no less commonplace intuition cuts in the other direction. It counsels against the having and holding of biases for which there is contrary evidence. But even where evidence is construed with suitable flexibility and vagueness — as something like reasons that have been or could be adduced — it is clear that the generalized injunction to avoid bias in the face of contrary evidence could not possibly be complied with by human beings, no matter how lavishly endowed. Any cognitive human agent seeking to comply would be faced with the task of considering reasons that have or could been adduced in support of any contrary belief he holds. But doing that would involve him in considering every contrary possibility with respect to everything he believes, which is already ruled out by the Clutter Avoidance Principle. In fact, that theorem can now be reissued as a

> **Revised Limitation Theorem:** The injunction to have and hold no belief except on consideration of evidence for any contrary of it violates the Clutter Avoidance Principle.

People might be unconvinced. Not all possible evidence for every possible contrary can or need be consulted, they will grant. But it is essential, if a belief is to resist the invasion of bias, that any contrary of it be considered for which the agent already possesses evidence which he considers to have some minimal probity. Sound advice, to be sure, but it has already been adequately provided for.[3]. Even granting that

[3]Perhaps this is too fast. For let T be any empirically adequate theory and E the total evidence adduced for T at time t. There are philosophers who hold — for example, Quine and van Fraassen — that E at t makes unboundedly many theories T_i, alternatives to and incompatible with T, are equally adequate empirically [Quine, 1975; van Fraassen, 1980]

some of the T_i are recursively specifiable from T, not all are, and a hefty multiplicity of them stand alone — contraries of T and just as good empirically. It is standard scientific practice to favour T or one of the T_i. However this is done, and whatever the justification of how it is done, there can be no question of the exhaustive consideration of each of the T_i. It comes to no less and no more than our earlier stricture against retaining with a bias under relevant challenge, whether of Nature or a dialogical adversary. This leaves us with two options, both merely semantic (and here we return to the point that definition 1 is counterintuitive). One is to reserve the name of bias for beliefs persisted with in violation of the present injunction. We can choose the option, if we like, but it is well to note that it makes ample provision for the rationality of holding and retaining unexamined beliefs. The second option is to reinstate the equivalence of "bias" and "unexamined belief" and to reserve regret for those cases of bias which fail the present rule. In the end, it matters little which we say, provided that we are clear in the saying of it.

3. Popular Appeals and Common Knowledge

A speaker addressing a crowd says at one point: "It is common knowledge that bad money drives out good". And having said so, he makes this without further ado a premiss in his evolving argument. In one of its modern meanings, this is an appeal *ad populum.* It pivots on what people know. It is offered on its own authority and without the supplementation of specific evidence. There is no need. How could there be, given that everyone knows it? Of course, the speaker may be wrong in his attribution of common knowledge and this we *could* call a fallacy. But the fallacy, here, is a "common misconception", the mistake of placing a false statement in the embrace of what is commonly known.

More interesting is an analysis of the conditions that obtain when the speaker's claim from common knowledge is not mistaken. Of more particular interest still, is an understanding of the following basic dynamic: When a speaker cites a proposition P on the basis of the common knowledge of it, he seeks for an authoritative positioning of P short of considering it himself. In fact, consider it he may not be able to do. How then is he entitled to P? The basis of a claim from common knowledge is a speaker, S's, membership in a community of knowers — an epistemic community. We can specify a proposition P's membership in a body of common knowledge CK for an epistemic community ec as follows:

Definition Four: $P \epsilon CK_{ec}$ if and only if for most $S \epsilon ec$,[4]

(i) S believes P,

(ii) S believes that P enjoys epistemic warrant,

(iii) S believes that he has access in principle to information implying that P has epistemic warrant,

(iv) the situation cited in (iii) is probative for S, and

(v) S has not accessed the information cited in (iii) and need not know how to do so.

That said, a further definition drops out.

Definition Five: S correctly claims that P is common knowledge for him and for an interlocuter S^* if and only if there is an *ec* such that $P \epsilon CK_{ec}$ and $S \epsilon ec$.

The definitions jointly provide that in the general case what is common knowledge is relative to a given epistemic community. What is common knowledge for members of the American Economics Association may not be for customers of Safeway in 1995. But, even so, at the limit, some propositions (for example, that there is something rather than nothing) will be common knowledge in every epistemic community.

It is notable that, by the evolving account, common knowledge is withheld from, say, mathematical propositions whose proofs are known to all mathematicians, or to members of the collective known as Bourbaki.[5] What is needed is a distinction between *tacit* common knowledge, which is what the definitions define, and *express* common knowledge, which is what the present example exemplifies. My interest here is in tacit common knowledge. Accordingly we need to correct our definition:

Definition Six: P is *tacit* common knowledge for S iff the conditions of definition 4 are fulfilled.

An aside: Hard-nosed epistemologists may grudge the name of knowledge for common knowledge, and skeptics will not give it so much as the time of day. I will not answer these complaints here, though perhaps it can be mentioned in passing that the hard-nosed objector is sometimes committing the utter folly of subscription to the KK-hypothesis, that is, to the hypothesis that for any P and S, S does not know that P unless he knows that he knows that P. Short of a considerable detour into

[4]But see footnote 10.

[5]There are other such cases. Let P="There are physical objects" or "There is something rather than nothing". Information implying P might include perceptual experiences, and to these S certainly does have access. Yet we would still want to say, surely, that P is part of S's background knowledge — that it is common knowledge for S. Definition 5 won't extend to P the status of common knowledge, short of fulfillment of condition (v). Perhaps, with some ingenuity, we could convince ourselves that perceptual background knowledge could in general be had independently of access to perceptual information, but I for one lack the ingenuity for such a case. Better that we revise our definition.

separate and difficult issues, it suffices to give our complainants modest sympathy. Let us concede, therefore, that the vulgar idioms of common knowledge are a trifle overblown. Perhaps the disclosures of common knowledge do not always, or routinely, measure up as knowledge. It is enough that they often measure up to warranted belief. This is enough for our purposes here. We could adjust ourselves to favour the more cautious idioms of common warranted belief, but the vernacular of common knowledge is entrenched, and we will stay with it as a loose equivalent of the other. In particular, the condition of epistemic warrantedness is re-invoked. *P* possesses an epistemic warrant for *S* iff apart from *P*'s being believed by *S*, *P* and *S* satisfy conditions for *S*'s knowing that *P*. Of course, it is a matter of lively contention as to what those conditions are. If we suppose that *P* has epistemic warrant for *S* when *S* has (in a certain way) evidence for *P*, then it might turn out that there is less to common knowledge than meets the eye (this would depend on how we were to understand "in a certain way"). If we thought of epistemic warrants as information about the reliability of certain processes and mechanisms, then it would seem that there is a good deal more to common knowledge than meets the eye. (I remind the reader that I am here speaking of tacit common knowledge). Suffice it to say that there is much that we say that we know that we don't, and much too that we are justified in saying that we know that we don't. It is the same way (only, perhaps, more so) with common knowledge. Depending on our respective epistemological perches, there will be less of it (or more) than we might have thought. For present purposes the idea of common knowledge does not require further refinement of a sort that would calm these issues down; and so we will let it be.

So far, the tacitness of tacit common knowledge has been a matter of the tacit knower not having achieved access to epistemic warrants for *P*, perhaps not even knowing how to do so. There is reason to think that this doesn't go far enough. If, for example, we have independent reason to think that a given proposition could itself be known tacitly or, less venturesomely, believed tacitly, then knowledge could be made tacit in two ways, one a matter of the inaccessibility of the epistemic warrant for something believed expressly, and the other a matter of the tacitness of the belief in the proposition known. In the first case we may say that

> Tacitness is the holding of belief without its pedigree, that is, unattended by epistemic warrants for it.

In the second case we may say,

> The knowledge that P is tacit when the belief that P is held in ways that *falsify* what we might call the *AB-hypothesis* for P. That is, if S believes that P, yet is not aware that he believes that P then, if S knows that P, S's knowledge of P is tacit.

Case two tacitness reminds us that it is very much worth noting how venturesome we sometimes are in attributing tacit knowledge (or belief). Consider the case of Cauchy who, in the *Cours d'Analyse d'Ecole Royale Polytechnique*, presupposes a real number system which includes the irrationals [Cauchy, 1821]. Now Cauchy would have had nothing to do with any theorems expressly about the irrationals, since the irrationals were not satisfactorily defined until 1840 by Weierstrass. Without the assumption of the irrationals, Cauchy's proofs of 1821 are no good. But there is no (classical) mathematician alive who would say that Cauchy's proofs were anything but good. So knowledge of the irrationals is attributed to Cauchy, never mind his own disavowals (see [Lakatos, 1976, p. 45, n. 3]).

How, then, is common knowledge got, and in virtue of what is it the real article, if it is? It is a long and interesting story but, in brief, common knowledge hinges on two interacting mechanisms:

(i) the mechanisms of epistemic trust

(ii) the mechanisms of deep memory.

Epistemic trust is a condition of there being epistemic communities at all. By its minimal provisions most members of epistemic communities must be truthful in their utterances. Not always, of course, but quite often. How often is a hard question. Its difficulty is evidenced by the attraction it has for vacuous answers. One is: truth is uttered with sufficient frequency when it underwrites epistemic trust sufficiently to make epistemic communities possible. Whatever the details of this complex dynamic, the mechanisms of epistemic trust help explain both the commonness of common knowledge and its reliability. They are not, of course, anything like the complete explanation of these things.[6]

The mechanisms of deep memory serve a different explanatory function. common knowledge — much of it anyhow — shares a feature with uncommon knowledge, the knowledge, say, that I alone could contribute to my autobiography. The feature that they have in common is that, often enough, they will be repositories of claims P which, when challenged, are attended by the wholly unadorned assurance, "I just know it" (to which is often added, "I don't know how I know it, I just do").

[6]Epistemic trust and the indispensability of truthfulness are discussed in [Hardwig, 1985; Blais, 1987; Woods, 1989a], the latter reprinted here as chapter 20.

This is knowledge at its most tacit. It is the output of deep memory, shorne of its explanatory and evidential provenance. What deep memory remembers, so far as common knowledge is concerned, are beliefs got from experience, beliefs acquired on the sayso of others, beliefs acquired by inferences therefrom and perhaps, too, beliefs sprung on us by innate mechanisms.

Human agents store this information and retain much of it without attending to it. Attending to it all, even if we could, would paralyze thought and action. If that were all there were to it, deep memory would be as useless as the appendix. It is spared this fate by the mechanisms of retrieval. Deep memories can be retrieved when needed. When this happens they are retrieved not *en gross*, but with great selectivity and considerable incompleteness. They are, under retrieval, memories whose pedigrees have often been eroded. The busy executive may need the Pythagorean Theorem and, if so, it will probably be vouchsafed to him from his deep memory of middle school Euclid. But it is much less likely that he will be able to retrieve its proof, assuming that he once had it. His memory of the Pythagorean Theorem is without pedigree, its epistemic provenance is shorne. He knows that the square of the hypotenuse of a right-angled triangle is equal to the sum of the squares of the their two sides; but he does not know how he knows this.

Though tacitness is not distinctively a feature of common knowledge as opposed to private knowledge, it is guaranteed to epistemic agents who possess rich and efficient memories. Tacitness is important and interesting in its own right. The importance, and the interest, are redoubled on re-summoning the concept of bias. bias, we said, is a matter of beliefs had and held without examination. The Limitation Theorem proposed that no rational agent, or none like us, could avoid having biases in just this sense. The theorem is now given striking additional confirmation. Tacit knowledge is knowledge got by retrieval from deep memory of recollections whose pedigrees have been erased. Possession by S, tacitly, of the knowledge that P entails that S knows that P but not how he knows it. As with the case of the businessman's recall of the Pythagorean Theorem, his not knowing how he knows it could encompass his not knowing in what the critical examination of it would consist. He "wouldn't know where to start". Tacit knowledge is, then, unavoidably biased. And to the extent that common knowledge comprehends tacit knowledge, common knowledge is biased knowledge and inescapably so.

If in making an *ad populum* remark a human reasoner is appealing to common knowledge, he may, with high probability, be appealing to bias. So the latencies of disapproval that the tradition reserves for appeals to popularity may themselves tacitly turn on the conviction that bias is inherently irrational and unreliable. As we have seen, nothing could be further from the truth.

At the intersection of common knowledge and bias something else of interest crops up in a rather natural way. It is *generic* knowledge. Generic propositions are propositions as general as they can be short of universal quantification. Generic statements will stand to general statements as "Tigers are four-legged" stands to "All tigers are four-legged". Generic statements often record what "everybody knows" about tigers, and what everybody knows about tigers, including that they are four-legged, is not something which in the general case needs the reinforcement of freshly reviewed evidence and the comforts of contrary possibilities considered and rejected.

Knowledge of generic truths resembles what Minsky has called a frame-system, briefly discussed in the previous chapter [Minsky, 1977].[7]

The notion of a frame-system is a partial elucidation of the concept of a prototype. A prototype can be thought of as a frame system in which all default values are specified. This is worth pausing over, since it is clear that bias and common knowledge alike involve, in a rather substantial way, the manipulation of prototypes. A rule suggests itself, though it too is slack and tentative.

> **ProtRule:** Try to organize your beliefs around prototypes.

ProtRule bids us to exploit frame-systems in the management of our cognitive affairs; it counsels us to be default reasoners whenever possible.[8] Default reasoning has much to recommend it for its remarkable efficiencies, but risks, too, trail along. The risk of organizing one's commitments around prototypes is that doing so may fly in the face of contrary indications, that is, indications that weigh against the assignment of default values. In ignoring or being culpably inattentive to contrary indications, one's default reasoning goes awry. We could say that *protoypical* reasoning goes over into *stereotypical* reasoning, in one established and disapproving sense of the word 'stereotype'. One makes a stereotypical attribution to a thing or event x of type K when K is a prototype whose default values are attributed to x in the face of contrary indications. K itself transforms from a prototype into a stereotype to the extent that a default value is contra-indicated for the K-things.

[7] Frames where foreshadowed some forty years earlier in Barlett's "scripts" [Bartlett, 1932]. See also [Schank and Abelson, 1977].

[8] Of course, what it *bids* is what beings like us already do, more or less automatically.

A prototype, then, would be a stereotype just in case for some default value Φ, the generic sentence "K-things are Φ", antecedently accepted as true, is in fact false and falsifying.[9]

4. Preconception and Dogmatism

For lots of cases (though not all) a belief of S is a bias iff it is a generic belief underwritten by a prototype or is the result of default reasoning therefrom. Should the prototype chance to be a stereotype, or the default reasoning from it chance to be stereotypical, S's bias can be said to be untenable. Similarly, if it is common knowledge for S that P, and P is a generic statement underwritten by a stereotype, or is the result of default reasoning therefrom, S's endorsement of P as common knowledge is likewise untenable.

All of this underscores what was previously said. There is nothing wrong with bias as such and there is nothing wrong with claims to common knowledge as such. Saying it this way is an understatement. For rich ranges of cases, the having of a bias, and the making of a biased inference, or the holding of something as common knowledge, and the making of an inference from common knowledge, is nothing more or less than conformity to the canons of default reasoning and, thus, compliance with **ProtRule**.

This makes of people's standing suspicion of bias and of appeals to common knowledge something to take particular note of. It suggests a significant failure of the burden of proof (as the saying has it). An accusation of bias is as such as groundless as would be an accusation of default reasoning. To accuse someone of default reasoning is in the general case to accuse him of what he should have done and what, in any case, he would have had difficulty trying not to do. Of course, the accusation can be made to stick in particular cases, as when the reasoning is stereotypical in the sense lately explored. It never suffices for a charge of stereotype that a prototype has been acceded to. The charge is groundless in the absence of evidence that the prototype has lapsed into a stereotype. And to show that, the burden of proof falls to the complainant. People who protest about putting 'labels' on things or placing things into 'pigeon holes' often appear to think that these practices are indefensible *per se*. This is the last thing that they are. Labels and pigeon holes are the standard currency of default reasoning and generic knowledge. They are to be regretted only in particular

[9] "Although [frame-systems] are valuable for chunking information together, they exact a toll in flexibility... [They work well] for highly regular and routine situations" [Holland et al., 1986, pp. 12–13].

circumstances, and the burden of proof for such cases more often than not is the complainant's.

The mechanisms of default reasoning cast a long shadow. As we saw in the previous chapter, some theorists are of the view that in interpreting the utterances of an alien community, it is best that the field linguist constrain his interpretations in such a way that maximize the yield of alien assertions that come out true. Quine and Davidson, among many others, dub this strategy the *Principle of Charity* (see [Quine, 1960, p. 59] and [Davidson, 1984, p. 196]).[10] It is a singularly unfortunate name. It suggests a favour volunteered to the alien community — a kindly gesture to strangers. This is unrealistic. If our anthropologist has identified his subjects as members of *homo sapiens* capable of speech, the Charity Principle bids him to do what he will do anyway, and it has nothing to do with charity. What he will do is reason prototypically about his subjects and their community. He will place them in the catchment of generic statements such as "Humans are truth-tellers" and "Human societies are epistemic communities" from which, in the absence of indications to the contrary, the inference that their assertions should not be interpreted so as to make them massively untrue, possess the virtue, not of charity, but of inductive competence. No more and no less, it is the anthropologist's fidelity to frame-systemic reasoning.

Still, the structure of common knowledge provides the critic of him who appeals to it with ample occasion for admonition. He can press his objections under three main rubrics, each of which invites the charge of fallacy. These are *ad verecundiam*, *conservatism* and *inconsistency*.

4.1 *Ad Verecundiam*

common knowledge is in part the output of communal epistemic trust. Epistemic trust is sometimes misplaced. It is misplaced if communal assurances fail to be authoritative, if they hinge on acceptance, underguided by collective memory of faded pedigree, of expert findings of debased or lapsed expertise. Communal confidence for a lay population in Newtonian platitudes about space are a case in point. So sometimes an *ad populum* appeal will be fallacious just by virtue of its commission of an *ad verecundiam* fallacy (see, for example, [Woods et al., 2000, ch. 11]. Whereupon a rule:

> **VerRule**: Try to make your popular appeals proof against the *ad verecundiam fallacy.*

---

[10] Attribution to Cauchy of tacit belief in the irrationals even when he would have expressly disavowed them presents Charity with an interesting datum. I leave it as something to think about.

4.2 Conservatism

Epistemic trust is *trust*, after all, and trust carries its inherent risks. That the risks aren't entirely foolhardy is secured by the fact that common knowledge is always common knowledge tentatively and for now. common knowledge constantly faces — much of it anyhow — what Quine calls the tribunal of recalcitrant experience, in the wake of which it is adjusted to a better fit. Some elements of common knowledge confront no such test, however, such is their distance from the periphery of sense. *In extremis*, some aspects of communal agreement are underwritten by no less than an inherited and unchallenged consensus that they are so (and, in so saying, bias recurs in our story).

In any event, it is always true that some of what we think we know we don't. True, and unavailing. It is the best that we can do. We transact our lives as best we can, on the basis of what we think we know and in the absence of guarantees. This comes to no more than the recognition that epistemic trust, like all trust, is risky. But it is differentially risky, and the competent reasoner will try to manage the differentials as realistically and efficiently as he can.

Our critic will grant this and still press an objection. It is a complaint about the commonness of common knowledge. common knowledge is, he will say, inherently conservative. Conservatism cuts in two directions, one a good thing and the other not. The good thing is that conservatism holds a contending knowledge claim to stiff tests of admissibility. The less good thing is that the very conservatism of those tests refuse admittance to new truths. So it is that, in any epistemic community of any complexity and interest, there are minority views that are denied standing and these views are true.

By Gresham's Law, bad money drives out good. There is an epistemic converse of this law that sometimes holds: unbiased belief drives out biased belief, thus the progressivity of science. A minority-belief can be defined as a belief that contradicts common knowledge. Minority-beliefs are pointful when they are effectively marshalled, that is, when they are attended by argument. Argument is effective, here as elsewhere, when it draws off supporting inferences from uncontested consensus. Sometimes that consensus will itself repose in tacit knowledge.[11]

Related considerations press for attention. If I am not mistaken, they show that, wrong though it has been in lots of ways, the tradition has

[11] Of related interest is Irwin Bross' approach to hypothesis-testing as a means of signalling new findings to the research community in ways that minimize errors in the signalling process [Bross, 1971].

been right in distrusting appeals to common knowledge. It has been right, that is to say, at least for an interesting range of cases which, it would seem, it never bothered to identify. These are cases in which, for certain kinds of phenomena K, knowledge of K-things begins as expert knowledge; it is the fruit of esoteric and specialized enquiry. As such, K-claims are unacknowledged in most epistemic communities and may even be implicitly contradicted there. Sometimes K-knowledge is subject to two significant dynamic functions. One, which I shall call '*pro*', takes account of the progressiveness of science (when it is progressive, that is to say), that is, its propensity for achieving rising levels of truthlikeness over time (see [Niiniluoto, 1984; Kuipers, 1987; Oddie, 1986; Zwart, 1998]). The other I shall call '*pop*'. '*Pop*' regulates the popularization of science, the passage of information from expert communities to non-expert communities which proclaim it common knowledge. Values of the *pro*-function are placed in expert epistemic communities which are temporal successors of those that initiated the arguments for *pro*. Values of *pop* are placed in non-expert communities under conditions, whatever they are is fine, that make of them there common knowledge.

Taken as dynamic functions, *pro* and *pop* can operate at different speeds. If *pop*-rates of change are slower that *pro*-rates of change, then there is a good chance that the popular conception of K-things will be wrong well beyond the usual margins of error. Where there is quite a lot of progressive science and quite a lot of popularization of it, and quite a lot, too, of *pro-pop* temporal asymmetry, an abundance of common misconception seems virtually guaranteed. A rule suggests itself:

> **ProPopRule:** The more *pro*-rates outpace *pop*-rates, the likelier that attributions of common knowledge will be wrong.

ProPopRule: teaches a useful moral. It reminds us that appeals to common knowledge are sometimes all right and sometimes aren't. It also reminds us that complaints against claims to common knowledge need to be principled complaints.

Our objector's hesitation began with the brainstorm that the conservatism of common knowledge guarantees that it is a repository of falsehood, of common misconception (and so of fallacy according to common usage). But short of repudiating common knowledge as such, which is an impossible folly, the bad objection pares down to a good one and withall a good rule to supplement **ProPopRule** with:

> **ChallRule:** Do not rely on or appeal to the common knowledge that P if P is under effective minority challenge.

The fallacy of the popular appeal is, frequently enough, just the violation of this rule.

4.3 Inconsistency

In its tacitness common knowledge pivots on deep memory. Deep though it is, something is known of deep memory. Deep memory is notoriously inconsistent (see [Cherniak, 1986, pp. 52–54] and *passim*). common knowledge is underwritten by inconsistency. So isn't it true that common knowledge is inherently unreliable?[12]

The saving feature of the inconsistency of deep memory is that its contents are retrievable and made serviceable for premissory and decisional action only bit by bit. If for some particular P, both P and not-P are nestled in deep memory they will, in the general case, occupy different "sectors" of memory which the memorial agent is unable to compare. Though there, the inconsistency is innocent and unavoidable.

So it won't do to press the complaint in its present form. The requirement of consistency is applicable at the level of retrieval. If P is retrieved from collective deep memory as common knowledge it is desirable that it contradict no Q previously retrieved or presented. It will be true, by hypothesis, that P will contradict something in deep memory, but P can be innocent of inconsistency with anything currently surfaced and available for premissory or decisional employment. Whence

CKRule: P is allowable as an item of common knowledge only if

(i) P contradicts no surface Q, and

(ii) no contrary of P has been retrieved from deep memory.

As we now see, bias is a natural concomitant of appeals to common knowledge. Where bias is a matter of holding preconceived opinions, this is often enough occasion for complaint. In an argument about abortion, S^* might be inspired to object that S's views are preconceptions, and that S's Catholicism suffices to show that this is so. What are we to make of this?

Definition Seven: P is a *preconception* for S at t_j, iff S believes P at t_j, Q implies P and Q is a bias for S at t_{j-k}, $k < j$.

Definition Seven is not terribly interesting as it stands. But it covers some interesting cases. If we were to think that preconception is made significant to the extent that it is in some way unawares — that preconception is something that takes a cognitive agent over — we could specify such cases as follows:

Definition Eight: P is a *tacit preconception* for S at t in context C if S believes P at t_j, in C, S's belief that P at t in C ensues from his belief that Q at t_{k-j},

[12] Or, for the grudging epistemologist, that what is *held* to be common knowledge isn't knowledge after all.

$k < j$, Q is a bias of S at t_j, in C and S's belief that Q is tacit for S at t_j, in C.

Definition Eight provides that a prior conviction of which S might not be aware and which, in any event, he is not prepared to reconsider, induces his current beliefs. It is left open whether the inducement of the present belief is rational or otherwise. It could be made rational, or cognitively competent, by addition of the clause,

(*) Q implies P;

and the over-takingness of the preconception, its seizure of S at t_j, in C, could be given additional emphasis by way of a further condition,

(**) S only tacitly believes at t that Q implies P_j.

In cases of maximal tacitness, S's preconception that P at t is something that j seizes him innocently, the causal output of a prior belief whose epistemic pedigree and causal provenance tacitness has wholly erased. For we say, "Poor fellow, he doesn't realize that his preposterous views on abortion are forced on him unacknowledged by his own latent Catholicism".

Preconceptions are not by any means always so opaque. Even where they are transparent for S they provide ample material for analytical attention. At the heart of the evolving account is the idea of causal thraldom: S would not believe P except for believing Q — his believing Q makes him believe P and moreover makes him believe P as a bias, that is, in ways that make him indifferent or unresponsive to the contrary case.

This is nothing, so far, to rail against. It is as entirely common phenomena whose blanket repudiation would constitute massive violation of principles such as Clutter Avoidance. The systematic removal of tacit preconception would lumber a cognitive agent with the undischargeable task of trying not to believe what his prior beliefs cause him to believe unawares. True, unawareness might give way to awareness, but this would be unavailing if it failed to disturb the causal arrangements. S might know that his Catholicism causally pre-empts any doxastic variation of his position on the abortion question, but knowing it does not in general suspend the causal connection. Harry now believes that Sarah is working at home, and believing this makes him believe that the house is not empty. Harry is aware of this, but knowing it neither disrupts the causal nexus nor should it.

Complaints against preconception are not in general well-made against the causal factor. Neither are they well-made against the factor of bias, for bias alone is no explanation of the preconceptiveness of pre-conceived opinion. Well-focussed complaints seem to require essential reference to the role of prior beliefs, though this may turn out not to be so. Once priority is given central recognition, reservations about the factor of bias and about the causal factor can be made to appear contextually appropriate. Preconceptions are thus eligible for regret in three broad ways. If we allow that Q is a *preconceiver* of P iff Q and P play the roles attributed to them in definition 8, then

Regret #1: Itself a bias for S and causally indispensable for S's belief that P, the preconceiver Q is either

(a) cognitively or dialogically inadmissible for S at t_j, in C

or

(b) is itself occasion of a threatening "standoff for" S and S^* at t_j, in C.[13]

In subcase (a), Q is either objectively disabled by contrary indications, as when Q is the layman's Newtonian conception of space; or it is, as in (b), a dialectically inadmissible unexpressed premiss in S's evolving argument; that is, it is *part of* S's commitment-store but *not of* S^*'s. S and S^* might belong to disjoint epistemic sub-communities in one of which P flourishes and in the other P's negation. Q can then be said to beg the question against S.

Regret #2: If Q's role is tacit, then Q is a rationally regrettable preconceiver for S to the extent that Q's availability to S is shorne of epistemic and causal pedigree, and yet S believes that a case for P needs to be made.

In *extremis*, Q 'gets' S to believe P innocent of that causal fact and of the epistemic whys-and-wherefores. S thus has a position with respect to P but knows not what it is. If that is all there is to it, S may claim that P is 'common knowledge' or that it is 'obvious', begging the question each time. Worse still, innocent of the actual 'case' for his position, S may be driven to invent one. This is an extremely interesting phenomenon, and far from rare. It is the situation in which one's best case for a deep conviction P does two things at once. It violates epistemic dependencies set forth by Aristotle in the *Posterior Analytics* at 71^b19 and 86^a21 [Barnes, 1984, pp. 141 and 115], and it fails the condition of causal inducement. Not only are the facts $Q_1, ..., Q_n$ of the 'invented' case not more certain for S than its putative conclusion P, neither is it the case that S wouldn't have believed P but for his believing $Q_1, ..., Q_n$. Of course, *these* are not things to regret as such. For one thing, Aristotle's epistemic dependencies are unconvincing. What's more, if our present

[13]A standoff is a particularly intractable sort of disagreement. Standoffs have an interesting logic, concerning which see [Woods, 1992a; Woods, 1996], reprinted here as chapters 12 and 13.

conditions were probative, inference to the best explanation, which fails them routinely, would be pole-axed.[14] No, the complaint should be this: that a case that you are driven to fabricate for P, innocent of P's place in your own causal-doxastic history, is a case vulnerable to *ad hoc*ness and irrelevance, among other things. In plainer terms, it will often be just a rotten case.

Regret #3: The causal dependency of S's belief that P upon Q as its preconceiver subjects S to cognitive misperformance. In clear cases, cognitively degraded belief-causation leads to a belief that is objectively false or unwarranted (a belief under the challenge of Nature) or to a belief (all right as it may be, just so) for which Q is inadequate epistemic warrant. In the first case the causal dependency misfires cognitively by making S believe "the wrong thing". In the second case, the right thing is believed but "for the wrong reasons".

5. Doxastic Loyalty

We wanted to know whether preconceived beliefs were a bad thing. We have seen that sometimes they are and sometimes they aren't. After a while, answers of this sort become tiresome. For here too, as with various matters earlier noted, there are no known algorithms. A sense of panic sets in. "Sometimes yes, and sometimes no" is not much of an answer to a good philosophical question. Something better is wanted, an answer to a "which is when" question. But there are no efficiently stateable routines for deciding.

The rising panic meets with two different responses from philosophers. One is to subdue the panic under the plea that the best that can be done has been done. We know the answer, they say, but only implicitly; it reposes in the coils of deep memory and tacit common knowledge. The other response is to give to panic free sway, to allow it to provoke the fabrication of efficient routines for deciding. Philosophers thus smitten are driven to invent idealizations and to define for them performance regulae that the idealizations make analytically true (recurrence of the *Can Do Principle*).

I dislike both options, never mind that each has its attractions. They are too easy and are made too easy, in the one case, by authorizing the premature throwing in of the towel and, in the other, by authorizing a change of subject. A mid-course is available. It is to seek for such generalities as we can unearth in prescribing for acceptable cognitive

[14]True, there are lots of critics of inference to the best explanation. But no one seriously proposes that its offence consists in the violation of these conditions. See [Cartwright, 983a, p. 87 ff], [van Fraassen, 1980, pp. 19–22], and [Hintikka, 1989, p. 4].

and dialogical practice, short of hasty normative enthusiasms for ideal agents. Mid-course is facilitated by an over-arching meta-rule.

> **FactRule:** In seeking to discover rules for the acceptability of cognitive and dialogical practice in human agents, seek always for a description of what human beings actually do. What is actually done makes a defeasible first claim on what should be done.

That said, **FactRule** suggests a further principle, close kin again to the **BiasRule** of pages ago.

> **PreCondRule:** Unavoidable in any case, preconceived opinions are not regrettable until they are admissibly challenged.

I have been trying to make the point that preconceptions are worth the analytical attention given them to the extent that they require us to extend the account of bias. Extended though the account now is, so that proper recognition might be given the factors of prior belief and its enduring causal output, it must be said that by and large, preconceptions are interesting because biases are interesting. Even so, there is an aspect of preconceptions which we have yet to take adequate notice of. As we will soon see, this too is an aspect of bias itself, apart from whether it fulfills conditions for preconceptiveness.

The missing ingredient is foreshadowed by the notion of causal thraldom, in the idea that S's being in a certain psychological state makes him, or strongly inclines him, to believe P and to hold it as a bias and, so, even in the face of contrary indications. "Of course, I'm biased", says the proud mother without a trace of apology. The object of her bias might be her son of whom she holds favourable opinions not shared by others. The mother might be in a special position to know what others fail to see, but that is not the case here; for here she is biased, and she knows it. It seems to be a matter of loyalty, a predisposition to think well of someone or something even when others don't (and here the passions re-enter our story). People, institutions and theories, even theories embraced by common knowledge, are given the hospitality of loyalty. The loyalist persists where others falter. Loyalty is the converse of hostility. Personal contempt and theoretical abhorrence are likewise part of the pathology of bias. What is true of loyalty will be approximately true of hostility, though in reverse, as we say. We will stay with loyalty.

> *Definition Nine:* S's bias with respect to P is *underwritten* by S's loyalty iff there is an x that is relevantly situated for P, and S's loyalty towards x induces or disposes S to hold P.
>
> *Definition Ten:* x is *relevantly situated with respect to P* iff
>
> (i) x is a subject of P

or

(ii) x is an institutional practice expressed by P

or

(iii) x is a theory of which P is a theorem.

Definition Eleven: P is a bias for S unacceptably underwritten by S's loyalty towards x iff

(a) P is under effective challenge

or

(b) S's *loyalty* towards x is misplaced, even apart from the challenge to P.

Definition Eleven covers two types of cases. In the one case, thinking that her son is a nice and law-abiding lad, that he wouldn't do that sort of thing, she holds to the belief that he didn't go joy-riding when he did in fact and there is plenty of evidence that he did ("Poor Sarah, she's blinded by her affection for that silly boy"). In the other case, her son didn't steal the car but it is not the exculpating evidence that convinces her that he didn't; she knew it anyway: "Ralph just wouldn't".

The accusation of Catholicism in our argument about abortion is frequently a complaint about loyalty-underwritten bias. It is thought that S's Catholicism gets in the way, that fidelity to it induces in S an indifference to contrary indications which he is epistemically or dialectically incompetent in ignoring. Or it is thought that his position on P is all right but that, out of blind loyalty, he holds it for the wrong reasons; that his loyalty is here misplaced. Sometimes the critic is right. He may sometimes be right about Catholic loyalism in the abortion debate. But he is certainly not right in the absence of contrary indications with regard to P or contrary indications with regard to Catholicism itself.

But in this latter, he will always be right, or so it will he said. Catholicism is a dogma, is it not?

I want to leave Catholicism (used in these pages in a lazily stereotypical way, as an expository convenience). Dogmatism is more interesting where it is less expected. Dogmatism underwrites all ideologies and is an abiding feature of common knowledge itself. It seems also to be a hallmark of our most basic normative commitments. If true, it would be terrible news if dogmatism were as such rationally disqualifying. Let us see.

Definition Twelve: P is an *unproblematic dogma* for S iff P is a bias for S and though P is underdetermined by the evidence, there is no contrary Q of P which is not equally undetermined by the evidence, and $P \epsilon CK_{ec}$ for some *ec* of which S is a member.

By some lights, there is not a single scientific theory that is not a dogma for its proponents. One associates the doctrine of the underdetermination of theory with Quine, [Quine, 1975]. Underdetermination seems to make one's subscription to theory arbitrary. Quine is aware of this. He invokes considerations of simpicity, predictive yield, and conservatism. Sometimes these are allowed, shabbily, on the standing of evidence; at other times evidence is restricted to sensory turbulence. We ourselves have picked up on conservatism. Of equally well-evidenced but incompatible theories, Definition Twelve bids us choose the one held in the epistemic community of which one is a member. Arbitrariness is evaded, but dogmatism is not.

It is interesting that dogma should be given such easy and routine accommodation in the modern life of reason. More interesting still is science's enduring embrace of dogmas that are not in the least unproblematic (Cf. Lakatosian science [Lakatos, 1976]). Thus

Definition Thirteen: P is a *problematic dogma* for S iff P is a bias for S, $P \epsilon CK_{ec}$ for some ec of which S is a member, but for some contrary Q of P, Q is evidenced to some non-trivial degree, and yet for the same CK and the same ec there exist procedural or methodological principles such that

(i) P's endurance is secured by cost-benefit considerations

or

(ii) a well-defined research programme is specified for reconciling P with Q.

Though it is not quite what I want, the situation described by clause (i) could be illustrated by our persistence with the calculus for its hefty fruitfulness, never mind its inconsistent treatment of infinitesimals.[15] Clause-(ii) situations abound. For several decades the special theory of relativity inspired enduring loyalties notwithstanding the nasty experimental discouragement of Kaufmann, the year following publication of the theory [Kaufmann, 1906, p. 487]. Wherewith a further rule:

[15] The inconsistency was pointed out by Berkeley. The calculus of Newton and Leibniz treated the infinitesimals as real numbers of a special kind. But their specialness was realized by way of inconsistent description: an infinitesmal was and was not greater than zero. Berkeley's complaint was acknowledged and continues to be to this day in standard texts. No one took the inconsistency to disable the calculus. Mathematicians just soldiered on. What explains this astonishing defection from mathematical rectitude? The Cauchy case suggests an answer. It is an enchanting possibility: just as we attribute to Cauchy tacit knowledge of the irrationals, so do we attribute to the history of mathematics tacit knowledge of Robinson's hyperreals (or, at a minimum, Weierstrass' limits). Ascriptions to reals, which result in inconsistency, evade the inconsistency, when made to hyperreals. Thus, since that very inconsistency did not disable the calculus, we assume that, all along, talk of infinitesimals was (implicitly) talk of hyperreals, never mind that nobody, until Robinson, would have avowed such. See [Robinson, 1966].

ProbDogRule: It is insufficient to disqualify a bias P for S for its undergirding theoretical loyalties, that is, just because P is a problematic dogma for S.

Up to now I have not shown much sensitivity to important differences between belief and acceptance [Cohen, 1992]. The span of cases considered so far provide ample occasion to mark some differences. A fundamental difference between belief and acceptance is that if S believes P then, for any Q that S recognizes as contrary to P, S is *unable* to treat Q in certain ways; and if S accepts P then, for any contrary Q, S *decides not* to treat Q in certain ways. It is a large question as to what these ways are and whether they are the same ways, and I shan't trouble with their further exploration here. It is for present purposes enough to note that belief is at bottom causal in a way in which acceptance is at bottom decisional. Thus acceptance is causal, too, but with a different phenomenological texture to it.

PART IV

PRAGMA-DIALECTICS

In Part 1, we considered some metatheoretical issues concerning fallacy theory and informal logic. This would be a good place to take up the meta-theoretic thread again. We briefly met with the Pragma-Dialectical Approach to argumentation in chapter 1. Enough was said there, even so, to carry the suggestion that in certain respects the Pragma-Dialectical approach is not only an alternative to the Woods-Walton Approach but a rival to it. The rivalry it produces is a fruitful one. It gives rise to important metatheoretical issues that will occupy us for the next three chapters.

Chapter 9

PRAGMA-DIALECTICS

In this chapter and the two that follow, I consider the contributions to fallacy theory made by the pragma-dialectical approach to argumentation theory. Pragma-dialectics possesses for me both a local and a global interest. The local interest is that, notwithstanding suggestions of reconcilability (discussed two chapters hence), the pragma-dialectical approach is a rival of my own perspective on fallacies. There was a time when Douglas Walton could have been supposed to share this interest. But he has, of late, succumbed, at least partially, to pragma-dialectic's charms. And that is a second reason for local interest. The point of global interest (and occasion for a pun), is the large international reputation that pragma-dialectics has won for itself. It is a widely reputed theory that has some distinctive things to say about the fallacies. It is grist for the mill of any would-be fallacy theorist.

1. The Pragma-dialectical Approach

Fallacy theory is in pretty doddering shape these days. Much of what is produced is not really up to the challenge, issued nearly thirty years ago by C.L. Hamblin, to take the logic of argumentation well beyond the superficialities and the sheer scholasticism of what he called the Standard Treatment [Hamblin, 1970]. It is altogether refreshing, therefore, to come upon a serious effort to bring to the theory of argumentation real innovation, an attempt to break new ground. Such is the work of van Eemeren and Grootendorst, and I mean to say something about its significance in the present chapter.

There is, I think, not the slightest doubt about the importance of the pragma-dialectical approach for the resolution of conflict by techniques of rational discussion. However, it is a measure of the ambitions of pragma-dialectics that it seeks not only for an account of rational conflict-resolution but also for an analysis of the fallacies which responds effectively and persuasively to Hamblin's admonition to take matters beyond the Standard Treatment. I intend here to limit myself to an examination of the extent to which the pragma-dialectical approach does constitute a step forward in the treatment of the fallacies, which, it must be emphasized, is only part of its agenda.

Van Eemeren and Grootendorst have three especially important things to say about the fallacies:

> *Thesis I*: The fallacies just are "any violation of the rules of conduct for rational discussants [in conversations of conflict resolution]" [Grootendorst, 1987, p. 337]. See also [van Eemeren and Grootendorst, 1984, p. 189].

Superficially this resembles other well-known approaches, such as Hintikka's information-seeking question games, in which certain of Aristotle's fallacies are said to involve breaches of dialectical rules [Hintikka, 1987, p. 211]. Cf. [Barth and Martens, 1977].[1] But in fact, with Thesis I we are met with a significant departure, a quantum leap, as we shall see more clearly as we proceed. For the present it suffices to emphasize that according to Thesis I, any breach of the procedural canon of conflict-relation is a fallacy.[2]

> *Thesis II*: Fallacies are not buttercups. Fallacies do not lead of life of their own independently of a properly articulated theory. Something is a fallacy only within a theory of fallacies (e.g., [Grootendorst, 1987, p. 335]).

Thus we are invited to be incredulous about any view in which "fallacies lead a more or less autonomous existence, like birds, trees or flowers ...[in which the] important job ...[is] to detect them and classify them correctly: ... 'Is this a buttercup or a dandelion?'" [Grootendorst, 1987, p. 335].

> *Thesis III*: It is a condition on the adequacy of the pragma-dialectical account of argumentation that it also account for the traditional fallacies.

Thus, though "*all* violations of the rules are incorrect moves in a critical discussion", it is necessary so to interpret these infelicities that they "correspond roughly to the various kinds of defects traditionally referred

---

[1] It is clear, by the way, that the work of van Eemeren and Grootendorst bears an ancestral link to Barth's pioneering dialectical approach, though there are in these respective accounts significant methodological and analytical differences. The debt to Barth and Martens is recognized in [van Eemeren and Grootendorst, 1987, p. 300].

[2] Hintikka's position, by the way, is an existentially quantified restriction of this view, transposed: *some* fallacies are (construable as) violations of dialectical rules, but not every violation of a rule is a fallacy.

to as *fallacies*" [van Eemeren and Grootendorst, 1987, p. 284], emphasis added in the first instance.

It seems to me quite clear that these theses, tugging as they do in different directions, make for an uneasy triad. Their troubled coherence is not dismissible out of hand as the relatively inconsequential result of hasty formulation; on the contrary, theses I to III, give the heart of the pragma-dialectical account of fallacyhood. So it will be necessary to pause and examine things more closely.

2. Difficulties

How is it that the three theses find themselves at cross-purposes? Consider, for example, Rule VII of the pragma-dialectical approach.

> *Rule VII*: A standpoint must be regarded as conclusively defended if the defense takes place by means of arguments in which a commonly accepted scheme of argumentation is correctly applied [van Eemeren and Grootendorst, 1987, pp. 289–290].[3]

Rule VII can be violated as follows:
(a) *Applying an unsuitable scheme of argumentation.*

-establishing the truth or acceptability of a standpoint by referring to some kind of authority (symptomatic argumentation)	'Professor X says so, so it must be true' (*argumentum ad verecundiam*) 'Everybody says so, so it must be true' (*argumentum ad populum*)
-testing the truth or acceptability of a standpoint by pointing out desirable or undesirable consequences (causal argumentation)	'This can't be true because it would destroy everything this country stands for' (*argumentum ad consequentiam*).

(b) *Inappropriately applying a scheme of argumentation*

-justifying a general conclusion on the basis of an insufficient number of (nonrepresentative) observations (symptomatic argumentation)	'The American medical system doesn't care what happens to the patient. I know of a man who was turned away by a hospital and then died' (*hasty generalization* or *secundum quid*)
-defending an opinion by comparing the matter in question with another, essentially different matter (argumentation by analogy)	'You can't have a compact disc, your father and I didn't have one when we were young' (*false analogy*)
-inferring a cause-effect relation from	'The fact that the shops are closed on

[3]It strikes me, by the way, that van Eemeren and Grootendorst may have been let down by their translator, for *Rule VIII* appears to have been misformulated. The condition which it forwards as sufficient seems at best to be necessary.

the mere observation that two events take place one after the other (causal argumentation)	Sundays is caused by the fact that they are open on Saturdays (*post hoc ergo propter hoc*)'
-rejecting a course of action because it is supposed to lead us from bad to worse, whereas it is not necessary for the alleged consequences to occur at all (causal)	'Tolerating euthanasia leads to genocide. That's why I'm opposed to it' (*slippery slope*)

Rule VII is violated just when, in the argumentation stage of a conflict-resolution discussion, a participant resorts to an argument schema not commonly accepted or, if commonly accepted, not correctly applied. As we see, among the indefinitely many argument schemata that are either not commonly accepted or are misapplicable (namely, all of them) we find some of our traditional friends: *ad verecundiam*, *ad populum*, *ad consequentiam*, *secundum quid*, false analogy, *post hoc* and slippery slope.

Now the oddity is that, by Thesis I and Rule VII, what makes for a fallacy is *not* that a *fallacious* argument is employed, but rather that the argument employed is an argument not commonly accepted, or incorrectly applied. And the analysis of the fallacy contra Rule VII owes nothing to the logical structures of the argument schemata resorted to in the infelicious move against the rule. The particular structures of *ad populum*, *post hoc*, *ad verecundiam* and the rest are not intrinsic to the fallacy with hose commission they are associated.

It is also important to note, that these argument schemata which bear the names of the (old) fallacies (but which are not as such fallacies in their own right) are briefly caricatured and exemplified, but at levels of sophistication no different from the most rudimentary forms of the Standard Treatment. To take just one example, false analogy is described as the error of "defending an opinion by comparing the matter in question with another, essentially different matter", [van Eemeren and Grootendorst, 1987, p. 290]. But this misses the heart of the matter. It is not enough to seize on essential differences as what goes wrong with bad analogies, for essential differences are also essential to *good* analogies. Cf. chapter 17.

We are now in a position to make good on the claim that the theses that give the heart of the pragma-dialectical analysis of the fallacies jointly exhibit an uneasy coherence. For I say again that the *ad populum* is not a fallacy in this context; resorting to it is. And the fallacy that is thus committed involves no more structure than what is expressed by "argument not commonly accepted or incorrectly applied". So in those contexts of discussion to which Rule VII is legitimately applicable, resorting to (say) the *ad populum* and resorting to (say) the *post hoc* is the very same fallacywhich, though efficient, is not otherwise a congenial result. Here we are met with a significant theoretical problem. In the spirit of Theses I and II, van Eemeren and Grootendorst extend to fallacies strikingly generous identity conditions, identity conditions which permit *de facto* pragma-dialectical similarities to overwhelm intuitively interesting and significant structural differences. In fact, a second intuition is lost at this juncture, namely, that a fallacious employment of an argument in context C is the employment of an argument that is fallacious in context C.

3. Unification

One sees what the pragma-dialectician is up to. He wants a *unified* account of the fallacies, a new and vigorous theoretical articulation of them that is free from the feeble scholasticism of the traditional accounts. But it may be thought that any account that identifies the infelicitous employment of the *ad populum* and the *post hoc* as the very same fallacy is an account that *over-achieves* the *desideratum* of a unified treatment, and that, in such surprising identities, we are granted too much unity for serious belief.

Thesis II plays an indispensable role in liberating the fallacies from the burden of their traditional analyses. Thesis II, that nothing is a fallacy except as articulated within a theory of the fallacies, clears the way for Thesis I to make no less than a radical breakthrough, concerning which the *nouvelle vague* theorist is encouraged to say: "Let us forget the old conventions about the traditional fallacies, and let us make a clean break: *any* violation of the rules of pragma-dialectics is a fallacy. Period". But the radicalism of Thesis I, and much of its excitement, is quickly muffled, for Thesis III demands that the old fallacies receive a theoretical reconstruction via the concept of dialectical infelicity. But, as we now see, this hasn't happened. Whatever the deficiencies of traditional accounts of the *ad populum* and the *post hoc*, there is no sense in which they acquire pragma-dialectical construal in any treatment in which committing them is committing the *same* fallacy. Treated that way, the traditional fallacies fast disappear. The infelicity that constitutes the fallacy contra Rule VII reconstructs only the most minimal feature of the seven traditional fallacies there considered ("argument schemata not commonly accepted or incorrectly applied"). In fact, those seven schemata receive two quite different treatments, one pragma-dialectical, driven by Thesis I, in which the fallacy is resorting to arguments having that minimal structure; and the other, in the Standard Treatment mode, driven by Thesis III. Thus, the *argumentum ad verecundiam*, which though it is not a fallacy in the context of Rule VII, is nevertheless described ("referring to some kind of authority"), and exemplified ("Professor X says so, so it must be true"), in a manner that the Standard Treatment in its most elementary versions reserves for those arguments precisely when they are considered as fallacies.

This said, it is crucial to notice that the old names crop up elsewhere in the pragma-dialectical account. Thus though the *ad baculum* is (implicitly), skewered as involved in a violation Rule VII, it is also expressly represented as a breach of Rule I. Notice: it is not a fallacy *vis-à-vis* Rule VII (*resorting* to it is), but it is a fallacy *vis-à-vis* Rule I. Rule I provides that discussants may not prevent or attempt to present one another from advancing or casting doubt upon standpoints [van Eemeren and Grootendorst, 1987, p. 284]. Thus, "Shut up" is on this account taken to be the fallacy of breaking Rule I; but so too is "Don't listen to this moron" the same fallacy, though I suppose a different version of it. Parity of reasoning requires that we extend the same recognition to a discussant's crossing his eyes and sticking his tongue out, or to his singing "I've got you, Babe" at the top of his lungs.

Here, too, the *pragma-dialectical* construal makes of *ad baculum*, *ad hominem* and *ad misericordiam* [van Eemeren and Grootendorst, 1987, p. 285], the same fallacy — infelicities all against Rule I whereas their Standard Treatment description recognizes that they are different fallacies, never mind that their descriptions, again, are scarcely more than caricatures. Thesis I thus confronts Thesis III. By Thesis I, these things are the same fallacy just because they disarrange conversation in a certain way. By the traditional account, these things are different fallacies, possessing significantly different structures which cannot find their way into the over-general pragma-dialectical construal. So it is doubtful whether at this juncture Theses I and III are co-satisfiable. An unsympathetic critic is almost certain to accuse the pragma-dialectical account of a substantial equivocation. It is not obvious that such a critic would be wrong, for any account which identifies the *ad baculum* both as a fallacy and as not, is an account which uses "*ad baculum*" equivocally.[4]

[4]Of course, not every equivocation is a fallacy, and it may be that the authors would wish to account for and justify their equivocation by appealing to "fallacies in the broad sense" and "fallacies in the narrow sense" [van Eemeren and Grootendorst, 1987, p. 300 n. 18]. It is clear, however, that the suggestion is, at this juncture, too programmatic to be of much use to us in the disambiguation of the pragma-dialectical treatment of fallacies.

4. Prospects for Pragma-dialectics

It is also tempting for the traditionalist to complain at this stage that the more serious shortcoming of the pragma-dialectical account is that they get (some of) the fallacies just plain *wrong*. The *ad baculum*, a traditionalist might say, is not a fallacy at all, since no prudential argument of the "Your money, or your life" type is ever incorrect by virtue of its turning on a threat or reference to dire consequence.

However, Thesis II again is called into play. If it is true that the fallacies have no conceptual purchase whatever, except within theory, it difficult to make stick any simple-minded charge that van Eemeren and Grootendorst have got things wrong, i.e., that they have misdescribed and misclassified the fallacies. By the provisions of Thesis II, it seems that one is rather free to make for the fallacies pretty much any theoretical accommodation as may suit the theorist's (serious) purposes. I shall not here explore the vexing difficulties of the idea that the fallacies are theoretical entities pure and simple. But a strategic admonition is, I think, called for. Theoretical entityhood is central in the approach of van Eemeren and Grootendorst, and it is a thorny bramble, perhaps no less prickly than the old idea of fallacy which their account tends, particularly at the join of Thesis I and Thesis II, to place into involuntary retirement. I discuss the question of the theoretical status of the fallacies in greeter detail in chapter 10. This suggestion is, if anything, reinforced by the pragma-dialectical claim that whether a Rule is violated is a matter of one or other discussant *thinking* so. Thus Grootendorst: Faulty argument schemata are not "seen as an objectively verifiable fact... [but are a matter of] the intersubjective agreement of the discussants" [Grootendorst, 1987, p. 339]. But, unrestrained, it is reinforcement at a huge price, for, unrestrained, such a view implies galloping subjectivism. If discussant A reasons, "G.B. Shaw thought that Stalin was the greatest statesman of the day, and that's good enough for me", and if discussant B replies, "Well, GBS was a perceptive and worldly fellow, so that's good enough for me, too", then by the present criterion of "intersubjective validity" no fallacy would have occurred. Not surprisingly, the pragma-dialectician will have none of this: The condition of intersubjective validity is a heuristic condition on discussive practicea discussion in which the parties cannot agree on whether the rules are being complied with cannot log proceed. Equally, a discussion in which incompetent participants fail to catch rule violations can proceed; but can it proceed properly? Not at all, say van Eemeren and Grootendorst. A further condition is required, which these authors call the requirement of *prob-*

lem validity [van Eemeren and Grootendorst, 1988, p. 280]. Problem validity would seem to come to this:

(i) The basic idea of a fallacy as a dialectical infelicity must answer to some kind of standard of objective correctness or adequacy, and in particular must appropriate into its service something like the old fallacies perhaps after (even fairly substantial) reconstrual.

(ii) Similarly, the rules of procedure for the rational resolution of conflict must be good rules, not just minimally heuristic ones. The manoeuvres that they prescribe must leave no room for the Shaw-Stalin kind of travesty.

So it is that the pragma-dialectician requires not only that the rules of conflict-resolution be accepted by the parties, but also that they be rationally satisfying rules; and that the arguments accepted by the parties be good arguments, and those they reject be bad. If, for example, we want, as surely we do, to discourage the acceptability of two dopes allowing bad arguments to pass muster, we must construct a dialectical linkage from the objective badness of bad arguments (and the objective impermissibility of bad questions, and so on) to the procedural canon that disenjoins using them. That is, for conflict-resolution to answer to the apprropiate requirements of objective adequacy, our dialecticians need to have been correctly briefed.

So I ask: What do we tell them? *Are the needed lessons to be found in the pragma-dialectical theory in its present state of development?* My own best answer is this:

(a) Though the pragma-dialectical rules are, on the whole,[5] intuitive and undisturbing, pragma-dialectics has not yet produced arguments which establish their own compliance with the requirement of objective validity.

(b) Though there is in the pragma-dialectical account an interesting effort to fashion a unified account of the fallacies (Thesis I), that effort is made to falter by the further requirement of Thesis III, that the (old), fallacies find an appropriate home in the new theory. Thus the complaint (which may be answerable), of equivocation.

(c) The assertion that the fallacies are theory-dependent (Thesis II), does not, just so, determine how radical a departure from the traditional accounts a new account of the fallacies is permitted to take. That is, Thesis II does not, just so, manage to adjudicate the tension between Thesis I, which clearly provides

[5]Though, as I've said, there is reason to think that certain of the rules are misformulated, citing as sufficient conditions what might more intuitively be thought as necessary conditions (see for example, Rules VI and VII). Then, too, there is Rule VIII, that arguments used in conflict resolution must be deductively valid or capable of being converted into a valid argument by the addition of missing premisses. Rule VIII is troublesome in a number of ways. First, it seems too strong in its affection for *deductive* validity. Second, it makes it absolutely essential to furnish an account of missing premisses which does not permit the vacuous satisfaction of Rule VIII by any argument whatever (by adding as premiss its corresponding conditional).

for a quite radical departure, and Thesis III, which nontrivially constrains this latitude.

(d) Thesis III is in other ways perplexing. It sets forth the traditional fallacies as things that require reasonably faithful representation in a pragma-dialectical reconstruction. But in providing for them brief caricatures straight out of the Standard Treatment, an overarching objective of the new theory — to advance well beyond the superficialities of the Standard Treatment — is stilled. Thus Thesis III is programmatic in two ways. It doesn't tell us which features of the (old) fallacies faithfully to represent pragma-dialectically; and it doesn't tell us in what such faithful representation would consist.

(e) Thus, as of now the pragma-dialectical theory does not tell us nearly enough about how to brief the would-be pagma-dialectical discussant. The concept of problem validity is imposed, not analyzed. If we turn to pragma-dialectics we do not (yet) find enough of what would enable us to answer, "But what would we tell them?"

Chapter 10

BUTTERCUPS, GNP'S AND QUARKS

One of the targets of critics of the Woods-Walton Approach to fallacy theory is the assumption that fallacies have a pre-theoretical conceptual identity, which any would-be theorist is obliged to honour. What this means is that, for example, there is something that the *ad hominem* actually is, independently of what any theory decides to say about it. What the critics are criticizing is the *conceptual realism* of the Woods-Walton Approach. In this second of the three chapters on pragma-dialectical criticisms, I want to try to take the measure of this objection.

1. Theory-dependency

Researchers who have been following the development of the pragma-dialectical approach to argumentation theory have been attracted by the promise that it shows for transacting fallacy in an innovative manner designed to take things well beyond the enfeebled scholasticism of what Hamblin has called the Standard Treatment. The principal sources are [van Eemeren and Grootendorst, 1984; van Eemeren et al., 1987]. As I tried to show in the preceding chapter, the extent to which it is possible for pragma-dialectics to furnish a new unified account of the fallacies turns out to be a function of the extent to which it is justifiable to free the fallacies from the burden of identities forged for them by the old traditional accounts. And so, for example, the pragma-dialectician is interested in being able to disarm, in a quite general way, objections in the form, "But that's not what the *ad baculum* really is!" Part of the methodological strategy which enables such a burden to be lightened, if not removed altogether, is reflected in the pragma-dialectician's espousal of what I called "Thesis II". Van Eemeren and Grootendorst make the point that fallacies are not buttercups. They are dubious of any approach in which

> fallacies lead a more or less autonomous existence, like birds, trees or flowers. The most important job concerning fallacies seems to be to detect and then classify them correctly. A 'theory' of fallacies seems to function like a flora in which the species are described: 'Is this a buttercup or a dandelion?' [Grootendorst, 1987, p. 335].

And so we have Thesis II: Fallacies do not lead a life of their own independently of a properly articulated theory. Something is a fallacy only within a theory of fallacies [Grootendorst, 1987, p. 335].

This is the crux of their approach to the fallacies, and I do not wish to do it the violence of a careless or fanciful interpretation. But an interpretation is needed all the same. The difficulty is this. If one came upon Thesis II equipped with a broadly Duhemian-Quinean (naturalized) epistemology of science, it is likely that it would be taken as trivial, something along the lines of "Buttercups do not (any more than buttercup stages) lead a life of their own, but only within a well-articulated physical theory". Since on a Quinean view of theories and things, every object is a theoretical entity, Thesis II could not be taken as anything particularly interesting or novel [Quine, 1981, p. 70]. So we see at once that van Eemeren and Grootendorst have a stake in preserving this central claim from the charge of triviality.

In fact, it is clear that van Eemeren and Grootendorst are entertaining a quite different notion of theory-dependency, because in forwarding the case of theory-dependency for fallacies they do intend to be attributing to them something striking and distinctive. Theirs is the idea, roughly, that (A) buttercups are not theoretical entities in the sense that they intend (for surely they have, also in the sense that they intend, a life of their own). (B) Gross National Products are to some degree theoretical entities, for, though the aggregate of goods and services that a country produces in a year is somehow pre-theoretically 'out there' in actual macroeconomic practice, the GNP numbers, in order to be useful to economists, are adjusted in Departments of Finance with the help of mathematical and statistical heuristics which have no natural macroeconomic counterparts. (C) Some of the entities of physics, for example, the permanently confined quarks of Quantum Field Theory, are theoretical entities par excellence; they have no life of their own independent of theory, and nothing is a permanently confined quark, except within QFT.[1]

[1]This is not to say, however, that Gell-Mann *coined* the word "quark". It crops up in Joyce's *Ulysses*: "Three quarks for Mister Mark".

Not only is the view of theory-dependency that van Eemeren and Grootendorst have in mind something rather like this degree-of-theoreticity approach, there is also reason for interpreting their Thesis II about the theory-dependency of fallacies rather extremely, that is, in the sense of example (C). Such is anyhow a sustainable interpretation initially, though it may have to be reconsidered in due course. I shall return to this point shortly, but for now I wish to concentrate on the theoretical gains that can be got from their account if we accept this extreme interpretation.

2. Fallacies

As is now well-known, van Eemeren and Grootendorst characterize as fallacies any violation whatever of their pragma-dialectical rules of procedure for rational conflict-resolution. This is their Thesis I. We can now confidently ask, "Does the van Eemeren and Grootendorst approach give at least the beginnings of a well-articulated theory of fallacies?" It does. A fallacy now is any appropriately disturbing procedural infelicity at any juncture of a critical discussion. The point is clear and it is also well-stocked with specifications of the procedural rules that van Eemeren and Grootendorst have worked out in their several writings and which they have amplified with illustrations of ways in which they can be violated. And this would appear partially to fulfill an important adequacy condition on any suitably comprehensive theory of such things.

The provisions of Thesis I and Thesis II do seem at first blush jointly to disarm any criticism in the form "But this is not what the fallacies really are". If the fallacies are nothing except within theory, then there exists considerable latitude to construct a theory, in which they cease to be nothing and genuinely become something, pretty much as one pleases. And it seems perfectly open to van Eemeren and Grootendorst to say, "No, they are not that. They are this, i.e., any infelicity contra the procedural canons of critical discussion". In doing so, not only do van Eemeren and Grootendorst attempt to simplify fallacy theory and to give it a generality that could hardly have been dreamt of previously, they also appear to have calmed the methodological waters. Endless wrangling about what precisely is the structure of the *ad hominem* or the dialectical character of complex question now lapses. It is unnecessary effort, as unrewarding as it is inconclusive.

Moreover, since the new theory provides that the fallacies (the infelicities of dialectical procedure), are copiously more numerous than the traditional gang of eighteen or so, we now have (part of), an answer to the boundary question: Why these? The answer: they are not these. They are as many as there are types of ways of offending against the dialectical canon. (The boundary question is discussed in chapter 1.)

It also emerges that though not decidable tout court, the van Eemeren and Grootendorst fallacies appear to be decidable in a more limited sense. Their fallacies are a recursive class for discussants A, B, ... in discussions D under the conditions:

If

1. recognize that R is a bona fide rule of D,

and

2. one or more of A, B... holds that move F of D violates R, then F is a fallacy.

What to make of it all? A masterly manoeuvre, one might think. As I have tried to say with some emphasis, none of this works without the linchpin that fallacies are nothing except within theory. Were it otherwise, it would invite, and deserve, the complaint that what van Eemeren and Grootendorst say the fallacies are they manifestly aren't, and what they say they aren't they are. But if fallacies have no conceptual purchase beyond their role in theory, then, again, the charge is stilled.

We are now at the crux of the business of this chapter. It is necessary to ask whether there are good reasons to think that fallacies are, in this way, wholly dependent on theory. If we take theoretical entityhood as coming in degrees, a matter of the distance, so to speak, from sensory stimulation, then "There are some cows over there" give us bovine theoretica of fairly low assay. The GNP of the Netherlands would command a loftier theoretical perch. The entities of QFT would be altogether out of sight. Negative-energy seas of electrons, permanently confined quarks, and spontaneous symmetry breakings would stand free of sensory signatures no matter how diffuse the linkage. Bereftness of sensory engagement would seem to vary proportionately with empirical *ad hocness.* Theoretical entities are posited for convenience, or for the further deep rationalization of theory, or for its filling in or smoothing out.

This is a large point. There is disagreement galore about the linkages of theory with the surface irritations of the theorist's sensory apparatus. Are theoretical sentences deficient in or empty of empirical content one by one, or is it theories entire that the data underdetermine, leaving individual sentences incapacitated for sensory kinship? It is not necessary to pursue these matter here, fortunately. But one admonition is called for, I think. Theoretical entityhood is central in the approach of van Eemeren and Grootendorst, and as I have said before, "it is a thorny bramble, perhaps no less prickly than the old idea of fallacy which their account tends to place into involuntary retirement" (see chapter 9). In the end, metatheory will have to judge the trade-off. I leave the question to the tender mercies of the ongoing research programme. However, I must say that it does not seem to me that the English word 'fallacy' stands in matters of sensory inducability more to 'permanently confined quarks' than to 'GNP'. 'Fallacy' is certainly somewhat sensitive to patterns of assent and dissent and dispositions thereto in the linguistic behaviour of human reasoners. But try as one might, one searches for like behavioural moorage for "Lo, a permanently confined quark!" in vain. It may be that the single best symptom of a word's having a purely theoretical denotation is that the word never before existed and had to be invented and conscripted for special baptismal duty. Whatever we make of it, 'fallacy' would fail this test hands down. Mind you, it is always possible to appropriate a word that currently lies about. If we can manage to semantically off-load it completely, this is tantamount to having a pristine name to do with as we theoretically please. If the off-loading is only partial, then we can try to make do with the ensuing ambiguity, the new tendril of which we invite to bear the new semantic weight. As things are, it takes time for semantic adjustments to take hold, if ever they will. For the time being, it is open to us, and necessary, to say that there isn't at present sufficient reason to think that fallacies stand more to quarks than to GNPs, and so no sufficient reason to accede to the central and most audacious feature of the van Eemeren and Grootendorst account on this, the extreme interpretation of it.

Of course, while waiting for the semantic facts to come to their senses one could bridge the process by stipulation or lexical invitation. Fallacy theory, old-style, is getting us nowhere, a pragma-dialectician might say. So why not realign some of the elderly and frail semantic connections and reconstrue the word 'fallacy' in this new way in quest of greater theoretical yield? This is fine, as far as it goes. But stipulations aren't authoritative except to those who freely adopt them, and even then only on pragmatic sufferance. For those who don't, and also those who do, the prudent course remains to wait and see how the ensuing costs and benefits shake out. I don't say these things polemically. Far from it; I am prepared to wait and interested to see.

3. Moderate Dependency

I turn now to a clarification and then a response to an anticipated objection from the pragma-dialectical side. First the clarification. Fallacies, so I have been saying, have no pre-theoretical purchase according to van Eemeren and Grootendorst. They are given a footing only in theory, and the place they occupy in theory can be in principle as distant from or near to sensory irritation as the theorist can arrange within the loose constraints of scientific method. Now I certainly have not meant to say that the fallacies, in the place in theory that van Eemeren and Grootendorst have presently given them, must exhibit the same level of theoretical ratification that QFT currently reserves for permanently confined quarks. Plainly they do not. I mean only that fallacies and quarks have the same pre-theoretical standing; namely none. That fallacies don't stand more to quarks than to GNPs is a fact, if fact it be, about pre-theoretical standing; and the fact is, or so I believe, that fallacies and quarks do not share the same pre-theoretical standing.

It should also be mentioned that nothing in the van Eemeren and Grootendorst account, as I am here construing it, requires them to ignore the existence of other theoretical attempts upon the (old) fallacies. They are not saying that the (old) fallacies have never had a theoretical home, no matter how humble or ramshackle — a hovel, as it were — and neither are they saying that until now fallacies have had no conceptual purchase whatever. All that they need say (and do), is [A] that all the conceptual purchase that the (old) fallacies have ever and could ever have had is intra-theoretic; and [B] that those theories have been sufficiently unimpressive to justify their abandonment and the semantic realignments that I have here been discussing.

As for the anticipated objection, perhaps, contrary to my interpretation, van Eemeren and Grootendorst don't quite mean what I say they mean, and so don't think that fallacies have just the same pre-theoretical vacuity as quarks. Fallacies probably don't, in fact, have the same standing as quarks, but admitting it is a nuisance for their account. The greater the degree of pre-theoretical autonomy that the (old) fallacies possess, that is, the more *buttercup-like* they are, the less they are dismissible out of hand. The more they are somehow 'there' with recognizable (I don't say transparent), buttercuppy structures to tell stories about, the more their abandonment is questionable. That they languish like eunuchs in the pragma-dialectical theory, unanalyzed and innocent of structural engagement with the fallacies that do occur there is a bit awkward (See chapter 9). It suggests a certain caprice.

It remains open, as I say, to stipulate. "Well, I don't want my word 'fallacy' to be about those things; I have no interest in those things. Let's try something else". But now the stipulation is doubly encumbered: (I) It puts pressure on the new approach to generate good and impressive results, the sooner the better; and (II) It requires that there be some systematic analysis of the old fallacies, if only that despairing of them is not seen as arbitrary. And so, in the end, you find yourself doing fallacy theory the old way, at least for the time-being.

As it happens, the requirement to attend to the (old), fallacies in something like their traditional identities, is expressly acknowledged by van Eemeren and Grootendorst. The acknowledgment is conveyed by their Thesis III which proclaims the necessity of construing pragma-dialectically those errors known as the traditional fallacies [van Eemeren and Grootendorst, 1984, p. 284]. This is striking and important. Thesis III straightaway restricts the latitude of the *nouvelle vague* theorist to make of the fallacies what he pretty well pleases, and thus diminishes the chances of the new account of evading debate about whether what it says the fallacies are the fallacies really are. Whereupon the radicalness of the departure announced in Thesis I and authorized by Thesis II, on an extreme interpretation of theory-dependency, is crimped. The theses tug in different directions, Thesis I in the direction of radical innovation, Thesis II in the direction of traditional connotations.

It seems, then, that the best way to understand the pragma-dialectical approach concerning the theoretical entityhood of fallacies is as one which provides for their transition from one theoretical environment to another, from the old traditional theories to the new pragma-dialectical theory. In describing the transition, Thesis III flatly disallows the supposition that the old fallacies have been wholly denatured, for at a minimum they can be taken as held over by the new theory for the time being and on sufferance, while possessing precisely the theoretical purchase they they've had all along in the old theories.

This can happen in two ways. In the first way, the fallacies of old are absorbed into the new theory with their natures more or less intact, and are given their more or less traditional place there along with the other things that the new theory decides to recognize as fallacies. This can be likened to the passage of the Ptolemic heavenly bodies to the Copernican theory, in which their Ptolemic natures are by no means wholly extinguished (though their behaviour is differently described).

The second mode of transition is one in which a new theory borrows from an old one some of its basic ontic commitments. The borrowed entities are accommodated in the new theory on sufferance, and there undergo an ever-increasing suppression of their natures, as genuinely new entities (often of the same name), emerge in the new theory and ultimately bring sufferance to an end. Here we might think of the transitional borrowings of the particles of atomic physics by subatomic physics. Experiment decreed that sufferance be short and it was. The transition was for the old particles a substantial change, as Aristotle would say. Of course, old particles still have a home in atomic theory, but they have none in quantum mechanics.

If you believe that Thesis I dominates over Thesis III, then it is possible to hold that the transition of the fallacies from the old theories to the new is like this second case; and thus that the fallacies do not survive the transition, which visits upon them a substantial change — a change that extinguishes their identities in the new theory. But if you give to Theses I and III, approximately the same weight, and so if the transition is of the former, more ecumenical kind, then as I have already indicated, the old fallacies persist and call out for an articulation of their natures, whatever else goes on in the new theory. To their credit, the spirit of wait-and-see is well appreciated by our pragma-dialecticians:

> The dialectic approach that van Eemeren and I have worked out for fallacies does not, of course, place the final seal on the subject. Within the framework that has been developed there are still a lot of things that need to he worked out, filled in and probably also put right [Grootendorst, 1987, p. 340].

So now we have it. I do not think that fallacies are theoretical in the same extreme way that quarks are. And contrary to my earlier conjecture, van Eemeren and Grootendorst don't think this either. If fallacies aren't buttercups, that is not to say that they can't be like GNP's, that is, like things having some degree of pre-theoretical conceptual identity which a new theory must not altogether trifle with. The task of the new theory is to characterize the fallacies in innovative and even startling ways, but not in ways that extinguish their traditional identities, or in ways that do so all at once.

The decision to accord to the fallacies a level of theoreticity that preserves non-trivially their pre-theoretical or traditional connotations is a significant decision. It encumbers pragma-dialectics with an old albatross, and it remains yet to be seen how it will be disposed of.

4. Theory and Analysis

I say again that it is a virtue of some recent writings on the fallacies that they are discussed in relative freedom from their traditional connotations, fixed by what is said to he an enfeebled scholastic tradition known, since Hamblin, as the Standard Treatment. The allure of the liberation is not hard to appreciate. One wants to be able to reexamine the fallacies with some prospect of revealing novel features of them which may contradict the tradition, yet without running into objections in the form, "But that's just not what we mean by the *ad baculum*". The freedom is desirable because it is the freedom to innovate theoretically in ways that improve upon our former effort.

As we have seen, the methodological instrument of this liberation is to be found in claims for the theory-dependency of the fallacies. In moderate versions, the theory-dependency of fallacies restrains the degree of pre-theoretical conceptual purchase which the fallacies need be assumed to have and thus leaves to theory at least some of the business of determining their character. This averts a procedural vexation which for ease of exposition I shall overstate to the point of parody: "difference from the Standard Treatment is allowable, because such difference would falsify the fallacies as they really are (in logical space, so to speak)". Thus the complaint against the Standard Treatment is for misunderstanding its own approach to the fallacies. The mistake it makes, on this parody, is that of supposing that in its own account of them, the fallacies are not theoretical entities, and thus that what it says about them realistically describes their true and antecendently fixed natures. Such is the confusion of theoretical postulation with identity conditions.

The virtue, then, of some of the new approaches is the prospect of proceeding with the investigation of the fallacious scientifically, that is, within the flexible limits of the scientific method under the loose encouragement of sensory stimulation. Doing fallacy theory scientifically stands in contrast to doing it analytically, that is, by way of the careful description of intuitions which are thought uniquely to determine the conditions of adequacy of the very enterprise itself. It is the way of conceptual analysis as opposed to the way of theoretical construction. To some extent, therefore, the doctrine of theory-dependency favours *ordo cognescendi* over *ordo essendi* and thus leaves certain features of the fallacies to be determined by conditions on theories, such as simplicity and comprehensiveness, that exceed the data whatever they are. Recall Grootendorst, in a charming phrase: "Fallacies aren't buttercups".

It is possible of course to identify more or less extreme versions of theory-dependency, for a domain of enquiry D, by way of the interplay between the mode of scientific postulation and the mode of conceptual analysis.

A *Extreme theory-dependency*: reserves the option that any and all putative truths in D may be overturned by postulation.

B *Moderate theory-dependency:* holds to a minimal core of truths in D as fixed by conceptual analysis, and permits postulational deviations from that core to occur only with great modesty — in dribbles, as it were — and only over large-ish intervals of our intellectual history

C *Minimal theory-dependency:*: reverses the dependencies of extreme theory-dependency, and then some. For the radical minimalist, nothing seriously said in D is subject to or secured by postulation. All D truths are held to standards of conceptual realism.

If, as in my parody of a few paragraphs back, the Standard Treatment is taken as a radical minimalist position, then the danger exists that it will enforce the at times silly things it says about a fallacy as part of its very concept — that is, it threatens to enshrine error in identity conditions — in consequence of which corrections and improvements risk dismissal *à priori.*

The virtue of moderate theory-dependency is twofold: first, it allows for theoretical innovations that genuinely improve fallacy theory without at the same time losing sight of the things that are generally recognizable as fallacies, as the same things, more or less, that our theoretical forebears were wrestling with.

Chapter 11

UNIFYING THE FALLACIES?

In the present chapter, I consider a further objection to the Woods-Walton Approach made by pragma-dialecticians. The complaint is that in the Woods-Walton Approach there is no hint of an underylying theoretical unity.

Although I have tried in chapter 1 to blunt the edge of this criticism, it seems to me that the issue would benefit from a more detailed examination.

1. Theory Reductions

Considered as an object of systematic study, fallacies have had a long and not especially distinguished history. Ensuing from Aristotle's notion of paralogism, fallacies have met with recurring attention and little theoretical consensus. A millennial failure twice-over, it is a wonder that there hasn't been a more broadly based call for the abandonment of fallacy theory as a bad job. Not that there hasn't been some crankiness. Contemporary readers will think at once of Hamblin, which scolds a long tradition for its lazy and uncritical subscription to a degenerate research programme [Hamblin, 1970]. Still, Hamblin is no Churchland. He does not see himself as calling for the outright retirement of fallacy theory in that way that Churchland calls for the abandonment of folk psychology (i.e., psychology of the common sense mentalistic kind). The nub of Churchland's complaint is that after two and a half millenia of effort, folk psychology has failed to produce stable, well-confirmed theories. Hamblin raises a similar objection to fallacy theory; but instead of arguing for its abandonment, Hamblin proposes is that we start doing fallacy theory *right* — to put it bluntly. On the other hand, what Churchland proposes is that the question of doing folk psychology right doesn't even arise.

Harsher critics than Hamblin there surely are. One thinks of [Lambert and Ulrich, 1980; Massey, 1975a; Massey, 1981] and their denial of the possibility of, or anyhow foreseeable prospects for, a *bona fide* theory of the fallacies. There is plenty of evidence that contemporary thinking has harkened to these objections. Writings on the fallacies do not routinely make their way into the leading journals.[1] All the same, largely in response to Hamblin's challenge to reform the Standard Treatment, a substantial literature has been produced in the thirty years since *Fallacies* appeared. It exists in quantities sufficient to sustain its own critical examination; and objections have come thick and fast. A goodly part of the post-1970 record derives from the work, jointly and severally, of Douglas Walton and myself. For reasons that will shortly become apparent, and which I trust will not be thought to be too self-serving, much of what I have to say in this chapter will focus on that body of work, in the abbreviating reference '*WW*'.

One complaint against *WW* is that it is too formal and context-insensitive [Finocchiaro, 1987]. Another is that although formal methods are admissible, *WW* sometimes does a bad job of motivating its own formal enthusiasms [Groarke, 1991]. Another still is that the formalities do insufficient explanatory work ([Scriven, 1989], *viva voce*, the Third International Symposium on Informal Logic), and that they are an expository and analytical impediment ([Scriven, 1989], again, and [Groarke, 1991]). An especially interesting complaint comes from van Eemeren and Grootendorst in several writings, most recently in their book *Argumentation, Communication and Fallacies.*

> The systematic exploration of advanced logical systems in order to analyze fallacies is characteristic of Woods and Walton's approach [van Eemeren and Grootendorst, 1992, p. 103].

WW employs inductive logic in its treatment of *secundum quid* and *post hoc*; a plausibility logic underlies *ad verecundiam*; dialectical game theory is plumbed for many questions and *petitio*; a relatedness logic is pressed into service for *ignoratio elenchi*; and so on. But this means that

> [e]very fallacy needs, so to speak, its own logic. For practical purposes this approach is not very realistic.... One only gets fragmentary descriptions of the

[1]There are welcome exceptions, of course. *Philosophy and Rhetoric*, *Argumentation* and *Informal Logic* stand out for their receptiveness to work in fallacy theory. There are recent indications that good academic presses are also lightening up, as evidenced by [Hansen and Pinto, 1994], and [Walton, 1992], not to overlook [Johnson, 1996].

> various fallacies.... Ideally one unified theory that is capable of dealing with all he different phenomena is to be preferred [van Eemeren and Grootendorst, 1992, p. 103].

I want in this chapter to see if I can take the measure of this objection.[2] Two questions press for attention. One is the question of what it is for a theory to be unified. The other considers prospects for unifying fallacy theory in compliance with our answer to this prior question.

Talk of unifying theories puts us in mind of reductionism. Logicism is a case in point, the reduction of mathematics to *FDL* and set theory (*FDL*, recall is Ralph Johnson's acroynm for formal deductive logic). Reductions of this sort interest philosophers largely for ontological reasons (though there is good reason to think that this has not been the central interest of mathematicians such as Frege). The reduction of chemistry to physics was considered important philosophically because it eliminated the ontological commitments of the reduced theory in favour of those of the reducing, or target, theory. Philosophically-minded logicists, such as Russell, felt the same way about the reduction of arithmetic to *FDL* and set theory. The basic idea is that for suitably constructed theories T and T^*, if there exist mappings and definitions such that the theorems of T are preserved in T^* and the ontology of T could consistently be considered as null under the preservation, then T reduces to T^*. Concerning the reduction of arithmetic to *FDL* and set theory, all the laws of arithmetic come out true in an ontology of pure sets.

Logistic reduction presupposes the meeting of certain technical requirements. One is that the theories in question, both the to-be-reduced and the reducing, be construed as the deductive closure of sets of axioms. It is striking that where a theory is also first order and extensional, the ontology of every true theory reduces to arithmetic, such is the providence of the Lowenheim-Skolem theorem in a strengthened version due to Hilbert and Bernays. Two things at once become apparent. One is that no theory announces itself *à priori* as the preferred reducing theory. So, for example, since set theory is true in the natural numbers and number theory is true in a universe of pure sets, the question of which, if either, we might favour as reducing must turn on considerations independent of the technical fact that the reduction can be brought off. Some theorists believe at times, what many others wouldn't; that set theory is conceptually and ontologically 'prior' to arithmetic in that the

[2]I shall not contrive a defence of *WW*'s formalisms beyond those offered by [Woods and Walton, 1989, ch. 17], [van Eemeren et al., 1996, ch. 8] and chapters 2 and 3 of this book. Responses to the unification challenge can also be found in chapter 1.

arithmetic may be right, but if so, its rightness does not reside solely in the fact that arithmetic reduces to set theory.

Readers will have seen long ago that those who recommend the unification of fallacy theory cannot have logicism in mind as their model or paradigm. Still less do they counsel any course which eventuates in the fallacies being artifacts of arithmetic. There is no chance that fallacy theory will ever answer to the technical requirements of such reductions. They lack the requisite metamathematical structures that make such reductions and such models possible. I return to this point in section 4 below.

2. Exemplar Theory

Ask a first year university student what a fallacy is, and there is a good chance that he will answer along the following lines: A fallacy is a mistake like begging the question or asking someone if he has stopped beating his dog. Our respondent might also cite hasty generalization as a further example, and illustrate with "Cynthia is a bad driver; so women are bad drivers". If you pressed with "What is begging the question?" you might be told that it is the sort of thing you get when you reason that Jesus loves me, this I know, for the Bible tells me so.

Suppose you asked for a characterization of hasty generalization. The answer might be that it is an induction from a single case. This is wrong of course. The Cynthia-case is a fallacious generalization and it is an induction from a single case. But not all single-case inductions are fallacious, as witness the fourteen month-old toddler, who, on reposing his hand on the element of his mother's stove, learns in just one encounter "Never to do that again" (see chapter 19 above).

I assume that for fairly large populations, say the population of university undergraduates, fallacy is a *concept-in-use*, that is, a concept with which there is natural familiarity evidenced by conversational ease and independently of its articulation in a developed theory. Another way of saying this is that 'fallacy' is a functioning item of the pre-scientific lexicon of this community. I also assume that this concept-in-use serves as a device of classification. Members of this conceptual community will have little difficulty in recognizing as a fallacy "VAT is a regressive tax, because VAT is a regressive tax." They might also be presumed to categorize it further as circular or question-begging. These assumptions are of a kind that one finds in psychological studies of concepts and their role in classification (see, for example, [Smith and Medin, 1981]).

What our imaginary case suggests (actually, it is far from imaginary; it is amply exemplified in the experience of lots of people I know), is that the concept-in-use of fallacy conforms to what psychologists know as the *exemplar theory*. If so, the concept 'fallacy' does not have a "summary representation". The concept of fallacy is not a unitary concept. This matters. Having rightly despaired of certain reductionist unifications for fallacy theory, it is natural to construe the unificationist challenge as calling for a treatment in which the unitary character of this concept of fallacy is explicitly recognized. But, as I say, there is reason to suppose that 'fallacy' is not a concept-in-use possessing a summary representation.

In the psychological literature a concept K is said to have a summary representation when there is a true sentence $\ulcorner K$s are Φ, Ψ, etc.$\urcorner$ of which the following three things are true.

1 The representation is an abstraction from K-phenomena.

2 The representation underdescribes actual instances of K (and so there will be K-things bearing features other than Φ, Ψ, etc.).

3 The representation is retrieved in situations of K-categorization; and so a possessor of the concept K would, whenever she were presented with a test item, retrieve the summary representation, against which the test item would be compared [Smith and Medin, 1981, p. 23].

If the exemplar theory of concepts is correct in general, or if it is correct for the concept-in-use of fallacy, then fallacy will not, I say again, be a unitary concept. On the exemplar model, everyday concepts are represented by users by "separate descriptions of some of their exemplars" [Smith and Medin, 1981, p. 23]. For the concept bird, the associated representation might consist of the exemplars robin, bluejay and sparrow, the sub-exemplar Carolina thrush, and the concrete instance "Fluffy", the concept-user's pet canary. In our population of university students, the representation of the concept of fallacy is similarly configured. Exemplars such as begging the question and hasty generalization are mentioned, as are particular instantiations — the Cynthia generalization, the spouse-beating question, and the Jesus-loves-me inference.

Consider now the mistake of characterizing an induction as the fallacy of hasty generalization when it is an induction from a single case. Experiments in cognitive psychology [Tversky, 1977], suggest a rather natural explanation of this kind of mistake. It is an explanation that takes seriously the exemplar theory of concepts and categorization. The error in question arises from retrieving a (concrete), exemplar, the Cynthia-case, for example, of which 'mistaken induction from a single case' was a correct characterization, and generalizing on that characterization.

Our particular example briefly to one side, it is interesting and a trifle ironic that on the exemplar model of categorization it is predicted that categorizers will routinely be hasty generalizers. If this is right and if generalizations from single cases were always fallacious, then the exemplar theory would make of us routine fallacy-mongers in the bulk of our categorizations. Though there is some experimental evidence that we are less pure in our everyday pursuits than we might have thought, it is too much to suppose that we get nearly everything wrong nearly all the time. At any rate, it is too much to suppose this in the absence of some hefty supplementary evidence that it is so. Actually we have a choice. Either single-case inductions are always fallacious or the exemplar theory is false because it implies that single-case inductions aren't always fallacious. I won't take the time to debate the options in detail. Suffice it to say that the exemplar theory is too good to be disabled by a *reductio* whose conclusion, in any event, is independently called into question by cases such as the toddler.

With that said, it is now apparent that if the target concept of fallacy theory is what I have been calling a concept-in-use, then we would expect a further three things to be true if the exemplar theory is correct in general or if it is correct for that concept-in-use. One is that the theory would be dominantly exemplary in its presentation of the fallacies. Another is that quite often the theory would correct standing or received characterizations of particular exemplars. That is, the theory would traffic in "theorems" (to speak tendentiously) in the form

(TH): So-and-so exemplar is not always (or ever) a fallacy.

Thirdly, the theory could be expected to attempt summary representations with regard to particular exemplars, but these would be motivated by *TH*-disclosures and various other pronouncements of theory. For example, in light of the 'theorem'

(TH1): Not every induction from a single case is a fallacy

it might be ventured that a generalization on an event E^* in the presence of a single correlation with an event E is justified by a risk-aversion strategy in those cases in which the costs attaching to recurrences of E^* are higher than the costs, if any, that attach to generalizing on it. Something like this fits our toddler case, but it clearly will not serve for women drivers, no matter what we might think of Cynthia.

It will not have escaped notice that the van Eemeren and Grootendorst (hereafter '*VEG*'), characterization of *WW* places it squarely in the ambit of these three features. *WW* is highly exemplary; it quite often corrects for misgeneralizations from exemplars; and such summary representations as it aspires to typically do not precede theory but are consequent upon it. And, as was attempted in chapter 1, a theory-driven summary representation of the concept of fallacy itself may be possible. What I have been saying of late is worth repeating. It is that the dis-unification of the fallacies is nothing to complain of. It is precisely as things should be. Or it is if the exemplar theory is correct.

It would be overdoing it to say that the correctness of the exemplar theory is by now a settled issue. No exemplar theorist sees it so. Exemplarism faces some non-trivial problems, [Smith and Medin, 1981, ch. 7], but there is considerable evidence that it does better than its natural rivals, the classical account and the probabilistic account, both of which are unitary theories [Smith and Medin, 1981, ch. 3, and 4–6]. To the extent that this is so, I conclude that if the challenge of *VEG* is to provide a framework in which the concept-in-use of fallacy is given unitary recognition, this is a challenge that cannot be met.

3. *VEG* as Theoretical Stipulation

Rival views are possible, of course. The theorist might hold that fallacy, considered as a concept-in-use, is unfit for the austere flights of theory. Theories, as Quine reminds us, are free for the thinking up; they are expressions of our conceptual sovereignty. The concept-in-use of collection was of no help in the reduction of number theory; *sets* had to be contrived. Why should it not be the same way with fallacy? Concepts-in-use resist abundant and deep generalizations; so why not dispense with them? Why not stipulate something more streamlined, and anchor the stipulation in the theoretical fruitfulness that it abets? In asking these things, a further interpretation of the unificationist's challenge becomes apparent. It is now seen as the challenge to displace fragmented concepts-in-use with theoretically powerful *stipulations*. *VEG* itself is interpretable as a positive response to this challenge. Whatever the antecedent nuances of the concept-in-use of fallacy, fallacies are now any violation of the rules of critical discussion in contexts of critical enquiry. "Fallacies are analyzed as... incorrect discussion moves in which a discussion rule has been violated" [van Eemeren and Grootendorst, 1992, p. 104]. The rules, ten in all, are held to be conditions necessary and sufficient to fix the extension of the predicate "is a rational critical discussion". To the extent that violations of the rules are efficiently (not: effectively) recognizable, so too is the extension of the predicate "is a fallacy". It is all rather tidy, and fallacy is now as unitary a notion as the idea of a rule-violation of pragma-dialectics will allow. (For an account of stipulated truth see [Woods, 2002b].)

If that were not progress enough, *VEG* proposes that the loose fragments of the concept-in-use need not be altogether given up on. They can be gathered up unanalyzed (after all they are, as such, analytically resistant) and given a place in the taxonomy of the stipulative theory. This is set out in [van Eemeren and Grootendorst, 1992, pp. 212–217], under the section heading, "Traditional Fallacies as Violations of Rules For Critical Discussion". "We think that all the traditional categories have their proper place in our system" [van Eemeren and Grootendorst, 1992, p. 104]. It is an opportune move. The pragma-dialectician can now say that his theory accounts for the traditional fallacies, never mind that it doesn't analyze them. (Their own use of 'analyzed' in the quotation cited earlier is a careless solecism, doing no damage in context.) Each 'traditional' fallacy — begging the question, complex question, *ad hominem*, *ad verecundiam*, hasty generalization — the whole gang of eighteen as I called them in chapter 1 — is pegged or triangulated by (a) the rule that it violates, (b) the stage of the critical discussion to which the rule applies, and (c) the party or parties to the discussion considered to be bound by the rule at that stage. Here is an example. Critical discussions come in stages, with various obligations on proponents of a thesis and those antagonistic to it to perform in certain ways at certain stages. The confrontation stage is the stage at which protagonist and antagonist acknowledge a disagreement, the resolution of which is the overall aim of the discussion. The object of the disagreement is called a 'standpoint'. Rule 1 pertains to this stage. It provides that "parties must not prevent each other from advancing standpoints or casting doubt upon standpoints" [van Eemeren and Grootendorst, 1992, p. 208]. The traditional fallacy ad baculum is now represented as a violation, either by the protagonist or the antagonist, of rule 1 at the confrontation stage. It is the violation of "putting pressure on the other party by threatening him with sanctions" [van Eemeren and Grootendorst, 1992, p. 212].

In this way, each traditional fallacy is recognized in *VEG*. Each is given a characterization in *VEG*. But there is a difficulty. *VEG*'s characterizations of the traditional fallacies are not much good. For example, the abusive *ad hominem* is said to consist in "doubting the expertise, intelligence, or good faith of the other party" [van Eemeren and Grootendorst, 1992, p. 212]. This is certainly not in general true, apart from a stipulation that makes it so. In other cases, the description is vacuous, as with the fallacy of ambiguity, which is the "misus[e] of referential, syntactic or semantic ambiguity" [van Eemeren and Grootendorst, 1992, p. 212]. This gives the appearance of a nasty setback. *VEG*'s descriptions of the traditional fallacies are just as callow and half-baked as anything one would find in the Standard Treatment against which Hamblin protested to such good effect. With *VEG* and the Standard Treatment alike the traditional fallacies come to the same thing. Even so, there is reason to think that, though they come to the same thing, they do not commit the same offence. For the Standard Treatment and its *WW*-like successors, too, it is essential that these descriptions be right. For a stipulator like *VEG*, getting them right is *optional.* The fallacy in question is not the *ad baculum* or the abusive *ad hominem* or the fallacy of ambiguity. It is any violation of the rules in their application at appropriate stages. In the case of *ad baculum*, the fallacy is doing anything that prevents people from advancing or challenging starting points. If making threats is one way of doing that, then it is an instance of that fallacy. (For a contrary view, see [Gabbay and Woods, 1998].) If *VEG*'s description of doing that is a vacuous characterization of the *ad baculum*, who cares? The only point of bothering with it in the first place was to give a general and loose indication of its tentative membership in a class of cases that exemplify the fallacy of violating rule 1 at the confrontational stage.

On the present interpretation of *VEG* as a stipulative rival of *WW* (and, of course, of the Standard Treatment too), *VEG* can afford being rather cavalier about descriptions of the traditional fallacies. If *VEG* gets them right, well and good. If not, it hardly matters. For, as the man said, "the traditional fallacies don't pay the rent in pragma-dialectics". They are house-guests only. They inhabit *VEG* on sufferance.

4. Reconciling *WW* and *VEG*

This, as I say, is the fate of the traditional fallacies on one natural interpretation of *VEG*'s programme. If it were the correct interpretation, it would leave the would-be fallacy theorist with a perfectly tenable option. One would be to stick with 'fallacy' as a concept-in-use and to work away at it piecemeal in something like the *WW* way. The other would be to make a radical departure and to contrive a unified notion *de novo* in something like the *VEG* way. The rivalry would persist and victory, if it ever came, would depend on future developments in the service of antecedent presumptions about what a good result would consist in. But there is reason to think that this is not the correct interpretation of *VEG*. In fact,

> ... we think the kind of approach which is carried out by Walton (and Woods) could be combined with a pragma-dialectical approach.... The combination could probably lead to a lot of progress in dealing with [outstanding problems]" [van Eemeren and Grootendorst, 1989, p. 109].

What is this combination? What would it look like? Would it constitute satisfactory fulfillment of yet another conception of unification?

In a celebrated paper, W.H. Stone followed up on some work of Tarski's and sought to unify logic and topology. (The rest of this paragraph may be a trifle technical for some readers. It doesn't matter;

only the gist is important.) Actually, he tried to unify Boolean algebra and topology, Tarski himself having established the link from logic to Boolean algebra. Like Tarski, Stone proposes a theory in which conjunction, alteration and negation correspond to (set theoretic), multiplication, addition and complementation. The closed-open sets under these operations form a Boolean algebra. Stone deals with a class of topological spaces, known as Boolean spaces, that display the following features [Stone, 1937]. They are totally disconnected, compact Hausdorff spaces. A particular case of these is the Cantor space which is got by according the pair set (0,1) the discrete topology and then assigning to the Cartesian product of countably many cases of it the product topology. The highlight of Stone is the Stone Representation Theorem [Stone, 1937]. It establishes a duality between any Boolean algebra and some or other Boolean space. Here is the important point for our present purposes. The Stone Representation Theorem is underivable in Boolean algebra and underivable in topology. It is got only from an arrangement that brings together and extends the two theories. It is a deep result. It identifies and characterizes certain structures which are inherently interesting for topology and Boolean algebra alike. It also facilitates the derivation of further results which provide new insights into each domain, with Boolean algebra illuminating the structure of Boolean space and Boolean space illuminating the structure of Boolean algebra where none of these results are derivable *prior to unification.* Stone's work unifies Boolean algebra, and topology in the way that Descartes unified algebra and geometry. In particular, Descartes' analytic geometry made it possible to give a good account of cubic curves, something that couldn't be done previously.

I apologize to readers for whom this brief interlude with Stone's Representation Theorem is unhelpful. Fortunately, we can now leave the mathematics and characterize its importance more abstractly and in ways clear to all. So let us say, to begin with, that

1. R is a r*epresentation relation* between a theory T and a theory T^* just in case there is a truth preserving interpretation of a nonempty subtheory of T in a nonempty subtheory of T^*, or vice versa.

Let us also put it that

2. X is an *R-extension* of Y iff R is a representation relation on a proper subtheory x of X and a subtheory y of Y and there are theorems provable in X that are unprovable in x (or in y).

With these things said, we have

3. The *two-way unification principle*: For theories T and T^*, there is a two way unification of them just in case R is a representation relation between T and T^* and there exists an R-extension of T and an R-extension of T^*.

Derivatively, a *one-way unification* of T and T^* would be an R-extension of T or an R-extension of T^*, but not of both.

We now have a precise question to ask. When *VEG* proposes the combination of *WW* and *VEG* in quest of further progress in dealing with outstanding problems, could these remarks fairly bear an interpretation according to which they are seen as calling for at least a one-way unification of *WW* and VEG? I think not. For one thing, as our discussion of logicism and the Lowenheim-Skolem theorem made clear, representation relations that stand any chance of constituting R-extensions will turn on structural features of the theories on which they are defined. There is little reason to think that *WW* or *VEG* possess, or could be made to possess meta-mathematical structures that will carry such representation relations in any literal sense. But the basic idea is attractive, and we may think that it could be approximated to intuitively by theories such as *WW* and *VEG*. The basic idea is that essential to a unification's success is an embedding of one 'area' in another in a way that produces new insights not achievable in either of the prior two. In this regard, the *WW* treatment of the *petitio* in Kripke's intuitionistic logic can be thought of as a one-way unification of *petitio*-theory in intuitionistic logic which produces new insights about the *petitio* but not necessarily new theorems in intuitionistic logic [Woods and Walton, 1989, ch. 10]. The present idea generalizes on this. Couldn't we try for at least a one-way theoretical unification between *WW* and *VEG*? I lack the space to explore this possibility in detail. Even so, some things are fairly clear. We might think that *WW* treatments of the dialectical fallacies would be especially amenable to unification. In that regard, [Walton, 1991], a substantial monograph on begging the question achieves, if only somewhat modestly, a unification with certain parts of *VEG* and in ways that facilitate some of Walton's own results.

Prospects are less encouraging, perhaps, for other fallacies, for certain of the statistical fallacies, or for the gambler's fallacy, the fallacy of division, or *post hoc*. But, bearing in mind that every mistake of reasoning that can be made solo, can be made in the company of others and can, with some artful contrivance, be made in our reasoned engagements with others, a certain openness of mind may be called for. Although it doesn't strike me as obvious that running what I think about hasty generalization through the sieve of pragma-dialectics will prompt new insights into hasty generalization, it doesn't strike me as obvious that it couldn't. It is something that I would be prepared to consider and to try.

At the end of the day one thing will be clear. The more that *WW* theories can be made to unify, one by one, with *VEG* the less it will be likely that the overall pattern of unification will be one-way. For it seems to me that under unification enough, the idea that a fallacy is just a violation of a rule of critical discussion will surely yield to something deeper and more complex. If so, the unification of *WW* and *VEG* would be genuinely two-way. And that would be interesting.

PART V

INTRACTABLE DISAGREEMENT

On the Goody Two-Shoes model, the role of argument is to produce agreement about some initially disputed matter. In those cases in which agreement cannot be found, the model requires that the participants cease their engagement and recognize its unfruitfulness in the case at hand. This is a version of an argumentative strategy known as "agreement to disagree". In the next two chapters we investigate modes of argument for which the agree-to-disagree strategy isn't appropriate.

Chapter 12

STANDOFFS IN PUBLIC POLICY

The Goody Two-Shoes Model of Argument serves as a kind of paradigm for dialogue logicians and argumentation theorists. In its most basic sense, arguments that conform to this model are exercises in free, open, and impartial enquiry, aimed at producing agreement or consensus. Habermas' Consensus Theory [Habermas, 1973] is perhaps the most faithful exemplar of the Goody Two-Shoes approach; but it is also discernible in the Pragma-Dialectical theory of critical discussions [van Eemeren and Grootendorst, 1992], and in Grice's logic of conversation [Grice, 1989, ch. 2 and 3]. In Habermas' approach, consensus is a regulative Principle of Rationality. This is an extraordinarily strong claim, and it rightly calls for a certain reflection before we pledge ourselves to it. In this chapter and the one to follow, I attempt to show that consensus should not be considered essential to dialogical rationality. I examine classes of arguments typified by their utter hostility to consensus. These I call *standoffs*. Standoffs are intimately tied to fallacy theory, given their propensity to trigger question-begging responses and other utterances sometimes said to be fallacious, such as the *ad hominem*.

1. Degrees of Standoff, Force I-III

Let me try to set my course. I begin with a disclaimer. Though I will be saying something about modern liberal pluralistic democracies, I shall venture nothing that is not already noticeable about how such governments and the ruling élites interact and communicate with their citizens in times of ordeal. It is no secret that governments sometimes behave badly, but what I wish to concentrate on here is a class of situations made distinctive by the limitations they impose on a government's doing well. The limitations that I am seeking are logical in character and they are interesting because they create, as I say, a paralysis in the public discourse. This paralysis arises from argumentational setbacks which I shall call *standoffs.*[1]

Standoffs are describable in that branch of logical theory which deals with the rational adjudication of conflict, itself, in one meaning of the word, a branch of dialectic. Dialectic is, among other things, the logic of negotiated rational acceptance. It lies open to a limitation theorem roughly as follows. Disagreements sometimes yield to ascending strata of intractability and so approach a limit at which they go into a dialectical black hole, so to speak, whereupon conditions for further negotiation lapse. The basic logic of such things receives early and incomplete codification in Book VIII if Aristotle's *Topics*, and more extensive and subtle *regulae* can be found in contemporary writings.[2] For my purposes it is unnecessary to summon up the full formal apparatus of any of these writers. I need to characterize standoffs and this can be done in a fairly intuitive way. I shall say that an argument is a basic standoff, or a standoff of force one, for participants P_i at time t with respect to a claim S iff

(ia) no consensus exists at t among the P_i which settles the matter with regard to S; and

(ib) no consensus exists at t among the P_i about procedures that would or might establish such a consensus.

Condition (ib) is important. It marks a distinction between an argument at pause and argument stopped cold, for the time being at least.

Standoffs of force one vary in their importance. Did the pretty woman in the window of the passing train wink at Charlie, or was it perhaps a cinder in the eye, as Harry thinks? Charlie and Harry are deadlocked on the issue, but given that her train rushes North and theirs South, they can't hit upon a practical way of settling their disagreement.

Standoffs of force two satisfy a further condition. I might imagine that Charlie and Harry, having had their fun, decide to let the matter drop. If so, they have reached the honorifically procedural consensus to agree to disagree. For standoffs of force two, I have the opposite of this:

(ii) there is no honorifically procedural consensus to let things drop.

---

[1]I apologize to purists for my failure to honour the plural paradigm of 'spoonsful'.

[2]See, for example, [Hamblin, 1971; Lorenzen and Lorenz, 1978; Barth and Martens, 1977; Hintikka, 1979; Hintikka and Saarinen, 1979; Manor, 1981; McKenzie, 1981; Barth and Krabbe, 1982; van Eemeren and Grootendorst, 1983; Walton, 1984; Krabbe, 1986; Blair and Johnson, 1987; Woods and Walton, 1989; Woods et al., 2002; Gabbay and Woods, 2001a; Gabbay and Woods, 2001c]. See also [Walton, 985a; Walton, 1988].

Standoffs of force two are, in any of a number of ways, too important, too pressing, too alluring to let go of.

The idea of not being able to leave the matter alone is entirely commonplace and it receives standard expression in a decision to put the issue to third-party settlement, whether to binding arbitration, final-offer selection, the civil courts or the legislature.[3] This is a procedural arrangement unlike that of condition (ib), for it is not a procedure designed to produce the consensus lost in condition (ia). Though it *could* do that, by happenstance as it were, the intent of the procedure is to produce not agreement but acquiescence. So now I have standoffs of force three, standoffs which fulfil condition

(iii) there is no explicit agreement to send the dispute to third-party determination.

Standoffs are bad things, logically speaking, never mind that they are also all too human. The natural mark of a standoff is its susceptibility to the fallacies of irrelevance and question-begging and the allied disposition to take these fallacies to the level of meta-argument. Thus, "Your argument begs the question". "It does like hell!" (Question-begging). Or, "Even St. Thomas Aquinas begged questions!" (Irrelevance). Standoffs of force three are, like the others, of shifting degrees of importance, and their importance will vary in turn with the interests of the protagonists. There are, however, cases in which the following things are true:

A The standoff is a standoff *in a community* and thus answers to certain demographic criteria. Whatever these might be in detail, they suffice to make it true not just that Harry and Charlie are in a standoff, but that the country[4] is;

B The standoff involves a claim about what to do or should be done, and so is of an essentially deontic character.

C The matter in dispute is of a type that places it within the legitimate authority of a third party to settle without the direct consent of the contenders. In many such cases, the standoff includes a disagreement about what the government, say, should do about the matter in dispute, concerning which it is empowered to act.

D The matter is frequently such as to present a third party — the government or the courts — with what William James called 'a forced option'. So not taking any action is equivalent to acting for one of the contending parties and against the other(s).

[3]Not to forget the O.E.D., Mum and Dad, Dear Abbey and Albert Einstein.

[4]Or the trade union, or the university, or the political party. The demographic conditions for the attribution of standoff to groups include group-percentages involved in the disagreement, the standing of the disputants in the group, statistical distribution o the issues in contention, the momentousness of the issue, and so on.

2. Extremism

It is time for an example. It is perhaps not an example of much interest for western European ears, but it currently engages in the two northern countries of North America. (For ease of exposition I shall concentrate on the state of things in Canada.) It is really two questions. First, is abortion ever morally permissible and, second, should abortion be subject to provisions of the criminal law? Whatever one thinks of the matter personally, it is clear that, for the present government of Canada, the criminal code aspect of this question is nothing short of a forced option. For, in present circumstances, to do nothing would leave it decriminalized — and thus would make it a lawful practice for any Canadian wishing to avail herself of it and for any legally competent practitioner willing to provide it.

Standoffs of force three, especially those that fall into the supplementary categories A to D, are nothing but derangements in communities in which they occur. In *extremis*, they are the occasion of civil war, but even in more composed circumstance they are the bearers of civil strife and not uncommonly the triggers of criminality. Prior to the decriminalization of abortion (by recent action of the Supreme Court), there were those who, unlawfully, operated abortion clinics, and those who, then as now, gathered to harass their would-be clients or, in some cases, who firebombed the premises.

I propose now to consider a way out of this black hole that is approximately within the bounds of dialectical effectiveness.[5] In so doing I well recognize that even if the rules of dialectic did show us a way out, it may be too late for the government, the passions in the country being what they appear to be. But at least the government would then have cogent means of resolution to offer its citizens, never mind that they might not avail themselves of them.

So the question is: are there, despite appearances to the contrary, dialectically tenable resolutions of force three standoffs of the category A to D sort? There are two remedies worth considering.

I Look for a heretofore unrecognized 'tie-breaking' methodological principle, one on which the protagonists might be expected to agree, and apply it.

[5]But can this be done consistently with the prior specification of standoff as a disagreement at a time t without the means of resolution at t? Though success cannot be guaranteed, to say the least, it is a move consistently available to our impacted dialecticians, for it is little more than a reconsideration of the issues involved and an attempt to identify relevant considerations heretofore unrecognized.

> II Try to determine whether one or the other of the theses in dispute admits of a certain kind of abstract re-description on which the protagonists can agree. Then award the laurel to the argument which instantiates the abstract re-description. That is, try to employ *analogical* arguments.

I shall here concentrate on remedy I. Remedy II I have discussed in another place ([Woods and Hudak, 1989] reprinted here as chapter 16).

As for strategy I, perhaps the most-used principle is *Ramsey's Maxim.* Regarding disputes about fundamental issues that seem insusceptible of consensus, F.P. Ramsey proposed that:

> in such cases it is a heuristic maxim that the truth lies not in one of the two disputed views, but in some third possibility which has not yet been thought of, which I can only discover by rejecting something assumed as obvious by both disputants [Ramsey, 1931, pp. 115–116].

Here is one way in which *Ramsey's Maxim* would appear to work. I might imagine two discussants who are deadlocked on the question of capital punishment. If they sought relief from their standoff by summoning up *Ramsey's Maxim* they might decide to abandon an assumption which, until now, they had shared. Namely: that the correct answer to the capital punishment question will be, in a sense to be elucidated, an *extreme* answer. That is, that the correct answer will be either that it is always morally permissible to invoke the death penalty in circumstances K or that it never is. If this assumption were abandoned, room would have been made for agreement that capital punishment is sometimes permissible in circumstances K, that is, always and only permissible in circumstances K*, where K* is, for example, the union of K and some further factors K†.

The example at hand turns on a conception of extremism about which it would be advisable to be a bit more exact. Let us say, then, that for a disputed question ?p, there exists a contention space, *CS*(?p), of its admissible answers. Membership in *CS*(?p) carries no implication of correctness or plausibility. *CS*(?p) contains all and only those responses that count as answers to ?p, not necessarily good answers.

For numerous questions ?p it is clear that they admit of answers in the form, 'p always', 'p never', and 'p sometimes'. Even where ?p admits of a bimodal response, Yes/No, frequently it is possible to provide answers of the 'always', 'never', or 'sometimes' sort which will imply Yes or No. For example, if the question were "Is the parallel postulate valid?" it would be admissibly answerable by "It holds for all cases", which implies "Yes", or by "It holds for no case", or "It holds for some cases (only)", both of which imply No.

Contention spaces whose admissible answers include responses of these kinds permit the following definitions.

Definition One: A position taken in regard to ?p is *extreme* iff it is fixed by an admissible answer, i.e., a response drawn from *CS*(?p), in the form 'Never' or 'Always'. Accordingly,

Definition Two: A position taken in regard to a question ?p is *moderate* iff it is fixed by an admissible answer in the form 'Sometimes (only)'.

To be sure, this is a rather crude interpretation of extremism and moderation, but it could easily enough be refined by introduction of the relations of more or less, with appropriate adjustment of quantificational idioms. Such refinements are not needed for present purposes.

Consider now the question, "Is the merely recreational torturing of babies wrong?" and imagine that someone replies that it is occasionally wrong. By the provisions of the present account, this is not an extreme answer, but a moderate one. In fact, however, it is nothing *but* an extreme answer for a further conception of extremism that clusters around such notions as excessiveness, outrageousness, preposterousness, incredibility, exorbitance and monstrousness. And it is obvious that for any question admitting answers that are extreme in this second sense, such answers are not wanted. So I might now speak of a second maxim, the *Maxim of Moderation*: Avoid extreme answers.

Now it is perfectly true that some questions are such that, by *Ramsey's Maxim* and the *Maxim of Moderation* alike, the same answers will draw the same verdicts of extremeness and moderateness. For many people, "Is spanking your child all right?" is just such a question. By *Ramsey's Maxim*, "Sometimes" counts as moderate, and the same is true for the *Maxim of Moderation.*

That said, it is easy to see that my two maxims do not coincide in the general case and that it would be a huge mistake to proceed as if they did. It is not an abstract folly, however, for some philosophers have either committed it or been drawn to it.[6] The two maxims are almost certainly jointly invoked by any suggestion that disputes should always be settled by taking a middle position — and this I might note is the *Fundamental Law* on collective bargaining. In fact, it is easy enough to take *Ramsey's Maxim* in this way, that is, as an expression of the *Fundamental Law,* though I must say that it is not entirely clear that this is what Ramsey intended.[7]

[6]There is plenty of textual evidence that J.S. Mill, among others, was at least drawn to this folly — perhaps in moments of uncharacteristic carelessness — and it may be that L.W. Sumner has committed it in [Sumner, 1981]. See [Woods, 1985]. I note in passing the equivocation on 'extremism', which undergirded the yowls directed against U.S. presidential candidate Barry Goldwater, following his speech to the Republican Convention in 1964: "Extremism in defence of liberty is no vice".

[7]Ramsey forwarded his maxim in the course of his discussion of a difference of opinion between Bertrand Russell and W.E. Johnson about whether terms designating or signifying

In any event, jointly resorted to, the maxims provide that for any disputed question, whose *CS* displays the appropriate quantificational grid, the best answer is the moderate answer (because least excessive and most credible), and this will be an answer (grossly), in the form 'Sometimes (only)'. Jointly resorted to, anyone locked in a standoff concerning the recreational torturing of babies could escape the standoff by agreeing on 'Sometimes'. But this is absurd. Answers moderate by *Ramsey's Maxim* (or for that matter by the *Fundamental Law* on collective bargaining) can be extreme by the *Maxim of Moderation* — extreme and quite mad. So there is no doubt that some questions that land their answerers in standoffs of force three will not brook resolution by way of *Ramsey's Maxim.*

We now have at hand a further characterization of force three standoffs in categories A to D. They are disagreements concerning which both disputants are of the view that the correct answer must be extreme in Ramsey's sense, that any Ramsey-moderate answer must violate the *Maxim of Moderation.* Such disagreements are non-negotiable in a now fairly precise sense.

Standoffs of this kind pose special problems for public policy formation. When the issue is of a type appropriate for government pronouncement, the government finds itself faced with a forced option, as I have shown. As a standoff of force three, the abortion question would seem to be rationally irresolvable with present and foreseeable dialectical resources. Being of force three and an issue of public concern involving substantial percentages of the whole population, anything that the government does, including nothing, meets with damaging consequences. Here are two:

1 At least one group is guaranteed to be infuriated and scandalized in special ways — the ways of memories that are bitter and long, and the ways of disapproval that are reactive rather than notional.

universals can function both as subject and as predicate in (atomic) propositions. Russell says No, Johnson Yes. The assumption which they share, according to Ramsey, is that the logical form of atomic propositions is knowable. And, siding with what he takes to be Wittgenstein's position (Ramsey is writing in 1925), Ramsey denies this: Since Russell's No and Johnson's Yes both imply this assumption, No and Yes are both wrong [Ramsey, 1931, p. 134]. *Ramsey's Maxim* seems always to provide for handling disputes by way of something like the *Fundamental law* on collective bargaining. All disputes concerning "fundamental questions" neither side of which they can be "disproved" [Ramsey, 1931, p. 115], must be *mistaken*, or anyhow, such that they should be given up. The trouble, of course, is that if Ramsey is right in the particular case, both Russell's and Johnson's case *can* be 'disproved'. So, as I say, it is not obvious what Ramsey's point is. Here I shall take the historical liberty of associating it with the *Fundamental Law.*

2 The government cannot truthfully (or anyhow convincingly) claim that its proposed actions are undergirded by factors that make for rational settlement.[8]

Thus such standoffs are precisely those on which the government founders. They are standoffs that will not yield to the words of Edmund Burke:

> Certainly Gentlemen, it might be to the happiness and glory of a representative to live in the strictest union, the closest correspondence with his constituents. Their wishes ought to have great weight with him; Their opinions high respect; their business unremitted attention.... But his unbiased opinion, his mature judgement, his enlightened conscience, he ought not sacrifice to you.... These he does not derive from your pleasure; not from the Law and the Constitution.... Your representative owes you, not his industry only, but his judgement; and he betrays, instead of serving you, if he sacrifices it to your opinion.
>
> [I]ssues which the member is blindly and implicitly to obey, to note and argue for, though contrary to the clearest conviction of his judgement and his conscience: these things are utterly unknown to the laws of this land, and which arise from a fundamental mistake of the whole order and tenor of... my constitution.

So I see that there are some legislative measures that significant portions of the electorate will not consent to,[9] and that this failure of consent is so utterly basic as to abridge any overriding notion of social contract. Such measures are an ultimate and radical insult to conscience.

It is no good to remind people that their consent is already implicitly pledged, for it lies in the nature of such disagreements that they resist the presumption of prior consent. Whereupon another distinction presents itself.

> *Definition Three:* A force three standoff fulfilling conditions A to D is a *political* standoff in a population **P** to the extent that **P** acknowledges the of prior consent to a legislated solution by a lawfully constituted government, recognized as such in **P**.
>
> *Definition Four:* A force three standoff fulfilling conditions A to D is a (purely) *moral* standoff in a population G to the extent that G refuses the presumption of prior consent to a legislated solution by a lawfully constituted government recognized as such in G.[10]

---

[8]Things are made more complicated, though not essentially different, if MPs are likewise divided over the issue at hand.

[9]This resistance would extend to a plebiscite, by the way.

[10]Cf. Dewey:

> Conduct as moral may thus be defined as activity called forth and directed by ideas of value or worth where the values concerned are so mutually incompatible [here Dewey will reveal a sunny optimism] as to require reconsideration

Let us now say that

(iv) a standoff is of force four to the extent that it satisfies Definition Four.

Even standoffs of force four admit of the prospect of *future* resolution within the bounds of dialectical rectitude. For such a possibility, I need a distinction between open-minded and closed-minded disagreement.

Definition Five: A disagreement is open-minded when its protagonists acknowledge that the opposite opinions are 'real possibilities'.

Definition Six: A disagreement is closed-minded just in case it is not open-minded.

Open-minded disagreements, even when they are standoffs of force four, are potentially resolvable by a Pascalian minimax strategy, which provides as follows: If the opposite positions are real possibilities for their respective advocates, then so constrain public policy that, of the options in question, it is best to back the one of least morally harmful consequence should it prove to have been the mistaken option. The manoeuvre presupposes, what might not always be the case, namely, that a coherent notion of least moral harm is specifiable for the options in question and that the disputants agree as to what it would be. If so, the following suggests itself for the abortion deadlock. For ease of exposition, I will assume that the standoff pivots on the question of fetal personhood. Thus, given that the contending positions are real possibilities for the respective disputants, there are two possible mistakes to take note of:

M1 This is the possible mistake of holding that the fetus is a person and on that basis supporting measures that would encumber countless women with unwanted pregnancies.

M2 This is the possible mistake of holding that fetuses aren't persons and on that basis supporting measures that would permit the killing of countless persons.

A Pascalian remedy would then attempt to adjudicate between M1 and M2 and to determine the lesser moral risk. Rough though it is,[11] the idea of the slighter risk turns on the prior notion of 'real possibility'. It is tempting to take real possibility as a theoretical primitive or to say, perhaps, that real possibilities are those for which Pascalian measures are (acknowledged to be) intelligible. But it would be best to be somewhat more forthcoming. So I shall propose

---

and selection before an overt action is entered upon [Dewey and Tufts, 1978, p. 194].

[11]And it is very rough. See, for example, [Woods, 1987; Woods, 1978]

Definition Seven: A proposition A is not a real possibility for a cognitive agent X iff

(a) X disbelieves that A

(b) X believes that he knows that not-A

(c) X acknowledges the possibility that he does not know that not-A

(d) X disbelieves that (c) is any reason for him to believe that he does not know that not-A.[12]

Definition Eight: A proposition A is a *real possibility* for X iff it fails conditions for being not a real possibility for X.

Whatever I make of my Pascalian ruminations of late, it is clear that the notion of a closed-minded disagreement gives rise to my final category of standoff. For

(v) a standoff of force five will be a force four standoff which is also closed-minded.

The idioms of closed and open-mindedness are perhaps misleading in a way that we should quickly correct. In particular, in a closed-minded disagreement in our sense there is no imputation of unfairness or of dialectical misconduct. Standoffs of force five resemble, by the way, what Robert Fogelin calls 'deep disagreements'. A deep disagreement about an issue, about reverse discrimination say, is one in which the argument is disabled by a disagreement about underlying 'framework principles', for example, about whether collective rights exist. I take Fogelin to be saying that so long as the framework dispute persists the disagreement about reverse discrimination is irresolvable in principle. My own view is that notwithstanding the persistence of the framework dispute, it might be possible to seek out and find common ground elsewhere which would tip the balance of the discrimination stalemate. If I am right about this, force five standoffs make agreement impossible only relative to the cognitive and procedural resources at hand for the disputants. So force five standoffs are not in the general case deep disagreements in Fogelin's sense [Fogelin, 1985].

3. Dealing with Standoffs of Force Five

It may be worthwhile to pause and consider whether the existence of force five standoffs is guaranteed by the Arrow Barrier Theorem [Arrow, 1963]. This celebrated theorem says that no social utility function answering to a conception of the general will can be got by aggregating individual preferences. The idea of *la volonté générale* is incoherent.

[12]Cf. the definition of chapter 5.

As it happens the answer to my question is No. Standoffs precede the behavioural basis of the Barrier Theorem, namely the expression of preferences by voting. Force-fivers won't put the issue at hand to a vote, not because they disbelieve that the vote will evince the general will, but rather because they are not prepared to lose it. A force-fiver, therefore, could be described as one who resists resolution of a conflict by way of the general will even assuming that there is such a thing.

What he *can* do, of course, is insist that his unwillingness to submit to a vote, apart from his force-five attachment to the issue in question, is not without rational foundation. Voting, he may say, is defective precisely because it fails to decide issues by way of a coherent social utility function. At the end of the day, it doesn't matter what force-fivers think of such things. For a standoff of force five is one that a protagonist is unwilling to surrender to any mechanism which might decide against him.

Governments, the dominant élites (to borrow Walter Lippmann's term) and the whole citizenry have a large stake in discouraging force five standoffs. Force five standoffs are the non-meteorological counterparts of force ten hurricanes. They require one to batten down the hatches and to head inland. And clearly it would have been better had they never cropped up.[13] But crop up they do, more in some societies than in others. It might be supposed that the propensity to standoff is greater the more a society is liberal and pluralistic in its dominant cultural arrangements. Whether such a culture will be beset by standoffs of force five, rather than by disagreements of lesser rank, will in large measure be function of what it is pluralistic *about* and so will turn on whether it fosters differences in value appropriate to the greater intractability of force five. It is sometimes apparent that certain kinds of multiracial pluralism are ripe occasions of force five turmoil, but it is by no means always so.

Optimists will seize upon the following possibility. Government and the dominant élites should marshall an argument something like this, and broadcast it widely and often:

[13]Some naifs celebrate pluralism about values as occasion for the promotion and intensification of normative conflict. Cf. Berlin, who counsels against attempts to eliminate normative conflict [Berlin, 1978]. Doing so, in the general case at least, incurs "the loss of a sense of loss" and a flattening of human experience [Williams, 1981, pp. 80 and 82]. This does seem rather precious. In fairness to Williams, he does say that normative conflict is not always pathological; and this leaves plenty of room to regret it when it is (as it clearly is in the case of force five standoffs). And Williams' metaphor of the flattening of experience is, in some ways profoundly apposite.

> *Last Gasp Dialectical Response, (LG):* "Citizens, the controversy which currently troubles my country seems impossible of resolution among the contending parties. Even the government's role is compromised. For although the raging dispute places at risk the public safety — certainly the public amity — anything the government might propose to do will, if it makes for any change at all, change things for the worse.[14]
>
> What is at risk is that minimal social harmony which is the very condition of a good and peaceful life [and, more to the point, of the general practice of dialectical felicity]. The government urges that the collective self-interest requires that tensions be lessened and that the most serious consideration be given by all to settling this matter [say, by a free vote in the House of Commons]".

Here the government is openly proposing the voluntary collective downgrading of the dispute from a force five standoff to a force three standoff which fulfils the conditions on being a (merely) political disagreement. In this proposal, it is not suggested that people change their values but that, out of consideration for the commonwealth, they not act on them intractably or unlawfully.

I shall not take the time to comment in any detail on the complexity of the issues raised by *LG*. It forces an attentive hearer into decisions about whether rational self-interest (ever) overrides moral certainties, and if it does, whether this represents the supremacy of non-moral value over moral value, or whether the overriding value is continuous in nature with the values it subordinates. But let me hastily make this point: If standoffs of force five push protagonists toward broadly utilitarian remedies even against their conscience, there is a moral theory that stands ready to admit all comers and to administer a palliative for the troubled heart. The theory could, and deserves to be, called Shifty Utilitarianism. It provides that all non-utilitarian and anti-utilitarian policies and practices which a society adopts as permissible have an ultimately utilitarian sanction. Thus it is utilitarianly justified to pronounce and enforce, as correct, policies and practices that contradict utilitarian precepts, if the costs and benefits pay out appropriately. Shifty Utilitarianism is thus tailor-made for morally sensitive policy-makers. It won't do, by the way, to suppose that various non-utilitarian principles are perfectly correct, morally speaking, just until their utilitarian costs get too high. For the utilitarianism here considered is altogether pre-emptive; it is Shifty Utilitarianism, the ethic that dare not speak its name.

Dialectically speaking, *LG* is not a bad idea, but how it fares as strategy is another matter. Recall that I said that a country caught in a force

[14] The government is landed in a quandary. Quandaries are possessed of considerable logical interest. See [van Fraassen, 1973; Hamblin, 1978]. Cf. [**A**quist, 1984].

five standoff is *in extremis* a country poised for civil war. True, a country *can* drift into a civil war unawares, and a friendly warning from a vigilant government might give the country pause. But if the country knows that its contentions are leading to civil strife, that is almost certainly because that the country intends such strife, thus the meaning of force five standoffs *in extremis*. In those cases, admonitions from the capital would be laughably inapposite. The same points also hold for the lesser derangements of polity: general strikes, riots, massive demoralization, and so on.

4. Education as Strategy: Persuasion and Thought Control

I conjecture that it is precisely for this reason that governments tend not to prefer the dialectical provisions of the *LG*-strategy. Something else lies ready to hand and it is available to the entire network of the country's rulers, its dominant élites, and it enters the spiritual marrow from a pervasive cultural mist. It is called *education*.

Education is here the education appropriate to the normative rationalization of a liberal, pluralistic population. It is, or involves, the inculcation or transformation of values by means of the social technology of persuasion. Dialectic now defers to rhetoric. Logicians tend to be dismissive of rhetoric, legatees all of Socrates' hostility to the Sophists of yore. I am not one of them. If concerning any issue there should happen to be a fact of the matter, then it would be oddly illogical not to want to present the truth persuasively, that is, in ways that maximize the chance of getting others to see that it is true. Most interesting are those cases concerning which no consensus exists with regard to the question of truth. For wide ranges of such issues, going fishing is not a realistic option, and we are left with the hard question of what to do in the wake of intractable disagreement. Some issues require the fixation of belief, never mind that the matter at hand is underdetermined by the agreed-upon evidence (thus every scientific theory worthy of the name). So a central task for rhetoric is to specify measures for the fixation of belief about matters are acceptable. It is a very hard question, as is evidenced by the frequency with which belief-fixation is accomplished by measures thought to be seriously unacceptable, as we will now see.

There is a huge literature regretting the social technology of persuasion. Much of the worry has to do with its concealedness, its manipulation and its sheer efficacy. People complain that the technology (i.e., the mass media), is concentrated in too few or in any case the wrong hands. Others fret about the domination of economic considerations over the selection, duration and mode of transmission of mass media signals.

Some say that the majority of such transmissions are continuous with mass advertising which, in turn, is continuous with propaganda, and that the continuity-relation is transitive. Still others are troubled by the monolecical inaccessibility of the media or by the fact, if it is a fact, that televisual argumentation is structurally illucid for the ordinary viewer and thus beyond his competence to assess.[15]

Of two things I have no doubt. One is that much of the mass media bashing of recent years is half-baked, paranoic, and politically self-serving. The other is that there is evidence enough of thought-control to warrant my concern and chagrin.

I want, however, to return to the business at hand and to concentrate on the rhetorical targets of the dominant élites in the course of the education in the 1980s and 90s of the Canadian public on the matter of abortion. I am no conspiracy theorist but if I were to invent a conspirator, say, a secret Ministry of Correct Thinking, I would attribute to it objectives consonant with my claims, and that, I suggest, lends them further support. But that said, I am *not* suggesting that agendas for the social technology of persuasion are entirely or principally the deliberate plans of individual technocrats. Such agendas are to some extent implicit in the culture itself.

The abortion issue gave the appearance of having become a nasty force five standoff in Canada. Objective: neutralize it, reduce it to nothing of higher rank than a political standoff. *Subsidiary objectives*:

I Discredit the leaders as extremists. Show them in disagreeable, noisy, uncivil and, if possible, criminal contention. Hope for early boredom.

II Demoralize the population. Restrict the national discussion to shrill set pieces from the *parti pris*. Do nothing to encourage the notion that this is an issue of complexity and perhaps of central national importance (merely saying that it is is another matter entirely).

III Marginalize the visible protagonists. Take care to identify Ms. X as a member of the Feminist Caucus and Mr. Y as a Mormon bishop. For all its dialectical limitations, endorse and promote *Ramsey's Maxim*: the most sensible solution is a compromise. It sounds so right.

IV Trivialize the contending values, not only as extreme but as *moral*. Thus moral choices are private, a matter of personal opinion only, are not to be taken up invasively ("forcing your personal values upon me"). Invoke the non-cognitivism of moral principles. Emphasize their relativity.

V Adjust the taxonomy. These extreme positions are also religious positions, fine as long as they don't intrude and, in any event, subject by implicit prior consent to the sanctity of the separation of Church and State. Since it certainly

[15] Here is a very small sample of literature. [Cantril, 1940; Innis, 1951; Grant, 1969; McLuhan, 1964; Ellul, 1965; Altheide, 1976; Geis, 1982; Govier, 1988; Herman and Chomsky, 1988; May, 1988; McMurtry, 1988; Langsdorf, 1988; Dolan, 1988].

isn't up to the Church to decide for all, the right of decision must be the State's to take or not. So then they really *are* political means, including governmental hands-off, if need be.

VI Saturate communications with euphemisms, the more vapid the better (thus 'pro-choice', 'pro-life'). Keep disclosure of clinical details to a minimum (no footage of an abortion; no footage of a thirteen year old's labour). Be 'tasteful'.

VII Guilt by association. President Ceauscescu ran a monstrous tyranny. He imposed absurd and burdensome reproduction targets on Romanian women, in an effort to increase the population substantially. Abortions were also forbidden. Now *all* that nonsense is over.[16]

There is reason to think that by the 1990s the Canadian population was indeed getting weary of this business, and that the drift of feeling was toward some kind of negotiated settlement by way of the *Fundamental Law* on collective bargaining. If this is right, then the overall objective was met, and a purely moral standoff was downgraded to a political standoff.

If my claim is right, it cannot be denied that the dominant élites' handling of the Canadian abortion mess was dialectically malodorous in all sorts of ways. But it must not be said of these dialectical felonies that they commit the further crime of displacing intellectually pure procedures for rational consensus. They are not capable of doing that. For recall, I have a limitation theorem. No such procedures are available. Such, in part, is the story of logic and politics.

[16] "Of all Nicolae Ceaucescu's legacies, perhaps the saddest can be found in Romania's maternity wards. Visitors are shocked at the sight of abandoned babies, sick pregnant women and doctors working 24-hour shifts to perform abortions — all because of a mad plan to increase Romania's population" [Economist, 1990a].

Chapter 13

STANDOFFS AND DEMORALIZATION

1. Demoralization

In the present chapter, I extend the analysis of standoffs to an examination of two important matters. One is the development of social policy in the absence of a substantive settlement of a standoff of high level. The other is the reaction of loser-groups created by the character assumed by such charges in social policy.

It is important to be clear about the normative character of standoffs of the sort we were examining in the previous chapter. I shall say that a subject S holds a *deontic* belief B *proactively*, if and only if,

(a) B is a belief in the form "Nobody should do action A"

and

(b) Holding B commits S to the truth of the proposition, P, "Everybody should take appropriate steps to prevent the performance of A by anyone".

The proactive commitments of S's subscription to a deontic belief B can, in turn, be characterized as follows: Anyone who satisfied (a) but who did not act in fulfillment of P is

pragmatically inconsistent[1]

and

guilty of wrongful omission by his or her own lights.

[1] Again, at a gently intuitive level, a charge of pragmatic inconsistency echoes the familiar complaint against "not practising what you preach". A deeper analysis reveals how surprisingly pernicious this breed of inconsistency actually is. See [Woods, 1993b], reprinted here as chapter 6.

Conspicuous by its imprecision is the requirement of the proactive deontic believer to take 'appropriate steps'. I shall not attempt to subdue that imprecision beyond pointing out that, in a rough and ready way, preventative measures are held to expectations of proportionality and procedural normalcy. If I think that students should not wear baseball caps in my classes, it may be enough of a discouragement that I merely say so. Or if I see that my neighbour's house is being burgled, phoning the police will suffice in the general case. If a toddler is being physically savaged by a drunken brute, and I am the only third party at the scene, it may be that I will have to intervene directly. But if I am ninety-three years old and much reduced by arthritis, my intervention may have to be more honorific than actual.

A further imprecision affects what logicians call the *quantifiers* embedded in clauses (a) and (b) of our definition of proactively-held deontic beliefs. These are the expressions 'nobody', 'everyone', and 'anyone'. Quantifiers have *scopes* or ranges of applicability. They 'range over' classes of individuals to whom their attached predicates are presumed to apply. If I think that everyone should try to discourage government corruption, whether I myself am bound to take such steps will vary as between my own government and, say, the government of France. Whatever the details of such variation we may take it that I myself am excluded from the range of quantifiers in the proposition "Everybody should try to prevent government corruption" in its application to France.

The Canadian parliament has abandoned all efforts to legislate the issue of abortion. Something else happened. A standoff of force five plummeted not to a political standoff of force three, but to a standoff of *force one*. With respect to abortion, Canadians have in effect agreed to disagree, a particularly striking outcome given the issues involved.

We see in this a consequential overdoing of the rhetorical technologies. The target was to downgrade the abortion controversy from the dangerous perch of force fivedom to a level on which Parliament could work; that is, the objective was to convert a closed-minded moral disagreement into a resolvable political one. But something else happened. Parliament tried to treat abortion as a political problem; it openly connived in the conversion of a moral irresolvability into a political resolvability. Parliament failed. And this, it may be said, was provocation of further degradation to level one. It is degradation in two senses. The standoff-grade is the lowest possible; and a sizeable percent of the Canadian electorate is seized of a fundamental matter of conscience on which it is now prepared to be inert. In the steep fall to force one, millions of Canadians have decided to do nothing about what they believe with unshaken conviction to be sixty thousand murders a year. In thinking so, they may well be wrong. But they do think so, and this means that the technology of persuasion has, under our very noses, transformed a noisy and even dangerously fractious population into a profoundly demoralized one. It is not a felicitous outcome.

There is plenty of evidence that the people of Canada face the new millennium with lowered spirits, but none that their lack of morale derives at all significantly from the present situation in the country as with regard to abortion.

By 'demoralization' I do *not* here mean a lowering of spirits, or any particular variety of lassitude. Such a meaning is comparatively new, ensuing from the past century. In the century before it, demoralization was the corruption of morals. *The Oxford Dictionary of English Etymology* (1966), reminds us that the French equivalent *demoraliser*, a word of the French Revolution, provoked LaHarpe's scorn: "*Si Demoraliser pourvait etre francais, il signifierait cesser parler de morale*". This, the eighteenth century sense, is closer to what I myself intend.

It is one thing to state the case you intend, it is another thing to make it. Doing so is the task of the present section. Its development will be facilitated if, from time to time, we replace the example of abortion in Canada of the 1990s with the example of the treatment of German Jews in the 1930s and 1940s. A great many people are infuriated by the abortion-holocaust analogy. In this they may be right. I myself intend no such analogy. I want rather to ask about a society in which the question of whether Jews should be exterminated is a question that divided opinion in ways that qualify it as a standoff of force five. I want then to consider, if the Canadian "solution" of the abortion question were applied to the Jewish question, how that state of affairs should best be understood.

Consider a community K of B-holders, where B is a proactive deontic belief. Suppose that in the broader community, U, of which K is a part, B is disbelieved and resisted by an opposite proactive belief. As it stands, in our definition of proactive deontic beliefs the quantifiers 'nobody', 'everybody', and 'anyone' range over the member of U. But if the disagreement between K and U with respect to B is downgraded from one of force five to one of force one then, in effect, the quantifier of B is itself restricted to K; that is, its scope is cutdown from U to K and is a perfect illustration of the ancient wisdom of "minding your own business" and "not imposing your own personal values on others". B, recall, is a proposition in the form "Nobody should do A". And since it is held proactively, it commits its holders to proposition P, "Everybody should take appropriate steps to prevent anyone's performing A". The quantifier cutdown already noted for B, may also apply to P. That is, if the issue in question has been downgraded from a standoff of force five to a standoff of force one, proactive holders of B are, in effect, committed to the proposition that everyone in K is obligated to take steps to prevent anyone in K from doing A. We then see that there are two subcases to take note of. One is the case in which

> I Members of K shrink the scope of the quantifier in B to K itself. (This is the case in which, for example, Dutch Reformed Christians come to believe that *they themselves* should not participate in abortion.)

In contrast is a second case, in which

> II the quantifier of B remains fixed over U, but the quantifier of P is restricted to K. (Thus, if a Dutch Reformed Christian believes that no one should participate in abortion, he is in a situation of type II if he also holds that *all Dutch Reformed Christians* should take steps to prevent *any Dutch Reformed Christian* from participating in abortions.)

In judging situations of types I and II, perspective is everything. From where a pro-abortionist stands, these are exactly the situations to be hoped for in a pluralistically tolerant society. But from the point of view of the anti-abortionist, both I and II are troubling. In situation I, the anti-abortionist decides to hold his strong belief against abortion in so 'cutdown' a form as to approximate contradiction. For although he believes that no one should participate in abortions, the claim of situation I is the lesser one, that no K-member should be thus involved. The anti-abortionist has settled for a good deal less than what he actually believes. Situations of type II avoid the problem with respect to the scope of the belief that abortion is wrong (*viz.*, for everyone), but they reduce the scope of preventative action. A B-holders' position becomes one in which, although he believes that no one should perform or receive abortions, he is not prepared to take preventative measures beyond those who already share that belief. We see, then, that a situation-II anti-abortionist is in violation of the definition of his own proactive belief. In restricting the range of preventative action in the ways here noted, he stands convicted of pragmatic inconsistency and moral wrongdoing, by his own lights.[2]

---

[2]It is worth noting how structurally intractable a type-II situation is. It is natural to suppose that members of K would aspire to a condition in which their role as "losers" of the contention in question is reversed, with their opponents now bearing that mantle. On the substance of the issue they would be quite right, against by their own lights, so to aspire, but structurally speaking it is no improvement. The pattern of alienation would merely have shifted from K to those who disagree with K. Either way there would be a significant chunk of the larger society which, in its own judgement, would be pragmatically inconsistent and guilty of non-trivial wrongdoing.

This being so, it cannot be an unmixed blessing for the pro-abortionist that his opponent's position has been downgraded from one of force five standing to one of force three standing or lower. Such is a transition that condemns his U-compatriot to pragmatic inconsistency and to the conviction that he is guilty of moral wrongdoing precisely for his having accommodated the pro-abortionists' own position. It is an unsatisfactory situation whatever the specific content of the disagreement at hand, provided that it was once a disagreement of higher than force three. It is not in general a good thing to have K's in U's that are false to their own beliefs in ways that convict them of inconsistency and serious wrongdoing.

A K-community in a situation of type-II is one that is alienated from the larger community U and whose membership in U is nontrivially compromised. People might, by and large, be prepared to tolerate such alienation where K is small and the issue in question sufficiently *recherché*, if not downright nutty. One might even venture to say that alienation and compromised U-citizenship (where U is a state), are the structurally inescapable constituents of normative pluralism in U, and (with Isaiah Berlin), that a multicultural U is in general a better thing to have than a normatively monolithic U.[3]

It might be proposed that to the extent that multiculturalism is a defensible ideal for polity, that situations of type II themselves must be defensible. Perspective being all, it depends on where the judges of this question stand on the issue whose situation-II 'resolution' constitutes the alienation. We have no hesitation in saying that a situation of type-II with respect to the Jewish question in Germany was monstrous. Germany's demoralization was radically complete owing to the collapse of resistance to genocide. No doubt it will be said that the policy towards Jews was self-evidently outrageous and that to think otherwise one has to be quite mad. This is the pathologization of normative disagreement, and from where I stand this is the right judgement in that case. But from where the racists of Germany stood any such judgement is dismissible out of hand. The relevant thing to say for present purposes has a peculiarly structured texture. It generalizes as follows.

> *demoralization rule:* For any group K, in a society U, and for any proactively held belief B with regard to an action A, which satisfy conditions on a situation of type II, U is in a state of demoralization approximating to that of Germany

[3] So we are recognizing that a normatively pluralistic society is not a matter of some people liking garlic and others hating the stuff, or of some people preferring to spend Saturday evenings at the Polish League's weekly dances and others opting for a night at the opera. Pluralism is a social condition with normative bite only to the extent that significant groups are significantly alienated from the whole.

in the 1930s and 1940s to the extent that members of K attach to action (or policy), A the same moral weight as opponents of the Jewish policy originally attached to it.

It is of little moment that members of K may be objectively wrong in their assessment of the moral gravamen of action (or policy), A. What signifies is that K, and U too, are in a situation of type II and that they could not be so unless, in particular, the K-community believed unshakably that its assessment of the moral gravamen of A *was* objectively correct. Thus the relevant question to ask is whether it is a good thing for societies to be in a state of affairs that models our demoralization rule? The answer is clearly 'No' to the extent to which it is a corollary of the demoralization rule that

Demoralization Corollary: For any K satisfying the structural conditions of the demoralization rule, members of K will judge themselves and their larger society to be guilty of moral cowardice and political impotence with regard to a matter of great moral significance.

That a group such as K could well satisfy the Demoralization Corollary in a merely tacit way only adds to its ills. For now it is a state of bad faith or self-deception.

What is to be done in a society[4] in which a demoralized group K takes the issue on which it has 'lost' to have a gravamen approximating to the killing of Jews in Germany? Clearly the broader society of which K is a part has a huge stake in promoting a shift from the type-II situation in which it finds itself. There are three options available for its consideration:

Option 1: It can try to move itself into a situation of type I. That is, it can try to convince members of K that its abhorrence of action (or policy), A is applicable to the performance of A only by members of K themselves.

Option 2: It can try to re-energize the members of K, and to encourage them to resume strategies of intervention with respect to the performance by A by anyone in the broader community U.

Option 3: It can try to reduce the standoff of force five with respect to the issue in question to a standoff of force four. That is, it can try to persuade disputants to be open-minded, rather than close-minded about its disagreements.

Of the three, the second option has the least to recommend it, in as much as it risks the encouragement of civil strife. When abstractly considered, option one has its attractions. For it is tantamount to attempting to use

[4]The idea of what a society might 'do' about a problem is, of course, a huge expository convenience. I shall leave this notion to function intuitively, beyond saying that as here applied it is the relevant totality of a society's varied devices for forming public opinion and conforming public behaviour to it.

technologies of persuasion to convince members of K that their harsh judgement of A is valid only as applied to themselves. But we need only remind ourselves that option one was as available to the Nazis as it is to our hypothetical state U to see that it is something that members of K would resist to the uttermost (and should).

This makes option three something of a standout. In it, disputants are encouraged to be open-minded about the object of their disagreement. For this to happen, each must recognize that his own position might be mistaken, that is, that it is a real possibility that what his opponent says is true. Recognizing that one's being mistaken is a real possibility is consistent with one's total confidence that one is not in fact mistaken. Acknowledging the real possibility that one is mistaken is not a probative matter, that is, it is no evidence that one is mistaken. Its significance, then, is procedural. Parties locked in a standoff of force four dispose themselves to the application of the resolution rule **RR**.

> **RR:** Settle the [force four] issue in such a way as minimizes the realization of the greater possible cost. The higher the cost the more it is mandatory to minimize the possibility of its exaction.

It is easy to see how option three and the **RR** rule which attends it would have settled things in Germany. Even the most ardent Nazi, if he recognized the real possibility that Jews have the full moral protections of human personhood could be made to see that if he were mistaken in thinking otherwise, the cost of the mistake would be mass murder; and, further, that this is a cost that outweighs any cost attaching clemency towards the Jews even if it chanced that they did not have the full moral protection of human personhood.

Although it has the look of a maximum rule in the manner of rational decision theory, *RR* is mute on the role, if any, played in its execution by the theorems of classical probability. *RR* is less a decision rule than a challenge. The challenge it makes is this: With regard to a dispute practice P, a rational agent will decide against P if he thinks that there is a serious enough possibility that P is wrong in a serious enough way. Since the part against whom the rule is deployed has not decided against P, *RR* has the force of requiring him to show that the possibility of error is not sufficiently high or the degree of mistake, if it were a mistake, is not sufficiently high.[5]

What is so deeply striking about **RR**-resolutions is that *they are arrangements that regulate behaviour without changing belief.* The Nazi who acquiesces in a **RR**-resolution of the Jewish question can go to

[5]Difficulties with taking *RR* as a strict maximum rule are discussed in [Stingl, 2004].

his grave with the untroubled conviction that Jews are subhuman and wholly without the moral protections of life, and yet in full consistency with that conviction quit killing Jews. It is a salutary arrangement.

Standoffs of force five are a structural embarrassment for any polity in which public policy is shaped significantly by any mechanism that cuts across the grain of popular disagreement. They are an embarrassment virtually guaranteed to eventuate in any polity that is significantly multicultural in its makeup.

In characterizing such embarrassments as structural, it is intended that standoffs of force five persist at instrumental cost. They are disagreements that disturb the workings of polities in which they occur enduringly; that is to say, they have the potential to threaten the very conditions under which a social arrangement even qualifies as a polity. And, as we have seen, there is no modern, technologically mature polity which lacks mechanisms of social persuasion whose function is to discourage, contain, marginalize, trivialize, or subdue these *crises de conscience.*

After consideration of the instrumental or structural debilities of standoffs of force five which are left standing, so to speak, it becomes clear that the central question is one of the quality of life in societies in which standoffs of force five are dealt with in the ways lately noted. But it is also a question of what the spinners will allow.

2. Spinning

I have not yet said what seems to me the more important thing about the linkage between standoffs of force five and the management by dominant élites of liberal pluralistic democracies. It is necessary to make it clear that the contemporary dominant élitedom embraces significantly more than is found in its ancestral moorage. Government, yes; industry, yes; the Church and universities, too; the Civil Service and the trade unions. And now the mass media, the advertising agencies and Hollywood; and MTV. But one must not forget that offspring of mass communications and the neighbourhood bully. This is the spin doctor, the media consultant, the 'spokesperson' and, at times, the head of government (see below), and they are the spearheads of single-issue politics. By its very nature, single-issue politics will in the general case be abjured by the citizenry of any complex and benign society. Nothing is worth fixing — except *in extremis* — independently of everything else. True, there are occasions, some majestic, some terrifying, others tragic, when a single issue does paralyze the mottled agenda of civil life. With luck they are a few and short-lived. Here a war, there a Black Plague, and yonder the collapse of an economy.

At the limit, such situations are disasters, and there is no élite in any society that will not spend up to its every effort in rallying the nation — telling lies, provoking countervailing hatreds, ransacking the Treasury, bullying the boy-conscript in an effort to preserve the commonweal. *In extremis*, force five standoffs tend to be pre-empted by such singularities. But they are pre-empted, when they are, not by a rational consensus which resolves them, but rather by consideration of their potentially lethal cost. People just quit their contentions or they enter into adjacent force five absorptions which oblige peaceability at home. Thus the splendid but bemusing solidarity of the British in World War II.

I return now to Canada. There was some years ago a major constitutional row underway (Meech Lake), and Canada has recently weathered an immense national effort to get clear about free-trade with the behemoth to the south. Canadians are familiar with aboriginal discontent and the usual disaffection with government and the big shots, unless perhaps they are the vulgarians of tabloid celebrity whose excesses fuel the fantasies and trigger muted jealousies and equivocal resentments. I say these things in order to suggest that, notwithstanding my earlier remarks, it is business as usual where I live — 'dull old Canada' as the *Economist* says in [Economist, 1990b]. So what *have* I been trying to say about the force five status of abortion in that country? I'll tell you; *it is a fraud.*

The sheer efficacy of mass media transmissions is seductive. The media are a standing encouragement to theatricality. People with access have discerned the untutored and all but insatiable appetite for *Sturm und Drang* of their consumers. Canadians, though not they alone, are vicarious and slovenly romantics. They love the fabricated prospect of disaster. The inexorable tug of sublimity is bred in the bone. I am going to say something further about abortion, but for now I want to introduce the paradigm case of my central point. The point is that people in control have noticed the allure of the *theatre* of force five behaviour and it is, for them, an irresistible strategic temptation.

The paradigm case is this. In the province of Québec, the late sixties and early seventies brought to the surface cultural tensions and, yes, even hatreds of approximately the kinds that Belgium has known and Spain is experiencing and that now faintly pullulate in parts of France, to say nothing of Azerbaijan. There were incidents of francophone hostility to the very idea of continuing geopolitical integration with a sprawling, incompetent, insensitive and *uninteresting* anglophone subcontinent. Some postboxes were blown up and some people died. A Minister of the Québec government — a coyly pro-confederation bunch

— was abducted and murdered. Soon after, the British Consul in Montreal was abducted and roughly used, and later released.

Things were bad in Québec then. Things were bad in many places. Rioting students nearly brought France to its knees; and the spoiled simpletons of Cornell University invaded the precincts of its president with such ferocity that he was stricken and died of heart failure (kids will be kids in Ithaca, New York). Some people in Québec wanted to separate from Canada. Others didn't. There is not the slightest doubt that some Québequois were locked in standoffs of force five over separatism. Some families broke apart and some marriages exploded. But there is no evidence that Québec itself was in a force five deadlock with Canada, or Canada with it. The necessary demographic conditions for group conflicts of such force and reach were not fulfilled. Single-issue provocateurs flooded the media with force five rhetoric. This was unsettling, but no more than that. Canadians were so far from head-thumping that the rebellion of the 1830s looked like a Disney flick. (Canadians thump their own heads very seldom. They await invitations from Europe.)

So Pierre Eliot Trudeau performed the paradigmatic act in the theatre of force five. *He invaded Québec.* Invoking the War Measures Act — since rescinded — the Prime Minister committed the armed forces of Canada to the task of putting down an 'apprehended insurrection'. The language of the Act was tailor-made for this, for the word 'apprehended' is thrice-wise ambiguous: *anticipated*; *alarming*; and *stopped.* Some people were detained and, in time, released. There was no loss of life. The bombings ceased. Things settled down. A few years later, in a provincial referendum, Québec decided to stay in Confederation, for the nonce.

Pierre Trudeau showed the marks of a great political leader. He manufactured reality. Canada was not in a force five standoff over Québec until the invasion made it so. In the end, Mr. Trudeau cashiered rhetoric for action. He caused the very calamity which his troops simultaneously were there to contain. In a single stroke he established that the nation was at force five and then dealt with it. Civil libertarians at the University of Toronto (where I was then) were furious with Trudeau. I thought that they largely missed the point.

The invasion of Québec is a limiting case of the point that I now come to. Single-issue politics are tailor-made for the theatre of force five. Nothing shakes a Minister more than the prospect that a conflicted issue is going to get out of hand in the press. It is an abiding inducement to the single-issue spin doctor, the spokesperson, the lousy pre-empters of civility, to represent disagreement as calamitous and to validate the representation by calling into play, or having one's cronies to do so, the appropriate symbols of débâcle (thus the televised 'riots' outside a Toronto abortion clinic). Put all of this on the telly and before long, if you know your rhetorical onions, the inert, slightly damp, national grey matter will shift its gears into collective suboptimality. And people will say, "Hey, this is bloody serious. This had better be settled".

And so, the common man and woman were bewitched, bothered and bewildered into rolling over on the abortion issue in Canada. He or she did so, not because the thing has been resolved, not because irresolution will blow the polity apart, but because the politics of force five theatre is subdued and numbed the national conscience with the fabricated, concealed, and vicious suggestion that the country had better cool it or else — precisely — the country will blow apart.

Some years later, I had occasion to ask Pierre Trudeau whether he saw the intervention in Québec in the terms that I have here described. He sweetly demurred, and talked of other things.

There is no intelligent and reflective democrat who does not hope that a dialectically felicitous and rationally honourable breakout from the black hole of the abortion contention could not have eventually been achieved, given time and good will and hard intellectual effort. But the very nature of the abortion problem implies that while wise people persist, stakeholders endure unacceptable, even momentous, indignities which their hearts will not allow.

3. Witnessing

A recent and important study in what I am calling the theatre of force five is that of Scott Jacobs in his chapter on witnessing and heckling, and limits on arguments resolution, in van Eemeren, Grootendorst, Jackson and Jacobs [Jacobs, 1993]. Jacobs reports on interactions at various American universities between fundamentalist preachers Jed Smock and Max Lynch, and campus audiences. These exchanges are routinely deadlocked, and the ensuring standoffs have have many of the features of force five. They are unresolvable and persistent, and there is no thought of third party resolution. The disagreements are close-minded, noisy, and abusive, and often they involve issues of public policy which neither party is willing to surrender to any resolution mechanism which might give what each party respectively would regard as an unacceptable result. But, intractable and abusive though they are, the case studies indicate a tendency to join and rejoin these standoffs mainly for the pure provocation of it. It is suggested that the preachers in particular have no interest whatever in persuading the apostates, but rather calling down their ire, since exposure to the abuse they incite serves as occasion for their purification and proof of their salvation. Clearly, then, the notion of a standoff of force five would need to be reworked if it were to capture the nuances of cases such as this.

4. Privatizing the Good

Radical pluralism rooted in what Ramsey would have called fundamental matters of conscience is a *de facto* political impossibility in a liberal democracy except under two broad constraints:

(i) for most issues concerning which there are fundamental contentions of conscience (i.e., contentions that are or are taken to be standoffs of force five) if there is a significant majority on one or other side of these issues, the dominant élites may validate that majority; or

(ii) concerning such matters, citizens may abstain from acting on their consciences in any way that carries non-trivial public consequences. They may 'speak out', but they do not impose, or encourage others to impose restrictions on the liberties of people having opposite opinions.

In briefer compass I can say that a polity can help itself to survive force five standoffs either by the device, as in (i), of the *tyranny of the majority* or, as in (ii), by the *radical privatization of conscience.*[6] Both are fragile arrangements, as we all know. A polity dedicated to strategy (i) risks the segregation of blacks in public places. A polity dedicated to strategy (ii) is a polity without a public conscience and it risks not having the resources to block the evolution of policies which lead to holocaust.

In any event, in cases galore the output of such arrangements is resisted from within and sometimes overthrown. The question of how much it is justified to persecute the Jews might move out of the moorage of conscionate inaction and receive a decisive answer by way of majority opinion. In plain English, there might arise a strong majority opinion that it's all right to persecute Jews to the limit. It is worth noting that what changes in these cases is the prior output of the strategies. The strategies themselves do not change. They are practically irresistible and this is accounted for, largely, by their enormous flexibility. In many countries today, strategy (i) suppresses minority conscience with regard to parental refusal of life-saving blood transfusion and also with regard to paedophilia, to take just two examples. Homosexuality in the western democracies was until recently also handled via strategy (i); it is

---

[6]Here again in Dewey:

> ... the severance of morals from human nature ends by driving morals inwards from the public open out-of-doors air and light of day in to the obscurities and privacies of an inner life. The significance of [the discussion of certain moral problems] is that it reflects precisely a separation of moral activity from nature and the public life of man [Dewey, 1976, p. 9].

now all but in the embrace of (ii). That, in fact, is one common way to evade the tyranny of the majority: try to place the issue in question within the provenance of strategy (ii). The civil rights phenomenon in America in the 1970s resisted the verdict of strategy (i) in a different way. It didn't change the strategy; it just changed the majority. The changes it sought to and did bring about not only did not change the strategic device of the tyranny of the majority, it presupposed its legitimacy throughout. And, under slightly more careful scrutiny, it is easy to see that homosexual activists have been transacting a joint campaign at once against the resentments now informing strategy (i) and in support of the accommodations offered by strategy (ii).

Whether it is the civil rights movement or the gay and lesbian liberation movement, or the incipient paedophilia rights initiatives in some parts of North America, I see that a more complex project has disclosed itself. It is a project right at home with the manufacture of consent. It is the effort to seek the protection of strategy (ii) while simultaneously assailing the present normative provisions of strategy (i). The efforts of homosexual activists over the past thirty years reveal a basic rhetorical pattern. In the beginning, homosexuality was handled by strategy (i) which suppressed it. Activists sought to loosen the grip of (i) by reducing its majority, so to speak, by showing or representing it that the majority opposed to homosexuality was significantly smaller than previously supposed. If the public buys this, the grip of strategy (i) does weaken, and if it weakens enough it releases the homosexual issue into the custody of strategy (ii). People who are disposed to live and let live are satisfied with the present arrangement. Those who go further, and think that homosexuality should not even be resented, take the additional step of attempting to return the issue to the provisions of strategy (i), but only after changing the majority.

It comes to the reformer as something of disappointment that changing the opinions of the majority, whether about the worth and natural entitlements of blacks, or about the essential sweetness of consensual pederasty, is an arduous and lengthy process, the sheer sluggishness of which guarantees the delay of targeted gains. Accordingly, a further manoeuvre suggests itself. Rather than wait until the hearts of scores of millions of individual citizens do change, employ instead the resources for the manufacture of consent so as to produce what can be produced relatively quickly, something that is called the *official* majority opinion or the *received opinion.* Wherewith another definition:

Definition Nine: An opinion on an issue is a *prime facie* received opinion in a polity to the extent to which (a) it is expressed and endorsed by dominant élites (b) resistance to that opinion is subject to effective sanctions either directly enforced or enforced *ad libitum* by sympathizers.

Prima facie received opinions are thus opinions dissent from which may get you fired from your job, or scorned by your associates, or your house vandalized. Under a lighter touch, you just become ludicrously *passé*.

Definition Nine leaves room, as it must, for the possibility of a issue giving rise to two or more mutually incompatible *prima facie* received opinions. The Church may press one way on abortion and the liberal press another. One way to flout a *prima facie* received opinion on an issue is to yield to another *prima facie* received opinion on the same issue. Sometimes, of course, a *prima facie* received opinion has no rivals. It can then represent itself as the received majority opinion. Thus

Definition Ten: An opinion is *the* received majority opinion with regard to an issue, A, iff it is a *prima facie* received opinion with regard to A and no incompatible opinion is a *prima facie* received opinion about A.

Polities in which strategy (i) is dominant, if they are organized in the way that most western countries are presently organized, will in a striking number of cases yield to the following variation of that strategy:

(i*) [Strategy (i), together with the rider that] there is in principle no issue on which the *bona fide* majority opinion could not be represented by the received majority opinion, never mind that they might be *different* opinions.

It is the first and most natural task of the manufacture of consent to produce and enforce received opinions. Some people resent the manufacture of consent for the reason that the received opinions do not comport with their own.[7] Others resent the manufacture of consent because it is pernicious as such, never mind that it gets some of the issues absolutely right; for it is an invasion of liberty. Such resentments are in some ways naive, even though the things complained of do happen. The whole apparatus of the manufacture of consent is a partnership between seduction and behavioural modification. This partnership is the classical *modus vivendi* of the normative socialization of human beings. It is the heart and soul of the education of children. It is indoctrination pure and simple.

5. Indoctrination

If indoctrination is the standard broadly political remedy for force five standoffs, it poses a problem for democracies. To locate that problem, it is necessary to see that the manufacture of consent is indoctrinational,

[7] I sometimes think that at its most visceral Chomsky's basic objection to the media is that it forwards and "legitimates" opinions which he abjures.

but it is not required to believe that indoctrination is wrong as such, or that it is contra-democratic as such, still less that it is avoidable.

A liberal democracy may be characterized by the provisions it makes for indoctrination. In democracies of the U.S. model it is not permitted that the State be a primary agent of indoctrination.[8] It may and will be expected to pass laws that give bite to some of the doctrines of the indoctrinated, but in the general case the production of the indoctrination itself is not a function allowed the State. Exceptions are provided for in this model, but only reluctantly. In critical situations the people might consent to laws establishing a Ministry of Propaganda in wartime, or where vital national interests seem to be at risk, citizens might concede to the State a right to manufacture some of the news some of the time. But such concessions are grudged, and they are tolerated only on sufferance. As originally conceived of, democratic governments assigned the business of indoctrination to others and enshrined its protection in constitutional arrangements or in custom. The constitutionally guaranteed separation of Church and State endowed the Church with the right to instruct, correct, and admonish its members. Families were similarly empowered. Such arrangements, properly handled, make for normatively stable populations. Such arrangements quickly become paradoxical for any polity which also pledges itself to normative pluralism which, as such, disposes the polity to instability. Public education presaged the first major default of the principle that excluded the State from primary agency in indoctrination. Once the State obliged children to attend school, into whose curriculum the State also intervened — the fundamental betrayal of academic freedom, by the way — the die was cast for further defections.

What the contemporary dominance of the mass media allows us clearly to see is that this basic restraint upon indoctrination has all but been reneged upon. In every western democracy the mass media are the indoctrinational surrogates of the State. In the marriage between the cathode ray tube and the potato chip, both parties owe their entitlement to operate, and the myriad other constraints upon how they do so, directly to the State. In saying this, I do not suggest that the media are the mere creatures of the government of the day. In many ways the media are a dangerous nuisance for governments, and it is not for nothing that every department of government now has its own substantial staff of media consultants.

[8]Britain is, largely honorifically, an exception, what with the establishment of its national Churches.

It used to be the case, and still is, that if you didn't like what you were being fed from the pulpit, you could demur quietly, or declare your apostasy and transfer your allegiance to the Church across the street, or to none at all. If things went badly at home, you could try some strategic disobedience, or you could decide to out-wait your parents, or you could just run away. If school offended, you might try the enchantments of truancy, or keep failing grade six until you were sixteen.[9] These defections bore their private costs, to be sure. But they were remedies of a sort.

It now becomes noticeable that similar sorts of defection from the indoctrinations of manufactured consent are not a realistic option. This is part of what is wrong with electronic thought control. We are in thrall to the opinion-makers, and knowing this lessens the thraldom not one whit. I know the techniques of the manufacture of consent direct energize the powerful engines of single issue politics, special interest politics[10] and the entire run of issues of public concern. I know further that those who control the techniques of manipulation serve without my approval, and usually without my knowledge. That is a second thing wrong with the manufacture of consent.

If what I have been saying is so, it will follow that the price that a polity pays for its pluralism is the price of its democracy, as earlier conceived of. Modern polities have reassigned the rights of indoctrination to the technocrats of social engineering and their bosses, and they are surrogates of the State. That is a third thing to regret. If what I have been saying is so, it should also follow that a mark of contemporary technological pluralism is the sheer speed and quantity of normative change on issues of public concern, on matters of no particular public concern, and in people's everyday behaviour. No one who has lived in such a polity for any length of time cannot but know that this is a mark of technological pluralism.[11]

[9]There is a story about a Toronto tycoon, lately dead. Ruminating about his early life, he acknowledged that he had left school at the age of thirteen. "That was a terrible mistake," he said. "I should have left at twelve".

[10]I mean here single-issue politics transacted more or less privately, but always attached to the threat of 'going public'. Of course these contentions are sometimes wages privately *and* publicly. So the contrast, as I have drawn it, between single issue politics and special interest politics is far from exact. It will do for present purposes, however.

[11]These and related issues are discussed in great detail in [Woods, 2000b; Woods, 2004b; Woods, 2004a].

PART VI

HOW TO INTERPRET ARGUMENTS

Having in the previous section taken pains to show that defections from the Goody Two-Shoes model of argument are not as such either irrational or argumentationally defective, it would do well to give some attention to two particular aspects of what might be called in a quite general way "the cooperative factor in argumentative discourse". For it certainly remains the case that much of argument is dominantly cooperative and all of its is cooperative to certain extent, even those cases that count as defections from Goody Two-Shoes standards of propriety. The issue that I shall here attempt to track has to do with how an arguer should interpret the contributions of his or her argumentative partner. More particularly, should these modes of interpretation have a predominantly cooperative character or not? Bearing on this issue are a well-received principle and a classic problem. The principle is the Principle of Charity. The problem is the Problem of Missing Premisses.

Chapter 14

CHARITY: CAN WE FIND IT A HOME?

Having taken some steps to promote the idea of non-cooperation arguments, we now examine a countervailing factor. It is the idea that in interpreting the argumentive practice of an interlocuter, an effort be made to see it as *not* non-cooperative. This in its most basic sense is one of the provisions of the Principle of Charity.

Charity is the most important of the theological virtues. Also translatable as 'love', charity is two things. It is empathy with fellow-creatures and it is sympathy towards them. The empathic element enables the charitable person to put himself in the shoes of his fellows and to see things from their point of view. The sympathetic factor induces the charitable person not only to understand his fellows but to accord them dignity and value.

The Charity Principle invoked by philosophers of language and argumentation theorists is not the same principle; and neither involves in its realization the theological virtue. Even so, there is in these principles an affinity to the theological virtue. Like the virtue, what the other two charity principles enjoin is empathy.

1. Slogans

Much of present-day argumentation theory is theory in name only. This is manifested in a number of ways, two of which seem to be important.

First example. Theories of argumentation frequently make use of talk about ideal types and normative models. This is loose talk and it needs some careful tightening. Here is one way in which it should not tighten. If we introduce the notion of an ideal reasoner or ideal interlocutor, we might mean (as some theorists have meant) a reasoner who, among other things, adjusts his deductive practice by conforming a proper subset of his inferences to the demonstrably valid rules of deduction. In particular, he will close his beliefs under consequence, and his beliefs will be transparent, i.e., he will believe that he believes what he believes, etc. If we now want to introduce the idea of a *normative model* of reasoning or argument, we might cash the notion of normativity as follows: reasoning in the normative model is reasoning done by ideal reasoners. The normativity of the normative model is secured by the ideality of its participants. (This is how reasoning should be transacted by ordinary reasoners because this is how reasoning is transacted by ideal reasoners, who, among other things, choose deductive strategies licensed by widely accepted rules of deduction.) If a skeptic asks, "Why these rules?", the answer to give him is that they are, in standard elementary formulations, sound and complete.

But this is wrong. Any real reasoner who even attempted to fulfill this deductive ideal would quickly paralyze thought and action [Woods, 1988], and, in any event, the rules of deduction are not good rules of inference (see [Harman, 1986, ch. 1] and [Woods and Walton, 1989, ch. 1]).

There are several respects in which this is so, provided that the logic that defines the ideal model in question is classical logic. To take just one example, the valid rules of logic are both *left* and *right* monotonic. A valid deduction is left monotonic if and only if the result of arbitrarily supplementing its premiss-set arbitrarily many times is itself a valid deduction. A valid deduction is right monotonic if and only if the result of disjoining to its conclusion any arbitrarily selected statement arbitrarily many times is likewise a valid deduction. Since belief is not closed under consequence, inference or belief-revision cannot be governed by monotonic rules, not even ideally.

So a condition on our normative model — a negative condition — is this: do not approximate to the behaviour of these ideal deducers. But what are the positive conditions that we should expect the normative model to honour? How do we know that the participants in our model are reasoning as they should? The short answer is that we won't know until we have a correct theory of reasoning. This we do not get by positing rules and procedures that hold in a model that we've decided to call a 'normative' model. The sentence "Response R to occasion O is correct in the normative model" does not, just so, imply that R is a correct response to O.

It is a large and difficult, and open, question as to how to warrant the normativity of normative models in argumentation theory. I am unclear about how to proceed, but one thing is clear. The normative model should describe the behaviour of beings who possess what Cherniak calls minimal rationality. Such descriptions should describe the optimal performance of beings endowed with intractable cognitive finitude.

Second Example. Theories of argumentation also bristle with slogans which not infrequently are given a prominent and load-bearing place in theory [Grice, 1975]. One example is Grice's Relation Maxim: "Be relevant". Grice is disarmingly candid about how inadequate his maxim is. He himself is chastened by the fact that he has no analysis of relevance to offer[1] and so, by implication, he evidently thinks that no one else does either. Grice pledges himself to that further task — unperformed, sadly, three decades later at his death in 1988. Still, references to relevance simply abound in all sorts of accounts of argumentation, but theories have been slow, very slow, in coming. This makes Douglas Walton's book on topical relevance something of an exception, though Walton is careful (and right) to say that his account hardly amounts to an analysis of relevance as such. Also important is the special number of *Argumentation* in 1992 devoted to the analysis of relevance. And, of course, there is the very good (but analytically flawed) book, *Relevance*, by Dan Sperber and Deirdre Wilson [Sperber and Wilson, 1986]. Even so, nothing has so far been done to produce a fully comprehensive theory of relevance. Not even Walton's *Relevance: A Dialectical Theory*, soon to be published, will rectify this omission since it treats of relevance only in dialogical contexts.[2]

A third example of sloganeering is the Principle of Charity. It seems to have found its place in the common currency argumentation theory by way of two sentences from Thomas' logic text. What I want to do in this chapter is to try for a reasonably deep analysis of this important maxim.

> When you encounter a discourse containing no inference indicators which you think may nevertheless contain an argument, stop and consider very carefully whether such interpretation is really justifiable... A good rule is "the Principle of Charity": If a passage contains no inference indicators or other explicit signs of reasoning and the only possible argument(s) you can locate in it would involve obviously bad reasoning, then characterize the discourse as "non-argument" [Thomas, 1977, p. 9].

Three years later, Scriven provided the principle with what has become a widely attended to and well-received characterization.

[1] I would say "theory" for Grice's "analysis" [Grice, 1975, pp. 46–49]. It is a big replacement, but it doesn't much matter for present purposes.

[2] It remains to be seen whether *Agenda Relevance* which I have written in collaboration with Dov Gabbay will fare any better in this regard [Gabbay and Woods, 2003].

> The Principle of Charity requires that we make the best, rather than the worst, possible interpretation of the material we are studying. The Principle of Charity is more than a mere ethical principle, but it is at least that. It requires that you be fair or just in your criticisms. They can be expressed in heated terms, if that is appropriate; they may involve conclusions about the competence, intellectual level, or conscientiousness of the person putting forward the argument, all of which may well be justified in certain cases. But your criticisms shouldn't be unfair; they shouldn't take advantage of a mere slip of the tongue or make a big point out of some irrelevant point that wasn't put quite right [Scriven, 1976, p. 71]. See also [Baum, 1975; Johnson and Blair, 1983; Hitchcock, 1983; Govier, 1988; Johnson, 1981; Rescher, 1964; Parret, 1978].

Charity, then, requires that we not be unfair in our disputes, and this seems all right by and large. Scriven's principle also implies

> (*Interp*): Interpret your opponent or your interlocutor correctly

which also seems all right by and large.

It appears to me unsurpassingly odd that in virtually none of the discussions of the Charity Principle in theories of argumentation is there any recognition that something called the Principle of Charity is at the centre of a brisk contention and a huge literature in the adjacent precincts of philosophy and cultural anthropology.[3] In the following section I shall try to repair this omission.

2. Radical Charity

"Chapter Two of Quine's *Word and Object* contain what may well be the most fascinating and most discussed philosophical argument since Kant's Transcendental Deduction of the Categories"; so says Hilary Putnam,[4] and he is right [Putnam, 1975, p. 159].

Quine is talking there about "radical translation and, as he proposes,

> There can be no doubt that rival systems of analytical hypotheses can fit the totality of dispositions to speech behaviour as well, and still specify mutually incompatible transations of countless sentences insusceptible of independent control [Quine, 1960, p. 72].

What is this all about?

Imagine that a linguist is making the first attempt to translate into English the language of a people deemed to be culturally very different

[3]Among the exceptions that come to mind would seem to be Rescher and Parret. According to F.H. van Eemeren and Rob Grootendorst, Rescher acknowledges a deviation in his own treatment of charity from Davidson's, whereas Parret's is a Davidsonian notion [van Eemeren and Grootendorst, 1984, p. 189, note 47]. Van Eemeren's and Grootendorst's own discussion of charity is a farther exception, to which I shall recur in due course.

[4]I closely follow Putnam's exposition, which is the best three-page account of indeterminacy thesis I know of.

from the translator's own. Our linguist will prepare a translation manual which Quine calls an "analytical hypothesis". Preparing such a manual in such circumstances is what Quine dubs "radical translation".

Quine assumes that the linguist knows how to recognize assent and dissent as speech acts in the host community. If this is right, he can locate truth functions in the host language and he can also discriminate occasion sentences, such as "Here's a cup", for which certain stimulations will produce assent and others dissent. Standing sentences, such as "Groningen is in the province of Groningen", are such that once a speaker has been prompted to assent (or dissent) he prevails in it without "further immediate stimulation from the environment" [Putnam, 1975, p. 159].

Quine identifies the stimulus meaning of a sentence with the set of stimulations of a host's nerve endings which would induce assent to that sentence. Two sentences will have the same stimulus meaning if the same stimulations would induce assent to each. A sentence is stimulus analytic for a speaker if he assents to it no matter what the stimulation, and a sentence is stimulus contradictory for a speaker if he dissents from it no matter what the stimulation. Observation sentences have intersubjective stimulus meaning.

Central to Quine's thesis is the notion of an analytical hypothesis. An analytical hypothesis is a general recursive function, *f*, from the set of all sentences of the host language to a subset of the set of all sentences of the visiting language. The function is constrained by three conditions.

> (i) if s is an observation sentence of the host language, $f(s)$ is a visiting observation sentence and $f(s)$ has the same stimulus meaning for visiting speakers as does s for host speakers; (ii) f commutes with the truth functions, i.e., $f(s \vee s')$ equals $f(s) \vee f(s')$, and so on; (iii) if f is stimulus analytic (or stimulus contradictory) in the host language, then $f(s)$ is stimulus analytic (or stimulus contradictory) in the visiting language.

Should our visitor chance to be bilingual in the host language, condition (i) is replaced by

> (i*) if s is an occasion sentence of the host language, then $f(s)$ is an occasion sentence in the visiting language, and the stimulus meaning of s for the linguist is the same as the stimulus meaning of $f(s)$ for the linguist.

Conditions (i)-(iii) or (i*)-(iii) are Putnam's paraphrase of conditions (1)-(4) or (1*)-(4) of chapter two of Word and Object. Translations made in conformity with these conditions are made in circumstances that collectively exhaust the evidence for the translation in question.

The thesis of chapter two is this. Suppose that our linguist entertains two rival analytical hypotheses f_1 and f_2. There is, says Quine, no "fact of the matter" as to whether the translations afforded by f_1 are correct or as to whether the translations afforded by f_2 are correct. Given that conditions (1)-(4) or (1*)-(4) exhaust all possible evidence for any translation, f_1 and f_2 could each fulfill (1)-(4) or (1*)-(4) and still "specify mutually incompatible translations of countless sentences insusceptible of independent control [i.e., beyond the sway of evidence]". So there is no such thing as an objectively correct translation.

But couldn't the translator become bilingual and from that position check to see which of f_1 or f_2 (or some other manual) is objectively correct? He can do the one, says Quine, but not the other. Becoming bilingual just means that the linguist's analytical hypothesis will be governed by (1*)-(4) rather than (1)-(4); but fulfilling (1*)-(4) will not block the specification of mutually incompatible translations of countless sentences beyond the sway of (1*)-(4). The fate of the bilingualist suggests to Quine that indeterminacy considerations apply equally to intralinguistic translation and interpretation.

How, then, is our translator (or interpreter) to choose among rival analytical hypotheses? Suppose he specifies some further condition (5), and that conformity with conditions (1)-(5) produced a unique translation, that is, that there were one and only one translation that fulfilled those conditions. Such translations would be rivals, but it could not be said to be objectively correct, for there is no fact of the matter as to whether there is a fact of the matter that (5) is an objectively correct constraint.

Charity is one way of specifying condition (5). It is a way of discriminating rationally among rival analytical hypotheses, never mind that none could be said to be objectively correct. If we take Charity in Davidson's way we will hold that "no sense can be made of the idea that the conceptual resources of different languages differ dramatically" [Davidson, 1984, p. xviii]. And

> Charity is forced on us; whether we like it or not, if we want to understand others, we must count them right in most matters [Davidson, 1984, p. 196].

Admittedly, this is a "confused ideal".

> The aim of interpretation is not agreement but understanding. My point has always been that understanding can be secured only by interpreting in a way that makes for the right sort of agreement. The 'right sort', however, is no easier to specify than to say what constitutes a good reason for holding a particular belief [Davidson, 1984, p. xvii].

The heart of Davidson's approach is that in interpreting another it is necessary, if we are to make sense of him at all, to assume that

> [*PrinChar*]: He is not noticeably deficient in holding true those things which we ourselves are rather adept at holding true. ("Oh, look, it's raining").

That said, a (non-deductive) corollary suggests itself.

> [*Cor*]: Concerning those things that we notice we ourselves are not particularly adept at holding intersubjectively-true (e.g., a theory in sharp contention), we should likewise suppose that the persons whom we are interpreting may not be significantly more adept about such matters than we. Where there is intersubjective disagreement at home it is not unreasonable to predict it elsewhere.

[*Cor*] carries interesting consequences. It permits us to preserve disagreement in our interpretations when two conditions are met: (A) When the interpreter takes heed of all available evidence and, thus, complies with (1)-(4) or (1*)-(4), and (B) when he has reason to believe that the subject matter of the interpretation has had an intersubjectively bad dialectical track record here at home. (The theory of being *qua* being; or the morality of abortion).

There is a great deal to puzzle over and to wrangle about in the modern history of, as I shall now say, Radical Charity, ensuing from Wilson's paper [Wilson, 1959]. I shall not join those wrangles here, for our prior question is what, if anything, does Radical Charity have to do with Charity as a dialogical maxim? The answer is "Not much".

If we consider the sorts of cases mentioned by Thomas and Scriven, it is plain that Dialogical Charity (as I shall now call it) is not the principle that fills for condition (5) on radical translation.

For if it does follow from Scriven's principle that the interpreter should furnish objectively correct interpretations, it is a presupposition of Radical Charity that it is precisely this that an interpreter cannot do for any sentence or discourse that outreaches the provenance of conditions (1)-(4) or (1*) to (4). Scriven's principle is not only not the same as Radical Charity but, if the implication I've been talking about holds, it is incompatible with it.

Suppose now that we have been radicalized by Quine's thesis about the possibility of objectively correct interpretation. In that case, we will say that our implication, *Int*, does not hold. Will Radical Charity now serve as an explication of Dialogical Charity? No. Here is why.

I. Radical Charity has nothing to do with fairness. It does nothing to disenjoin the attribution of embarrassing or loony views to natives or others on grounds that this is a nasty, culturocentric thing to do.

II. Radical Charity, applied at home, holds us to the view that the community of my co-linguists is not in general to be counted as massively in error on those matters concerning which I am in relative semantic serenity, that is, intellectually sincere and reflectively untroubled serenity. So Radical Charity gives no instructions for the individual case, except as noted below.

III. Radical Charity does have a kind of negative provenance for various kinds of case, contextually specifiable in individual situations. But here, often as not, the principle is suspended, not applied. So, for example, the very fact that I am locked in a dispute with a co-linguist counts as some evidence that he is wrong, and my interpretation of what he says must preserve this fact or make a nonsense of our disagreement.

I can hear the charitable dialectician exclaim, "Look here, you've got things muddled. Neither Radical Charity nor Dialogical Charity is a carte blanche. Neither requires the suppression of disagreement in individual cases". This is quite right. Radical Charity posits uniformities in human nature that would justify the reflection of agreement in interpretation in the general case. But I want to suggest that not only does Dialogical Charity permit disagreement-preserving interpretations in particular cases, it also permits them and counsels them in the general case as well or, more carefully, for the general run of cases for which the Dialogical Principle has apparently been designed. If this is right, the two Principles remain significantly different. The question is: Is this right? (I return to this in section 7.4.)

3. Dialogical Charity

I want now to turn our attention to a discussion of Dialogical Charity which stands out for its careful attention to detail. It is in other ways a limited account, but there is a reason for this, as we will see. In the work cited van Eemeren and Grootendorst invoke a Charity Principle in the course of trying to solve a problem that might be said to be the classical motivation of all dialectical invocations of charity. (Henceforth Charity is Dialogical Charity unless otherwise indicated.) The problem is that of reconstructing an interlocutor's argument in a way that captures a premiss heretofore unexpressed. Van Eemeren and Grootendorst locate the problem in a theory of the co-operative resolution of disputes, for the procedural canons of which they rely heavily on features of Gricean co-operation. They propose that the unexpressed premiss be associated with a conversational implicature, for whose specification they provide essentially Gricean instructions. We suppose that *X* and *Y* are in the process of co-operative conflict resolution. *X* produces what van Eemeren and Grootendorst call a "single-argumentation", that is (roughly), an argument with just one conclusion. The argument is invalid as it stands, is recognized to be so by *Y*, and may be recognized to be so by *X* as well. The task for *Y* is to attribute to *X* a belief or a commitment which makes the expansion of the original argument valid (the expansion just is the result of adding the new or unexpressed commitment to the premiss-set of the original argument), and, moreover, which is held to the condition that expansion remains a single argumentation (more of this anon) [van Eemeren and Grootendorst, 1983, pp. 134–35]. Their account requires the following assumptions:

1 The conflict resolution mechanisms available to the disputants are understood by each to be Grice-co-operative.

2 It is a violation of the procedural canon to attribute violation of the procedural canon unless (counterfactually, here), there is reason to do so; [from (1)].

3 It is mutually understood that with regard to *X*'s utterance of his argument, *A*, it is a condition of adequacy that *A* be valid.

4 *Y* knows that *A* is not valid and, moreover, that taken at face value it violates the Relation maxim; that is, at face value there is reason to question the relevance of the premiss-set to the conclusion of *A*.

5 To take *A* at face value would be a violation of procedure for *Y*; [from (2)].

6 to avoid violation, *Y* must attribute to *X* some commitment *P* which validates an expansion of *A*.

Since any number, in fact an infinity, of sentences will serve to validate an expansion of *A*, the question now is for which value of '*P*' will a Grice co-operator produce a valid expansion of *X*'s argument *A*? For ease of exposition let *A* be the lustreless argument,

1 John is English.

2 Therefore, John is brave.

Earlier writers, Scriven, for example, propose that the specification of the missing premiss might be held to a condition of minimality or a stronger condition of optimality [Scriven, 1976, p. 43].

By the minimality requirement, *Y* would opt for "If John is English then John is brave", which has two significant features. One is that it does validate an expansion of *A* and the other is that it is not "independently well-supported" [Scriven, 1976]. This can be understood in the following way: although *Y* will see that it produces a valid expansion of *A*, he may doubt that it is true. If he supposes that *X* tacitly holds this premiss as true on no grounds other than the validity of *A*, the original argument, he attributes to *X* an appalling error and has placed himself at risk of unco-operation. Or if he attributes to *X* tacit resort to this premiss just because *X* holds-true the premiss and conclusion of the original argument, then although this makes "If John is English then John is brave" truth functionally true and although, so taken, this will generate a valid expansion of *A*, *Y* nevertheless attributes a question-begging manoeuvre to *X*, in violation of procedural assumption (2), above. We may say, then, that fulfillment of the minimality condition can't be an option for Y without procedural default. Optimality can now be understood as the attribution to X of a commitment which (a) produces a valid expansion of A, and yet (b) does not land Y in procedural defection.

Can we give instructions to Y as to how he is to fulfill the optimality requirement? Van Eemeren and Grootendorst introduce a pair of terms which preserve the distinction, as I have explained it, between Scriven's minimality and optimality. They speak of producing a premiss which does "the logical minimum" for the repair of X's defective argument, and a premiss which will do the "conversational minimum" for it. They propose that the condition of conversational minimality will turn the trick and that it does a better job than optimality.

Let us renew our bearings. Van Eemeren and Grootendorst are grappling with the problem of how to produce a valid expansion of A in the most Grice-co-operative way. They point out that inasmuch as their "analytical reconstruction" (as they call it) of the interplay between X and Y is Gricean, anyone who Grice-co-operates is one "who observes the co-operative principle [and who] must also observe the conversational maxims". So the co-operator is held to the maxim of Relation (which they mistakenly say commits him to produce a valid expansion of A[5]) and also to the maxims of Quantity and of Quality. The Quality maxim holds a co-operator to providing just the information needed, and no more, for the task at hand; and the Quality maxim holds him to say what is true and, negatively, not to say what he doesn't believe to be true, or that for which he (realizes he) lacks adequate evidence.

We are now at the nub of the van Eemeren and Grootendorst solution of the missing premiss problem. Various candidates are now easily dismissed:

* If John is English then John is brave

is dismissed because it gives only the logical minimum, which is insufficient, for the reason already sketched. There is in fact an algorithm that stands in wait. If A is any argument "$p_1, p_2 \ldots, p_n, \ldots$, so q", the corresponding conditional of A is "If(p_1, and p_2 and $\ldots$ and p_n, $\ldots$.) then q", and the conditional expansion of A is the argument that results from adding the corresponding conditional of A to the premiss-set of A, and adding no other premiss. The mechanical rule is: Do not solve the problem of the specification of the missing premiss of an argument A in

[5] Y could just as well attribute to X the *generic* sentence, "The English are brave" which, while it won't produce a valid expansion, certainly takes care of any worry about the relevance of the premiss to the conclusion in the original argument.

such a way as to generate the conditional expansion of *A*.
Another candidate is:

> * All island dwellers are brave

It is excluded since it doesn't produce a valid expansion of *A*.
A third:

> * Since all island dwellers are brave and all Englishmen are island dwellers, then all English are brave.

won't do since it violates the maxim of Quantity; it says more than is needed for the task at hand. In fact, it is easy to see that any proferred premiss which has the structure of an argument, such as in the present example, will violate the maxim. So another algorithm presents itself.

> *Mechanical Rule 1*: reject any premiss that produces an expansion of *A* that is not itself a single-argumentation, that is, an argument with one conclusion only.

Another possibility,

> * All Englishmen are brave and island dwellers and subjects of the Queen are eligible for the National Health Service

while true, is also disqualified by the Quantity Maxim. Here too another mechanical rule drops out.

> *Mechanical Rule 2*: admit as a premiss no conjunction containing one or more conjuncts validating the expansion and at least one that does not. Call such conjunctions "distributively inert".

We can see that the tight and admittedly dialectically unrealistic constraints that van Eemeren and Grootendorst impose on their Grice-co-operators make possible a mechanical solution of the problem of the missing premiss, as follows:

> *P* is an admissible choice in the generation of a valid expansion of *A* iff
>
> (a) *P* generates a valid expansion;
>
> (b) the valid expansion produced by *P* is not the conditional expansion of *A*;
>
> (c) *P* does not produce a valid expansion of *A* which fails to be a single-argumentation (or as van Eemeren and Grootendorst say, is a compound argumentation);
>
> (d) if *P* is a conjunction fulfilling (a), *P* must not be a distributively inert conjunction.

Applying this rule will get us just what van Eemeren and Grootendorst are after:

> * All Englishmen are brave.

It is well to note that people will complain. The example here discussed is naively simple and unrealistic, and the co-operative constraints imposed, for all their benign and collegial charm, are as untrue to real life as a Bowdlerized fairy tale. The story told throughout about Charity is descriptively inadequate, and earlier admonitions about careless attributions of 'normativity' to 'analytical reconstructions' are so many chickens now come home to roost. All of this is true, I fear, but explicable.

For one thing, the appeal of the van Eemeren and Grootendorst solution bears an undeniable ancestral link to the solution of a classical problem of which their own is close kin. The classical problem is that of converting a defective argument into a valid categorical syllogism. Given the syntactic devices for the specification of propositional type and valid syllogistic form then, if the theory of the syllogism is good for anything, it is certainly good enough for the solution of the problem of the enthymeme as it there occurs. It is not for nothing that the classical problem endures to this day in logic textbooks: it is solvable mechanically.

It is a virtue of their approach that Gricean-co-operative strictures can be mapped onto mechanical resolution procedures. True, once you start dismantling the structural assumptions of the case in question, once you abandon the unrealistic requirement that the expanded argument be valid, those mappings break down, and the Gricean presumptions are left to do their work all but unguided. But they do work for the restricted kind of case. But they work in the wrong kind of way. the attractiveness of the solution is that it is mechanical, not that it is right.[6] It is not obvious that missing-premiss problems have algorithmic solutions. Perhaps they do; but it is no good to assume that they do without argument. The attractiveness of the van Eemeren and Grootendorst solution also exemplifies what we might call the *Can Do Principle.* According to the *Can Do Principle* the work that a problem-solver does on a given problem is justified by two considerations: First, that it is work that he knows how to do; and second, that it is work that bears on the problem at hand. The *Can Do Principle* applies illicitly when only the first condition is met. A celebrated example of the illicit application was the decision by neoclassical economists to pretend that utilities are infinitely divisible, in order that the theory could appropriate the fire-power of calculus (which even neoclassical economists are good at). In the present case, provided you pile on the requisite assumptions, it

[6]For a discussion of why this is the wrong solution to the problem of enthymematic reconstruction see [Gabbay and Woods, 2004; Woods, 2002c].

is easy to solve missing-premiss problems. The trouble is, they are the wrong assumptions.

This brings me to a second point. I said at the beginning that slogans in argumentation theory are in want of theory. Normativity and relevance are, certainly, and perhaps Charity as well. I have already examined and rejected the suggestion that Radical Charity is any particular illumination of Dialogical Charity. If this is true, we must seek for a theory of Dialogical Charity elsewhere. It may seem to some that I stand ready to despair of Gricean Charity once the mappings collapse. For what, in the realistic and structurally unfettered case, is it to say or attribute too much, and what is it for one thing to be relevant to another?

It is a fact that in some rather basic and primordial ways we are co-operators. We all bring to reflection and social intercourse our dark-side commitment, as Douglas Walton calls them ([Walton, 1987, pp. 125, 142], and [Walton, 1984, pp. 247–254]), those fuzzy sets of background knowledge, common knowledge, unnoticed doxastic predilections, ceteris paribus understandings, facility with generic assertion, which sometimes lurk unnoticed and resist effortless articulation. Somehow there is something that intelligent beings are able to do routinely, if not perfectly, and it is to figure one another out, and themselves too, to penetrate into one another's dark-side commitments. Of course we want to know how we do this and in virtue of what it is done right. For all I know, it may be that these accomplishments are no better explicable than by recognition of what Chomsky and Fodor call our general intelligence. "Describing the nature of general intelligence is a hopeless problem, according to Fodor, and the whole point of Fodor's 'moduality hypothesis' is to separate the problem of understanding the 'language organ' from the problem of understanding general intelligence" [Putnam, 1988, pp. 11–12]. Perhaps it is this way, too, with figuring one another out, with my knowing where you are coming from, as the vulgate has it. If so, it is unjustified to be wistful about Gricean Charity uncoupled from its mappings, and it is unfounded to demand of van Eemeren and Grootendorst, "Where is your theory for uncoupled Charity?" And, if so, other things become intelligible as well, including why so many informal logicians, self-called, simply abjure theory.

Now, appropriating a different but related thought of Putnam, I want to say that lots of problems, problems collected under such labels as "meaning", "sameness of reference", "relevance" and even "knowing where you're coming from", are every bit as daunting as the problem lurking under the label "general intelligence". All of these are manageable in lots of cases (see above) but to determine the intrinsic complexity of a task is to ask, How hard can it be in the hardest case [Putnam, 1988, p. 21]? Just so.

4. Radical Charity Again

A word about Charity, now 'radical'. It is possible to be a charitist about radical translation without embracing either Quine's indeterminacy thesis or Davidson's defence of Radical Charity. David Lewis is one such [Lewis, 1983a, pp. 108–115]. Lewis, too, considers Charity in the context of radical translation. We suppose that our linguist wants to make sense of Karl, his host. Lewis' Charity is restricted to attributions we should be prepared make to Karl, knowing only his physical situation and his physical history, but not knowing anything whatever about his psychology and not having (for now) any linguistic access to him. Lewis' Charity directs us to attribute to Karl the beliefs and desires that we ourselves would have in his physical situation. Karl holds his hand out and upwards. Facing the sun, Karl now stops squinting. Charity instructs the linguistic to assume that Karl had the sun in his eyes and wanted to shield his eyes and did. Lewis' Charity directs us to make uniformity assumptions. The human animal is approximately the same wherever you find him. Charity is thus a basic inductive strategy regarding the doxastic and conative properties of human beings. Charity for Lewis is but one of six principles of which the linguist is invited to avail himself.

The Principle of Rationality extends the inductive strategy to Karl's inferences. Karl is taken to be an inferer rather like the linguist. The Principle of Truthfulness attributes to Karl (tacit) subscription to a convention of truthfulness in his language just as the linguist so subscribes in his language community.

The further three principles tell the Lewisian linguist how to approach Karl's language. The Principle of Generativity posits that Karl's language is much like the visitor's own and, in particular, that the truth conditions of sentences in Karl's language are simply and finitely specifiable and uniformly so. How else could Karl learn his language? The Manifestation Principle provides that the linguist should assume that the beliefs expressible in Karl's language should ordinarily be manifest in his dispositions-to-utter: So he assumes that Karl is not a radical concealer of his beliefs. And, finally, the Triangle Principle bids the linguist to constrain his translations of Karl's language in such a way that sentences expressing beliefs and desires express the same beliefs and desires.

Lewis believes that, armed with his six principles, a linguist can overcome indeterminacy latencies in radical translation. But the more important point, for our current purposes, is that Lewis' principles constitute a certain view about uniformities in human nature. Each licenses a generic sentence in the form "Human beings are or do so-and-so". The principles can be linked to what Marvin Minsky has called a frame-system.

> A frame-system is intended to be a structured piece of knowledge about some typical entity or state affairs. It contains variables onto which go specific values... [Johnson-Laird, 1988, p. 107].

For example, following [Putnam, 1975], if I possess a frame-system for tigers and a tiger that I am inspecting is three-legged, then the variable NUMBER-OF-LEGS will take the value THREE. But in the absence of such information, the frame-system will specify the characteristic value FOUR. FOUR is said to be the "default value" for this variable. What, then, is a characteristic property of a thing or event? It is the default value with regard to a given variable furnished by, or which would be furnished by, its frame-system in the absence of contrary information. The notion of a frame-system is a partial elucidation of the concept of a prototype, for a prototype of something just is a frame-system in which all default values are specified.

Lewis' broadly charitist heuristic comes to this. The linguist should treat Karl as exemplifying the prototype human being, and unless he does, radical interpretation of him is impossible. Lewisean charity (broadly speaking — I now mean all six principles), posits a dialectical demeanour not greatly unlike that of a Grice-co-operator. Even so, Radical Charity, now taken in Lewis' sense, diverges from Dialogical Charity in a fundamental way. For although I may approach a co-lingual disputant prototypically, the whole point of disputation is the systematic displacement of default values with contextual determined "real" ones. Our fellow disputant is measured against prototype with regard to his rationality and his truthfulness. Dialogical Charity commands no more than that I not attribute to my co-linguist defections from rationality and truthfulness that are not manifested in his dialectical performance and evidenced by my beliefs about the world.

5. How Reliable is Charity?

So, then, Charity is an heuristic which presupposes the wide and even distribution of certain traits in human beings. In the form in which Donald Davidson presents it, the Charity Principle presupposes that in the general case people hold-true beliefs that are true; that people have and act on desires objectively appropriate to their interests; and moreover that these facts are manifested in their linguistic exchanges and other forms of behaviour. Massive doxastic and decisional error can be attributed to creatures that look like us only at the cost of our ability to count them persons, only at the cost of our not being able to regard them as "just like us". Charity, then, is the principle which counsels us to accept the presupposition. It is available in different strengths (and degrees of interest). Undiluted charity provides that the presupposition is true. Watered down charity advises that we provisionally approach others as if it were true.

Running against the nap of Charity are, as we have seen, recurring and forceful convictions about human incompetence, the output of our intractable cognitive finitude. Experimental evidence suggests — though it is not uncontested — that we are bad at all sorts of things. (This is the whole nub of the traditional conception of fallacies.) We seem to be especially maladroit in dealing with complexity: nested negations, quantifier iterations,[7] the compounding of probabilities, the routines of induction, memory storage and retrieval, and eyewitness testimony. There is, in this, plenty of room for the conjecture that people tend to get things wrong rather that right, though, providentially, this seems not to matter much. Perhaps our errors are self-cancelling, but in any case they don't kill us.

Plucky Charitists are undeterred by this. Such errors, they say, may be frequent and widely distributed, but (a) they could not be intelligible as errors except for the existence of non-errors with which they contrast; and moreover, (b) the distribution of non-error in the human population must outrun the distribution of error if attributions of wide-scale error are to make sense. What lies at the heart of the Charitists' bonhomie is the fact (a fact, certainly, if anything could be said to be) that our cognitive and decisional practices are not in the general case fatal, that human beings have had, so far, a good evolutionary track record.

It is tempting to say, therefore, that what an interpreter owes his interpretee is the assumption that he possesses beliefs and desires and that he compounds them, acts on them and expresses them to others in ways necessary for his survival as a being of a certain sort on this planet. To acknowledge this is but to recognize him as a human being. This is not politesse, this is *induction.*

The induction ends where theory begins. It is overtaken by a further induction: Human beings are attracted to theories, and some of these theories are *painless*; they are, as we say, "just theories". The whole human population could believe them without much evolutionary risk.

---

[7]It is well-known that nested functions are a problem for human information-processing. The same problem extends to existential quantifiers via the Skolem functions that replace them. See [von Neumann, 1958].

Utterly painless theories are in general behaviour-neutral, e.g., Berkeley's idealism. Another induction to consider is that all such theories are wrong or, more soberly, that they are no more right than other theories with which they are incompatible. The less deniable induction is that theory is variously and inconsistently distributed over the human population. Painful theories on the other hand attach huge costs to behavioural defections. Thus if faced with an on-rushing grizzly, experience, it is ill-advised to pause and review arguments for and against the reality of the external world.

It will seem to those who are disposed to find a theory in any systematic filter of a subject's nerve end hits, that the posited distinction is unpopulated at one end, and so is a distinction that fails to draw a contrast. I have no quarrel with this. The distinction can be re-drawn. There are theories that people take in through their pores, so to speak, just by virtue of being functioning human beings; these are the theories one learns in (or along with) the learning of one's mother tongue. It is here that one can expect to find painful theories. Other theories require radically different modes of inculcation; they are those you go off to the university for, or the pages of the 11th edition of the Encyclopaedia Britannica. It is here that painlessness often lurks.

All that I really need or want is the idea of painless theories and the inductions pertinent thereto and, in particular, the induction that painless theories are variously and inconsistently distributed in the species home sapiens. Painful theories are different.

The Charity Principle is nothing but the induction that my painful theories are also your own: Charity is needed for pain. Charity-governed theories may, for all I know, be hardwired into our cognitive and decisional devices. But this is far from clear.

If a linguist finds himself in a host population previously unknown to him and his ilk, could he ever have evidence that it was in the throes of massive infidelity to its painful theories? To ask this is just to ask whether evidence could present itself which would warrant the hypothesis that the host population had gone collectively and primitively quite mad. People might refuse nourishment, swim out to sea beyond the point of return, set fire to their dwellings, throw their children to the crocodiles, and so on. Must it be supposed that such a population has lost the power of speech and the linguist the power to interpret it? No. The linguist might learn from an informant that God has commanded her children to cast themselves out of their earthly confines and to join her in paradise. This is just Jonesville writ large.

What would this case show? It would show no more than that human beings who have mastered the rudiments of survival are likewise knowledgeable about strategies of self-destruction. It would also show that human beings might adopt theories, painless as such, that take an occasionally painful turn ("This is what God now commands"). So the induction needs refinement.

Suppose now that our linguist were confronted with a population in considerably greater disarray. Suppose that it were like this. For every pair of men and crocodiles at any time of their interaction, every physically possible variation of pairwise behaviour as is consistent with those numbers and those beings is actually evinced; and so, too, infant-crocodile pairs; young women-pomegranate pairs; elder women-young men pairs; question asking-answer giving pairs, and assertion making-information receiving pairs. (The device of pairs is for expository convenience only). Now randomize all such interactions in the entire collectivity; that is, as events turn, assign to similar stimulations randomly different responses. So at t_1, in response to the question "When is dinner?" Lrak's wife replies "Two is the only even prime"; at time t_2 she says "Isn't that Sherlock Holmes over there in the bushes?". At t_3 she replies "If you knew Susie like I know Susie".

It would be unpleasant for the visiting linguist. He would do well to stay out of the way. If he plucked up his courage, he might try to persuade Lrak to serve as his linguistic *entré* to the hapless group. It could not be done. This is a community now defeating all the helpful inductions. The concept of native assent and dissent is too fine-grained for any interpreter's grasp and, we may suppose, its own. The host community has lost its capacity for intersubjective stimulus meaning. This is the truth of Charity.

It is an interesting feature of painless theories that they sometimes contradict — and if true, semantically obliterate — painful ones. More interesting still, subscribers to a painless rival, though they believe it to be true and believe it to falsify the painful theory, do not in the general case behave as if the contradicted theory were false. There is good reason for this. If they honoured the painless theory in their everyday transactions they would perish. Substantial millions of people believe — know in their hearts — that their material existence is but an illusion and that all their (first order) beliefs and desires pertaining thereto are false. But ask a thirsty Buddhist if he wants a cold beer, he'll thank you for it or turn it down for some fruit juice if he's devout.

How could a Charitable linguist come to know that his Buddhist interlocutor believes false all the (painful) things that he himself believes true? If our linguist is a raw and slovenly Davidsonian, he will assume what is false, namely, that devout Buddhists hold true approximately the same painful sentences that the linguist holds true. Charity will not cash semantically. It cannot be the principle that all humans hold true approximately the same painful theory; that is, it cannot be this principle and be true. It is, rather, the principle that all humans behave for the most part as if, and give appropriate manifestations of assent and dissent, they hold true what we behave as if, etc. We have all learned to stay out of traffic and to feed the kids, and have tailored our languages thereto.

Once a community has been Charitably penetrated, a further induction could be invoked. It is more problematic, but it has an appeal nonetheless. It is, as Lewis has said, that human languages are (approximately) generative, that a linguist should assign truth conditions to host sentences in a finite, uniform and fairly simple way, that is, recursively. That way, once the linguist can figure out the language of staying out of traffic and feeding the kids, he can also figure out the more painless bits. He might even cotton onto to the fact that his host community disbelieves every sentence of his own naive painful theory. Suppose that he did discover this, would he have discovered that the host's language contained conceptual resources radically different from his own? Not in the least.

Would this show that the host community was not a culture radically different from his own? That is less clear. No one there believes anything in the linguist's own painful theory. That would appear to be a powerful difference. Yet everyone there stays out of traffic and feeds the kids. And that's a powerful similarity. One might say that having a radically different culture is different from having radically different conceptual resources. Culture is a matter of how those resources are deployed.

On what, then, does the hypothesis of cultural difference turn? In the present case, it turns on how the host employs the device of negation from the safe perch of his painful theory. In the case before us, if T is the host's painful theory and T^* is the contradicting painless one, then the native reveals something of his cultural distinctiveness by mapping each sentence of T onto its negation in T^* and by adjusting for tidiness. (For example, he might posit truth value ambiguities: 'true', 'really true', 'not true' and 'not really true').

The principle which enjoins an interpreter to count most things true that his interpreter assents to nevertheless allows that the linguist could find all of his science false, his moral codes, his politics, his philosophy and his religion, too. (And, yes, his logic. The host might be a dialethicist in T^*, but he'll not be so-minded in T).[8] The principle itself is,

[8]A dialethic logic has two distinguishing features. It sanctions the truth of some contradictions; and it denies that from a true contradiction everything follows [Priest, 1987].

charitably construed, both overstated and under-applied. Its defence of cultural invariance is hardly more than the discovery that people don't resist the instinct to breathe.

Charity is itself a theoretical principle. Is it a painful or painless principle? It is no threat to our place in the evolutionary scheme that we find lots of painless theories to be false. Its sole chance of being a painful principle is this. If an interpreter fails to apply it in his quotidian social engagements he will ruin his life. And so he will, provided that he keys his life to successful linguistic exchanges. Of course, he may take to the Gobi desert and live alone, but he could not do this without prior resort to the principle at home, without being able to manage a "Where is the Gobi desert?"-exchange. Once there, his hermitage (solitary, by definition) could prove to be lengthy. But it will also be nasty and brutish. Charity, as is said, begins at home.

Earlier I ventured that Dialogical Charity permits disagreement-preserving interpretations in a quite general way. We see now what that claim comes to: The linguist could interpret all of Lrak's theories in such a way that they disagree with his own. Not only could he do it, he should do it.

Chapter 15

MISSING PREMISSES

In the preceding three chapters we examined the contributions of pragma-dialectics to fallacy theory. In chapter 7 we considered how pragma-dialectics handles the problem of missing premisses. I want here to re-engage this second question in the light of more recent pragma-dialectical work.

Neither enthymematic as arguments nor their explicit interpretations are considered either as fallacies or occasions of fallacy. They are, so far as I am aware, in no one's gang of eighteen. Even so, human communicators routinely understate their messages, and yet their interlocuters are routinely capable of filling in the missing bits. They do so in an especially interesting way — confidently, speedily, automatically (largely), and correctly (largely). Our habit of filling in another's message is so ingrained, so effortlessly executable in the general case, that it deserves the name of seduction. We are seduced by the incompletions of our partner-discussant to make his message complete.

Like most things that we do in this way, i.e., confidently, speedily, automatically and correctly, the conditions under which we fill in a discussant's message are enormously difficult to specify. Solving enthymeme-problems is something we are practically adept at, albeit not perfectly, but defeasibly. But we have got nowhere with a *theory* that accounts for such successes.

1. Cooperative Conditions

The problem of missing premisses is notorious for its knottiness. That it has endured unsettled, in one form or another, since classical times attests to its allure and its difficulty alike.

The pragma-dialectical approach to missing premisses is examined in contexts of critical discussions ([van Eemeren and Grootendorst, 1992, ch. 6]; cf. [van Eemeren and Grootendorst, 1984, pp. 124-127, 129]). Such conversations are located in an ideal model. The idealizations of the model are designed to suppress or minimize features of the din and swirl of actual conversation in ways that make reconstructed speech amendable to efficient theoretical treatment. The authors have no interest in, or time for, caprice in the selection of idealizing constraints. What goes on in the ideal model must approximate to what goes on in life, and the model must portend, if it does not actually produce, application procedures by virtue of which theory can be said to have practical significance. So the idealizations are intended o be modest and circumspect.

The conditions that are cashed in in the ideal model, include those that in a rather central way are tied to a broad Principle of Charity, which we have already discussed in chapter 14. Protagonists and antagonists are assumed to be sincere and serious. What they assert they are held to believe, and both are imagined to proceed with the discussion in ways that are timely, relevant, efficient and sober-minded. Of the several conditions with which the model's participants analytically comply are these two:

R(1): *Honesty:* participants will perform no insincere speech acts.

R(2): *Efficiency:* participants will perform no superfluous or futile speech acts.

Efficiency subsumes two distinct sub-cases. For expository ease it would be well to divide R(2) into R(2(a)), which disenjoins superfluousness, (and is reminiscent of Harman's Clutter Avoidance Principle, discussed in chapter 8), and R(2(b)), which disenjoins futility. Neither superfluousness nor futility is defined. But clear examples are available. One could be said to have performed a superfluous speech act in uttering Q if Q is an immediate consequence of a proposition affirmed in that speaker's most immediate predecessor speech act. As for futile speech acts, a paradigm of futility is the forwarding of an argument which is transparently invalid.

It is very much in the spirit of its Charity Principles that the model's participants not carelessly attribute to one another the employment of or subscription to rotten arguments. When actual speech is reconstructed for its various roles in the ideal model, the principle carries a particular message. To see how the principle works here, it is useful to bear in mind that a premiss can be identified as 'missing' in argument A if and only if (i) it does not occur in the surface structure of A and (ii) without it A is invalid on its face, that is, invalid by way of factors attaching to its surface structure alone. It follows that a proposition P counts as a missing premiss for argument *A* if the addition of P to A's premiss set produces a valid argument, or is a conjunctive part of such a proposition.

Everyone who writes about missing premisses seems ready to buy the idea that some arguments are bad beyond the fact that they miss a premiss, and some arguments are bad arguments on a mere technicality, where the technicality is just that they miss a premiss. If the purported distinction between the two kinds of case held up, then we could venture an allied distinction between remediable and irremediable arguments. Remediable arguments would be those whose invalidity is removed just by adding a missing premiss (one or more, as needed). Irremediable arguments would be those whose invalidity is not removed just by adding missing premiss(es). With that distinction in hand, our present principle drops out. It, too, is a Charity Principle.

R(3): *Minimization of Stupidity Assumptions:* participants will restrict their accusations of invalidity to irremediable arguments.

R(3) gives a corollary of some interest.

Corollary: In reconstructing an opponent's argument for representation in the ideal model, if it is invalid on its face, find a missing premiss, if you can, and thus validate it. In such cases, arguments are represented in the ideal model as valid.

Of course it won't work. In what has become the classical bug-a-boo of deductivism, it is notorious that every invalid argument is remediable. For every argument has its corresponding conditional, that is, a conditional sentence whose antecedent is the conjunction of the argument's premisses and whose consequent is the argument's conclusion. By the prior specification of 'missing', the corresponding conditional of any invalid argument qualifies as a premiss missed by the argument and, so, as one the addition of which wholly rehabilitates the original argument *sub specie validitates.* Our distinction of late, between remediable and irremediable arguments, is now seen to implode, thanks to the emptiness of its second term. *Corollary* now bids us never to attribute invalidity in the ideal model. But if in the ideal model no one ever forwards an invalid argument or takes another as having advanced one, an earlier caution is trashed. For we said that the van Eemeren-Grootendorst idealizations were held to conditions of modesty, circumspection and ease of empirical re-entry. *Corollary* encumbers the ideal model with far too much logical optimism for serious theoretical belief. So *Corollary* needs to be abandoned or refurbished. That task is a central preoccupation of the deductivist research programme, concerning which it can fairly be said that the jury has yet to return a verdict. Whatever the verdict will eventually be — and whatever we chance to think of it — it will turn on an appraisal of attempts to re-define the notion of missing premiss in ways that avert the promiscuous levels of validity occasioned by the present definition of it. Readers interested in the current fortunes of deductivism should consult [Hitchcock, 1985; Hitchcock, 1987]. Also of interest is [Hitchcock, 1994].

2. Validity

It must be said with some emphasis that van Eemeren and Grootendorst are aware of these difficulties. In saying that they propose no solution to them, there is no risk that the absent solutions are for problems that van Eemeren and Grootendorst fail or refuse to recognize. On the face of it, this wrecks their programme — certainly that part of it having to do with the problem. Once the distinction between valid and invalid arguments collapses in the ideal model, the ideal model ceases to justify our having an interest in it, except negatively.

Though it has its evident attractions, the objection is too harsh. It overlooks an alternative which, while far from perfect, it is well worth reflecting on. It is this: Take it as *fait accompli* that the distinction between remediable and irremediable arguments has found the welcome of a disciplined analysis under which it does not implode. For present purposes we need to press for details. It is assumed, *pro tem* and pending further effort, that the class of remediable arguments is both well-specified and presented to us as a theoretical given. It is a lavish presumption, right enough. But it still presents the theorist with challenges which he will have difficulty in meeting. It is to these challenges that I wish to direct my attention and to which I shall turn after a brief preliminary.

The preliminary is about validity. Van Eemeren and Grootendorst know well the folly of holding all good arguments to the requirement that they be deductively valid, that is, that their premisses conjoined entail their conclusions. Evidently theirs would be an approach, if only they had contrived it, in which the validity of good arguments is made to answer to a certain criterial multiplicity. There are more ways than one for an argument to be valid, and some of these ways are *strictly* disjoint. No such account of validity is produced in *Argumentation, Communication and Fallacies.* It is an issue not overlooked, but foregone or postponed. The authors say that they do not want to take a specific and definitive stance on the question [of] exactly what kind of logical validity is preferred.

Here is another place in which there is room for severe judgement. By prior (and generous) stipulation, theory is presented with a subclass of invalid arguments made distinctive by the fact that they both can and deserve to be reconstructed as valid. But forebearing to tell us what validity is, the authors conceptually destabilize the whole project of rehabilitation.

This, too, is over-severe. At a minimum, the authors see it coming and have an answer for it. For the sake of simplicity, we shall restrict ourselves in our present exposition of unexpressed premisses to the well-known and ready-made instruments of propositional logic and first-order predicate logic, or *FDL*.

At first glance the proposal is not very impressive, but it can be lived with, or so I believe. Van Eemeren and Grootendorst are saying, in effect, that even if there was a good way of specifying remediable arguments and even if, contrary to fact, the validity required by the goodness of any good argument were just first order validity, there would still be difficulties in solving the problem of missing premisses. Moreover, these are important difficulties; important enough to justify our trying to deal with them even under the dubious and costly assumptions lately discussed. This, anyhow, is the spirit in which chapter 6 of *Argumentation, Communication and Fallacies* needs to be read.

We begin, then, with an assumed class of arguments invalid on their face and remediable by way of a choice of premisses which the argument misses. The target is the remediation of such arguments, the reconstructing of them in ways that make them valid. And for simplicity, validity is *FDL*-validity, plain and simple.

What, we might ask, goes on when someone makes a transparently invalid argument? Van Eemeren and Grootendorst think that the argument is made under conditions that seem to violate rule R(2(b)) — the futility rule. Making a transparently invalid argument is a paradigm of futility. It also appears, they say, that R(1) — the honesty rule may have been breached. For how can one sincerely propose an argument that is transparently invalid, that is, sincerely advance as valid an argument that is transparently not valid?

3. The Logical Minimum and Pragmatic Optimum

Because the ideal model is governed by strong charity presumptions, it must be supposed that in participants comply with these rules, not breech them. Accordingly, the apparently offensive argument needs to be reconstructed into regulatory compliance. It needs to be made out to be valid. This is done by postulating a validating missing premiss, one or more. One way of doing this is to identify as the missing premiss the argument's corresponding conditional. It is not only a way of proceeding, it is mandatory. In so doing, the argument's reconstruction furnishes what van Eemeren and Grootendorst call the 'logical minimum'. A further rule emerges.

R(4): *The Logical Minimum Rule:* In reconstructing a target invalid argument it is mandatory to add to its premiss set its corresponding conditional.

It is advisable to be clear. We have already seen that any invalid argument is validated by the premissory augmentation of its corresponding conditional. It is not to be tolerated that an R(4) — manoeuvre extinguishes the class of 'genuinely' that is, irremediably, invalid arguments. We have spoken of a class of arguments whose invalidity derives from nothing but a technicality, from the fact that it leaves a missing premiss unexpressed. Nothing I have said here establishes that the class is non-empty. I have only assumed that it is and the assumption, contrary to fact or to possibility as may be, is made with those risks identified and assumed. It is precisely this class, taken now to be non-empty, that contains all and only the 'target' arguments of R(4). R(4) is relativized to that class of target arguments and this spares us the embarrassment of having to concede that no argument whatever is genuinely invalid, namely, the untargeted ones.

Mandatory though it is to supply the logical minimum for reconstructions of target arguments, R(4) produces an unwelcome result. It appears to violate R(2(a)), the superfluousness rule. *A* speaker is convicted of superfluousness in affirming an immediate logical consequence of something he has committed himself to at the nearest predecessor stage of the discussion. If at a given point, our speaker proclaims his original argument *O*:

P1. Angie is a real woman.

C. Therefore, Angie is nosy.[1]

and if, in the next breath, he proclaims argument *LM*: (logical minimum)

P1. Angie is a real woman.

P2. If Angie is a real woman then Angie is nosy.

C. Therefore, Angie is nosy.

his assertion of P2 is an immediate consequence of his prior subscription to *O*, an instance nearly enough of the Deduction Theorem:

(DT): $A \vdash B \rightarrow \vdash (A \supset B)$. (That is, if B is deducible from A, then it is a theorem that if A, then B.)

[1]Some readers, sensitive to issues of gender-specificity, may regret the present example, never mind that there is not reason to think that the authors take it as a serious sociological claim.

So R(2a) is breeched.

This is troublesome. It lands the pragma-dialectician in a dilemma. R(2a) and R(4) are mandatory rules, but there is no way of complying with R(4) that doesn't violate R(2a).

In what is arguably the most interesting and imaginative feature of the pragma-dialectical treatment of the missing premiss problem, van Eemeren and Grootendorst seek to evade the dilemma by stipulating a further rule.

R(5): *The Pragmatic Optimum Rule:* In reconstructing an original argument, it is mandatory to furnish as a premiss some proposition P the denial of which would be pragmatically inconsistent with espousal of the original argument prior to premissory argumentation.

Here is what R(5) attempts to codify. Suppose a speaker espoused *O* and there went on to deny that real women are nosy. That, say van Eemeren and Grootendorst, would land the speaker in a 'pragmatic inconsistency'. R(5) requires that the denied proposition be added as a premiss. Thus we have argument *LMPO* (logical minimum and pragmatic optimum).

LMPO:
P1. Angie is a real woman.
P2. Real women are nosy.
P3. If Angie is a real woman then Angie is nosy.
C. Therefore, Angie is nosy.

Van Eemeren and Grootendorst leave the notion of pragmatic inconsistency undefined. That is regrettable for so load-bearing an idea. I myself am prepared to concede that there is something decidedly odd about saying "Angie is a real woman, so Angie is nosy; but it's not the case that real women are nosy". The issue is tangled by a couple of factors. One is whether "Real women are nosy" is to be taken as a full-blown generalization, that is, as a universally quantified conditional, or as a generic claim. Another is, if a universally quantified conditional, whether the denial of it by an *O*-espouser convicts him of anything seriously deserving of the name of pragmatic inconsistency. I shall come back to these points, but for now I want to concede that the inconsistency is pragmatic in some reasonably intelligible sense, and to attend to how our dilemma is thought to be slipped.

Our trouble arose if "If Angie is a real woman then Angie is nosy" were added as a premiss as an immediate consequence of *O*. This is the situation represented by *LM*. The preferred solution is to apply R(5) first, which in the case at hand gives us premiss 2 of *LMPO*, that is, "Real women are nosy". If the statement is a universally quantified conditional, then premiss 3, that is, "If Angie is a real woman then Angie is nosy", comes by way of immediate inference from it, not from the original argument *O*; in fact it comes to us by the device known as universal instantiation. Even so, premiss 3 we do have and, with it, fulfilment of R(4), the logical minimum rule. Given that R(5) and R(4) are invoked sequentially, and in just that order, it is possible to say that P3 of *LMPO* does not arise by immediate inference from *O* and hence that there is no transition from *O* to P3 that qualifies as a violation of the superfluousness rule. Moreover, since the negation of P2 is not logically inconsistent with the espousal of *O*, but only pragmatically inconsistent with it, it cannot be said that P2 arises from *O* as an immediate (logical) consequence of it. So the transition from *O* to P2 fails to qualify as a violation of the superfluousness rule. The dilemma is escaped, apparently.

4. Genericity

It now becomes necessary to deal with the status of P2 — "Real women are nosy". If it is a universally quantified conditional, then P3 is indeed an immediate consequence of it, as we have just seen. This presents us with two further difficulties. One is that since P3 arises from P2 by immediate consequence, and if P3 is asserted on that basis, then although the assertion of P3 is no longer a superfluous move from *O* itself, it is nevertheless a superfluous assertion from P2; and since P2 and P3 are essential premisses of *LMPO* and *LMPO* is the mandatory reconstruction of *O*, the reconstruction of *O* inescapably involves a superfluous speech act, in violation of R(2a). Our dilemma now is repositioned, not averted. The other difficulty with construing P2 as a universally quantified conditional, is that intuitive damage is done to the notion of pragmatic inconsistency, which is indispensable in furnishing the pragmatic optimum. Recall that the pragmatically optimal premiss is any proposition whose denial is pragmatically inconsistent with the espousal of the original target-argument. In the case at hand, we selected as premiss 2 the statement "Real women are nosy", because we were prepared to concede that the espousal of *O* and the denial of P2 was sufficiently odd to qualify as pragmatically inconsistent. Of course, this was largely stipulation: it's odd, so let's bite the analytical bullet and *call* it pragmatic inconsistency.

But if, as we are presently supposing, P2 is to be taken as "For all x, if x is a real woman then x is nosy", then I quickly begin to lose my former confidence that its denial is pragmatically inconsistent with *O*. Let me say it again: "pragmatic inconsistency" was a term stipulated for a kind of intuitive oddness. Neither the inconsistency nor the oddness was specified, but the basic idea was clear. *A* speaker has no business espousing *O* unless he is also ready to accept P2. With P2 bearing its present construal, we would have to be satisfied that unless a speaker were prepared to asert that every real woman whatsoever and without exception is nosy, he has no business saying that Angie is nosy, since she is a real woman. My intuitions resist this. Couldn't there be just one real woman who managed not to be nosy (she took a course in curiosity-suppression at the YWCA), without depriving a speaker of entitlement to say this of Angie (who obviously didn't take the course)? Those who share my unease have dealt the reconstruction of *O* a further blow, for P2 is no longer available to us as a premiss that satisfies pragmatic optimality. And with P2 gone, so is P3 in our reconstruction of *O* as *LMPO*.

No doubt it will occur to the reader that these difficulties call for a different construal of P2, that it not be taken as a universally quantified conditional. An alternative presents itself: Construe P2 as a generic statement. Generic statements make characteristic assertions. If "Real women are nosy", like "Tigers have four legs", counts as a generic statement then it is as general as may be, short of universal quantification. Counterexamples are tolerated by generic statements, whereas universally quantified sentences are toppled by them. The use of generic statements is linked to what Minsky calls a frame-system, as we have seen in preceding chapters.

> A frame-system is intended to be a structured piece of knowledge about some typical entity or state of affairs. It contains variables onto which go specific values... [Johnson-Laird, 1988, p. 107].

What then, is a typical property of a thing or event? It is a default value with regard to a given variable furnished by its frame-system in the absence of contrary information. The notion of a frame-system in a partial elucidation of a *prototype*; for a prototype of something just is a frame-system in which all default values are specified. The notion of default values gives rise to the kindred notion of default reasoning. If, for example, G is a generic sentence, one is allowed to infer its instances except where excluded or over-ridden by specific information to the contrary. A rule comes to mind, the default reasoning rule as we might say.

> R(7): *Default Rule:* If *G* is a generic statement then so long as the speaker takes it to be true, apply *UI* (i.e., universal instantiation) defeasibly, unless contra-indicated by specific information.

Like generic assertion itself, generic or default reasoning is defeasible. In those cases in which instantiation to a generic statement's default value is contra-indicated, the inference must be corrected, but that fact does not falsify the generic statement, it falsifies only that particular default instance. It is a nice question, of course, as to what does falsify a generic claim. *A* further rule comes to mind, vague though it surely is; it is a rule of generic falsification.

R(8): *Generic Falsification Rule:* A generic claim is falsified when its default instantiations are false in sufficient numbers to defeat the presumption that the default value in question is *characteristic* of the subject of the generic claim.

We now have the means to revive the van Eemeren-Grootendorst notion of pragmatic inconsistency. Let *O* be, as before, our original argument; and let P2 be its *genericization*, that is, its corresponding generic claim. Then it is pragmatically inconsistent to allow *O* and yet to deny P2, its genericization. And the rule for finding the pragmatically optimal premiss in the reconstruction of *O* is

R(9): *Generic Pragmatic Optimum:* Where *A* is the original argument, the missing pragmatically optimal premiss is the genericization of *A*, *G*.

Notice, then, that if we decided to introduce P2 by way of rule R(9), P3 would not arise from P2 by unqualified universal instantiation. It would arise by default instantiation and, so, would not be an immediate logical consequence of it. This, it might be said, is a sufficient weakening of the consequence relation to justify our thinking that superfluousness is now evaded. My own view is that this is a promising enough possibility with respect to the generic treatment of P2. What makes it promising is precisely the weakening of the consequence relation. "Real women are nosy" could be true even where "If Angie is a real woman then Angie is nosy" is false. And yet the default reasoning from the former to the latter is still all right for any reasoner not in possession of the relevant contra-indicating information.

Further problems also find solution. One is the problem of the non-uniqueness of an argument's optimal premiss. Certainly our earlier rule that any proposition whose denial would be pragmatically inconsistent with *O* was ludicrously permissive. For one thing, it admits as our pragmatic optimum the premiss, "But I don't believe it". Van Eemeren and Grootendorst suggest (and then abandon) a syntactic rule for specifying pragmatic optimality. The pragmatically optimal premiss of an original argument is the universal generalization of its logical minimum. But, since this requires us to construe P2 as a universally quantified conditional, it revives our earlier dilemma. Better, perhaps, to amend the syntactic rule and have it call for the *genericization* of the argument's logical minimum.

It won't work. The present course presupposes the prior availability of P3 — *O*'s logical minimum — and that lands us back in superfluousness. But a corrective is now at hand, as we have lately seen. Let, instead, the rule for the specification of *O*'s pragmatic optimum call for the genericization of *O* itself. Of course, this requires the prior availability of *O*, but since without *O* we would have nothing to reconstruct in the first place, it can hardly be supposed that genericizing *O* catapults us into superfluousness.

5. Find the Premisses

Within the generous and problematic constraints that the pragma-dialectical approach imposes on the treatment of missing premisses, it is possible to reconstruct the reasoning of chapter 7 in the following way. The principal virtue of *this* reconstruction is that it solves or makes some headway with solving problems that arise even under those liberal constraints. So, in reconstructing an original target argument, *first* genericize the argument. Since espousing the argument would be pragmatically inconsistent with denying its genericization, and since genericization affords a syntactic criterion of it, call the genericization the pragmatically optimal premiss in the reconstruction of the target argument. Since the generic premiss is not an immediate logical consequence of the original argument, superfluousness is avoided.

Second, apply the default-reasoning rule R6 to our generic statement and default-derive the further premiss which, as it happens, is the corresponding conditional of the original argument. This further premiss satisfies the syntactic criterion of logical minimality. Though it is an immediate consequence of the original argument it is not immediately inferred from it. It is not immediately inferred from anything. It is, rather, default-inferred from the pragmatically optimal premiss, P2. Once again superfluousness is evaded. At the same time, the requirement of validity is met. For P3 surely does validate *O*.

Still, it must be said there is a great deal to be dissatisfied with in this account. Various of its constraints are over-generous and unrealistic; others are over-severe. Key concepts are vague and unanalyzed. The present treatment will work only for certain classes of arguments, for example, those whose imbedded syntactic structures will tolerate the operation of genericization. There are arguments galore, ones with premisses obviously omitted, which either resist the genericization-test of pragmatic optimality or whose genericization no one would be ready to commit to. The later case can be illustrated by the argument

A: It's raining.
So Eveline won't be driving to Calgary.

Assuming as its generalization "People don't drive in the rain", two points require attention. One is that the generic statement is untrue; the other is that it doesn't fulfill the same syntactic criterion for genericity as does "Real women are nosy". Perhaps a different syntactic criterion could be supplied for "People don't drive in the rain". But now the problem is that selection of pragmatically optimal premisses is subject to multiply-syntactic criteria, and the multiplicity is left unspecified. The former kind of case is suggest by the argument

A*: Fred is President.
So, it is Fred who will be in Denver on Friday.

Concerning *A** it is doubtful that anything like "Presidents are in Denver on Friday" even counts as a generic statement. The implied generality, qualified though it would be by the inputed genericity, seems entirely specious. Generic statements thrive in contexts giving appropriate purchase to factors of typicality; they thrive, that is to say, in contexts of nomicity (or lowlikeness), of natural kinds, of counterfactual instantiation. No such context is available for *A**. The missing premiss which most interpreters would furnish is surely, "The President will be in Denver on Friday"; genericity does no work here at all. That said, there is room to regret the Angie example on grounds other than rudeness to women. For there is reason to think that the 'genericization' of *O* — "Real women are nosy" is specious, for it lacks that factor of nomicity.

Even so, limited in several ways though it is, the present account does solve some problems in an imaginative and credible way. The same cannot be said for all treatments of the missing premisses issue. Van Eemeren and Grootendorst leave us with an open question and a challenge. Can we build on and refine their account in ways that tolerate the removal of its more unrealistic assumptions? It is a good question and an engaging challenge, and is discussed further in [Gabbay and Woods, 2004, sect. 4.4].

PART VII

ANALOGY

In most approaches to fallacy theory, some attention is given to the fallacy of false (or faulty) analogy. There is a prior question, needless to say. What is the nature of analogy, and what aspects of it are compromised by the fallacy? As the two next chapters attempt to show, there are two different concepts of analogy, each having a quite distinctive logical structure. The one concept is that of *analogical argument.* The other is *analogical predication.*

Chapter 16

BY PARITY OF REASONING[1]

"analogy is an unstable form of argument".

[Perelman and Olbrechts-Tyteca, 1971, p. 393]

False analogy is on many people's list of standard fallacies. It is a particularly good example of a claim made in chapter 1: that the fallacies are entire research programmes. In this chapter and the next I want to try to develop this claim. I will suggest that a disciplined account of when an analogy is fallacious requires a good general theory of analogies, to which the account of fallacies bears an unbreakable linkage.

Analogies come in two different forms. There are analogical arguments and analogical predications. Arguments are the business of the present chapter; predications will be dealt with in the next.

1. Parity of Reason

analogy is rife in the lives of certain species. It scarcely seems possible that we ourselves could do without it. We have analogical predication, analogical pattern recognition, analogical explanation, analogical inference, and analogical argument. analogy is one of those ideas as old as logic itself, and yet, perhaps more than the others, it has resisted analytical reconstruction to the point of theoretical impotence. In this respect, analogy is like relevance, another momentous discouragement to theory, and yet another of those ancient and indispensable guides to rational life. This is ironic, since, as we shall soon see, our own account

[1]Co-authored with Brent Hudak.

of analogical argument will press relevance into theoretical service in a non-trivial way. So at the outset there may be cause to wonder whether ours is an account in which the blind lead the blind.

We will here concentrate on analogical argument. It is not the same as analogical inference, if only because inference is not the same as argument. It doesn't follow, of course, that a good theory of analogical argument will be of no use for accounts of analogical inference; it is just that they cannot be the same account. The argument that argument isn't inference is well-known and we will not take the time to reproduce it here [Harman, 1986, pp. 3–7].

Analogical argument must also be distinguished from analogical explanation, if only as a reflection of the fact that argument also differs from explanation. Even if it were true that explanations can always be made out to be deductive-nomological arguments, we would not have lost our distinction, for not every argument is a deductive-nomological argument. Many writers propose a sharper divide between argument and explanation than is drawn by the deductive-nomological account of explanation. Mill is one such and it was he who reserved explanation — as opposed to argument — for analogy's pride of place: "[A]nalogy's sole value is that it makes it possible to formulate a hypothesis for verification by induction".[2] Though analogical explanation plays a large role in rational life, we do not share Mill's restrictive view of it.[3] We are rather more drawn to a remark of Perelman and Olbrechts-Tyteca: "Any complete study of argumentation must... give [analogy] a place as an element of proof" [Perelman and Olbrechts-Tyteca, 1971, p. 372], emphasis added.

Some may find it surprising that analogical argument differs from analogical inference. More surprising still is the suggestion that analogical predication differs from analogical argument, not just in the way that predication differs from argument, but in rather more striking ways that suggest the unlikelihood that they exemplify a single, unified idea of analogy. It is almost as if the structural elements that make for analogousness in argument are not those that deploy to make for analogousness in predication; at a minimum, anyhow, they deploy quite differently. This we will attempt to demonstrate in due course, but it is well to flag

[2][Perelman and Olbrechts-Tyteca, 1971, p. 372]), paraphrasing J.S. Mill at [Mill, 1959, pp. 365–368].

[3]In fact, there is reason to wonder whether this is Mill's considered view, given his predilection for contradicting it.

the point here, to sound the admonition that predication and argument do not draw down the same concept of analogy.

Other things are said about analogical argument which we think untrue or, if not untrue, imperspicuous from the point of view of theory. These we will take up in passing. There is, however, one matter which should be addressed at the outset. In particular, it is the claim that arguments are analogical because they conclude from the fact that things are similar in certain respects that they are also similar in certain other respects.[4]

This is not so much wrong as unhelpful. For if we say that such an argument is good just when the conclusion (that the things are similar in certain other respects) cannot consistently be resisted in light of similarities already noted, this leaves it unexplained as to how the similarities interact with the inconsistency. That is, why should it be, and what is it about them, that the noted similarities give rise to the concluded ones? Without further elucidation, we trade in one obscurity for another. We trade in "and so by analogy such and such" for "and so, because they are similar in certain respects, they must be similar in these other respects".

Thus, a characterization of analogical argument cannot be theoretically illuminating until the appropriate connection is made between the

[4]For example

> ... as this resemblance admits of many different degrees, the reasoning becomes proportionably more or less firm and certain. An experiment loses its force, when transfer'd to instances which are not exactly resembling; 'tis evident it may still retain as much as may be the foundation of probability, as long as there is any resemblance remaining [Hume, 1888, p. 142].

We also have [Mill, 1865, p. 402]. Here Mill says that there is no *basic* difference between inductive and analogical inference. Also:

> If we can discover, for example, an unknown animal or plant, resembling closely some known one in the greater number of the properties we observe in it, but differing in some few, we may reasonably expect to find in the unobserved remainder of properties a general agreement with those of the former, but also a difference corresponding proportionately to the amount of the observed diversity [Mill, 1959, p. 367].

For a more recent example:

> The basis of arguments by analogy is that when two things are similar in a number of respects, we have some basis for thinking that they may well be similar in further respects also [Govier, 1988, p. 100].

Cf. Quine and Ullian: "The common core of all the uses [of 'analogy'] is fairly well covered by 'similarity' or 'parallelism'" [Quine and Ullian, 1970, p. 62]. Cf. also Shaw and Ashley: "An argument from analogy is, in sum, an inference from some points of resemblance between two or more objects to other such points" [Shaw and Ashley, 1983, p. 419].

factors of similarity and the inconsistency of resisting the conclusion.[5] If we ask whether there are clear cases in which factors of similarity admit of descriptions that genuinely elucidate such verdicts of inconsistency, the answer is Yes. Consider, for example, the following two rather simple-minded arguments, X and Y.

Argument X

1 The cat is on the mat.

2 If the cat is on the mat, the dog is in the manger.

3 Therefore, the dog is in the manger.

Argument Y

1 Bill loves Sue if and only if the market does well.

2 Bill loves Sue if and only the market does well, only if the total memory capacity of the human brain is $4x10^{16}$ bytes.

3 Therefore, the total memory capacity of the human brain is $4x10^{16}$ bytes.

Now it is evident upon inspection that although X and Y differ substantially in their surface structure (appropriating a term from elsewhere), they possess a common deep structure. The 'similarity' (an understatement) of deep structure just is identity of form, in this case, of truth functional form. It is this identity of form that accounts for the fact that one could not consistently hold that X is a correct argument and yet that Y is not. Here the interaction between similarity and inconsistency is elucidated. For identity of form is such that properties of a given argument which obtain on its account cannot consistently be withheld from arguments possessing the identical form. That said, it is now clear that the similarity-from-similarity notion of analogy admits of an interpretation under which it comes out true. Since X and Y are similar in a certain respect (viz., sameness of logical form), they are similar in a certain further respect (viz., in this case, deductive validity).

It is important to see that it is not a question of X's and Y's both being such that their respective conclusions cannot, with deductive con-

[5]That said, it is clear at once that we must separate ourselves from the position of Shaw and Ashley — among others — that the basic logical form of an argument from analogy is this:

1 Objects O_1, O_2, O_3, ..., O_n have Properties P_2, P_3, P_4, ..., P_k in common.

2 Objects O_2, O_3, ..., O_n have Property P_1 is common. Therefore, it is probable that

3 Object O has Property 1 [Shaw and Ashley, 1983, p. 420].

What this account loses is the factor of a charge of inconsistency, of the idea that an analogy threatens with a kind of inconsistency someone who acknowledges the relevant similarities but who refuses or fails to act in ways deemed appropriate to the similarities. It is certainly going too far to claim that someone who assented to 1 and 2 but not to 3 is guilty of an inconsistency. But if not there, then where?

sistency, be resisted in the light of their respective premisses (though this too is so in the present case). The point, rather, is a good deal more general. Whatever verdict — whether of deductive validity, inductive strength or whatnot — that is conferred upon a given argument by virtue of the logical form of its deep structure is also conferred upon any argument sharing that structure.[6]

It is our principal contention that it is precisely this way with arguments from analogy. Arguments from analogy are arguments by parity of reasoning, so-called.[7] They are arguments about arguments, meta-arguments. They argue that two or more target arguments stand or fall together and that they do so because they are relevantly at parity, that they possess similar deep structures by virtue of which they coincide in logical form. The target arguments of the meta-arguments are thus analogues of each other. This basic idea, or something similar, can be found in pragma-dialectical writings on argumentation: "Argumentation in which there is an ordering based on a similarity in the structure of things connected with the thesis being defended and a structure of things that is not subject to doubt in the mind of the audience is used in argumentation by analogy" [van Eemeren et al., 1987, p. 242]. That said, it is still doubtful that the pragma-dialectician has our analysis in mind.[8]

If the meta-argument account of analogy is true, it is important in another respect. For it shows that the analogy-relation defined over arguments is symmetrical. But, as we have shown analogical predication is not symmetrical. We return to this point below.

2. Standoffs

It now seems appropriate to speak of the basic structure of an analogical argument, somewhat as follows:

1 Argument A possesses a deep structure whose logical form provides that the premisses of A bear relation R to its conclusion.

2 Argument B shares with A the same deep structure.

[6]This we take to be Govier's intent in saying that argument analogues need not "depend on a deductive connection" [Govier, 1985, p. 32].

[7]Strictly speaking, the appellation is off the mark, given the difference between argument and reasoning (inference). However, with the point noted, we will allow the expression to stand.

[8]In particular, though they provide little theoretical elaboration in the Handbook, van Eemeren, Grootendorst, and Kruiger seem drawn to the proportionality account of Perelman and Olbrechts-Tyteca [van Eemeren et al., 1987, 243 ff.]. Cf. [Perelman and Olbrechts-Tyteca, 1971, 375 ff]. In contrast, Govier adumbrates part of our position in the memorable phrase "logical core" which leaves "the essence of the argument [analogues] intact" [Govier, 1985, p. 31].

3 Therefore, B possesses a deep structure whose logical form provides that its premisses likewise bear R to its conclusion.

4 Hence, B is an analogue of A. A and B are good or bad arguments, by parity of reasoning, so-called.[9]

On the face of it, the basic structure is something of a letdown, reminiscent of Peggy Lee's "is that all there is?" Where, it might be asked, are the brilliance and ingenuity that characterize analogical arguments at their best? Where, in particular, is what would account for the sheer cleverness of some such arguments? Defending argument Y by analogy with argument X is a humdrum enterprise, perhaps of help only for the spectacularly dim. The use of certain analogical arguments, by contrast, is arresting and mind opening. How are we to explain this?

The more useful thing to do at this point is to produce an analogical argument and to plumb it for its theoretical yield, if any. Ours is a good but manageable example; good in as much as it has won the admiration of many, if not most, who have studied it; and manageable inasmuch as it seems to admit of analysis in something less than a whole lifetime; unlike, say, the Design Argument. We should also add that it is deliciously controversial.

The example is drawn from Judith Jarvis Thomson's celebrated paper, "A Defense of Abortion". In it Jarvis Thomson produces an argument designed to show that the termination of a rape-induced pregnancy is morally justified.

> You wake up one morning and find yourself back in bed with an unconscious violinist. He has been found to have a fatal kidney ailment, and the Society of Music Lovers has canvassed all the available records and found that you alone have the right blood type to help. They have therefore kidnapped you, and last night the violinist's circulatory system was plugged into yours, so that your kidney can be used to extract poisons from his blood as well as your own. The director of the hospital now tells you, "Look, we're sorry the Society of Music Lovers did this to you — we would never have permitted it if we had known. But still, they did and the violinist now is plugged into you". Is it morally incumbent on you to accept this violation? No doubt it would be very nice if you did, a great kindness. But do you have to accede to it? What if it were not nine months, but nine years? Or longer still? What if the director of the hospital says, "Tough luck, I agree, but you've now got to stay in bed with the violinist plugged into you, for the rest of your life. Because remember this: All persons have a right to life, and violinists are persons. Granted you

[9]It is worth noticing that on the present account, it is a serious mistake to try to distinguish between analogical arguments on grounds of deductive or inductive considerations; that is, to try to distinguish between deductive analogical arguments and inductive analogical arguments. Arguments Z and W may be deductively correct, inductively strong, and so on. But whatever their logical character, it is not recapitulated in the general case, in the analogical meta-argument. Its structure is the same either way.

have a right to decide what happens in and to your body, but a person's right to life outweighs your right to decide what happens in and to your body. So you cannot ever be unplugged from him... [Thomson, 1971, p. 49].

How, then, are we to account for the sheer cleverness of Thomson's use of the Violinist? Part of the answer is that the clever, ingenious or breathtaking analogy turns on the disclosure or postulation of deep structure that may have been obscured by surface structure. Clever arguments from analogy are somehow striking in their structural perceptiveness.[10] That, as we say, is part of the story. The other parts are more easily told dialectically. So let us imagine that two dialecticians, Bill and Sue, are trying to determine the moral standing of abortion in the case of a rape-induced pregnancy. As might be expected, Bill and Sue are getting nowhere; in fact, they have landed themselves in a *standoff*.[11]

A standoff is a kind of dialectical black hole, afflicted by a paralyzing dissensus, depriving the contenders of a basis for rational settlement. Slightly over-simplified, an argument counts as a standoff within argumentative parameters P_1, ..., P_n to the extent that arguments and counter-arguments, claims and counter-claims, couched in terms of the P_i tend to be either irrelevant or question-begging. Thus, if Bill and Sue were transacting their argument in terms of foetal rights, maternal reproductive control rights, quality of life, and so on, then if their argument threatens to become a standoff, their only practicable chance of escaping from it is by finding new parameters; by finding factors relevant to the issue at hand on which they do agree.

Here is one way in which the escape from a standoff might go: The juncture at which there is a dialectical shift from the P_i that make the dispute a standoff to other parameters that might be agreed upon and considered relevant, is the point at which the original argument is abandoned and replaced by an analogical argument. The analogical argument, let us repeat, is a meta-argument, an argument to the effect (somewhat over-stylized) that another argument — let's call it a 'comparison' argument — shares an identical form with the original argument. Thus the analogical argument both makes an argument and presents a (comparison) argument. The argument it presents (e.g., the Violinist) is not the argument it makes. The argument it makes is analogical, that is, it holds that the comparison argument is identical or

[10]Thus, "[a]nalogies are important in invention and argument fundamentally because they facilitate the development and extension of thought" [Perelman and Olbrechts-Tyteca, 1971, p. 385], emphasis added.

[11]The idea of the standoff is developed in chapters 12 and 13. A standoff resembles what Robert Fogelin calls a deep disagreement [Fogelin, 1985, p. 5].

— at a minimum — relevantly similar in form with the original, and therefore that the original stands or falls with it.

The dialectical breakout from a standoff that an analogical argument tries to achieve will settle the original dispute if Bill and Sue agree

(i) that the original argument and the comparison argument are indeed analogues of one another;

and

(ii) that the comparison argument is good or, as the case may be, bad.

3. The Structure of Analogy

We said, just above, that this account of the escape from a standoff by way of analogical argument is somewhat over-stylized; and so it is. There are, in fact, features of actual cases that call into question our characterization of analogy as a dialectical manoeuvre. In such cases (the abortion case is a good example), the original argument is not fully formulated. The brunt of the disagreement falls upon the conclusion, upon whether abortion is morally permissible in certain circumstances. Since the disagreement is also a standoff, there is bound to be serious difficulty not only in agreeing on which premisses to count as true, but also about which statements to recognize as premisses in the first place. An abiding feature of a standoff is, as we have seen, that considerations advanced by the one disputant have a way of being found irrelevant or question-begging by the other, and so not worth stating, and hence, in a functional sense, not premisses at all.

Though nothing in principle prevents a disputant from giving a complete articulation of his side of the issue, we admit that in actual dialectical practice, prospects for complete articulation are often dashed by dialectical dissonance. Yet it is quite clear that, frequent or not, the fact of the half-expressed original argument need not be a barrier to eventual analogical settlement. For if the comparison argument does provide a basis for settlement, it is usually possible to reconstruct from the original, half-expressed argument a fully articulated version of it, in which the original parameters are integrated with parameters exhibited in the logical form of the comparison argument.[12]

[12]But see below, section 7. Also, there is another respect in which our account is over-stylized. It is of course sometimes true that the comparison argument is presented ("Consider the following argument, which you'll see at once to be valid") as opposed to waged ("If I keep hammering away, I think I can get you to buy this other argument"). In conditions of dialectical serenity, this would be the norm. In actual practice, often enough, the comparison argument must be aggressively prosecuted, and may itself seek relief from supplementary comparison arguments. Still the bold distinction between presented versus waged is upheld

We can see this all coming together in the Violinist. The Violinist is a comparison argument. It is offered as more or less obviously correct. If it is correct, this is because it yields to a structural re-description that lays bare a logical form in virtue of which it is correct. If it is also a dialectically tenable comparison argument, it will be possible structurally to re-describe the original argument in such a way as to lay bare the same logical form, which is obscured by the surface structure of the original.[13] If the comparison argument and the original argument turn out to be analogues of one another, it will be because they share that logical form and thus collect the same verdicts on their success or failure as arguments. The Violinist yields the following structural re-description:

Human beings H_1, H_2, and H_3 are so related that

1. without H_2's consent, H_1 has placed H_3 in a state of vital dependency on H_2;
2. the period of dependency is indeterminate (perhaps nine months, perhaps nine years, perhaps forever);
3. the dependency is a grievous impediment both to locomotion and to (stationary) mobility;
4. the dependency constitutes a grievous invasion of privacy;
5. it is an invitation to social disaster, for H_2 (and in some cases for H_3 as well) is a laughing stock;

and

6. it threatens H_2's economic self-sufficiency.

This suffices to set up the basic question: does H_2 owe to H_3 the hospitality of this arrangement, unconsented to and at the cost of locomotion, mobility, privacy, income, self-respect, and prospects of determinate relief? Let us say, for now, that he does not. The comparison argument finishes, then, with the conclusion

7. therefore, it would be morally permissible for H_2 to terminate the vital dependency.

It is perhaps odd to call (1) to (7) the logical form of the comparison argument of Jarvis Thomson's example. But it is certainly (part of) its deep structure, for it is, obviously enough, an abstraction; an abstract re-description of essential factors. In the earlier example of arguments

---

by the fact that the dialectically less hazardous comparison argument is one that can be just presented, one that stands without need of vigorous supplementary advocacy.

[13]Thus we recognize that "[a]lthough 'reduction to the familiar' is not necessary for successful understanding, it is an aid to comprehension, and analogy is frequently employed in this way" [Shaw and Ashley, 1983, p. 416]. This, however, has to do with surface structure and psychology. People may well grasp the logical form of some arguments more readily than others. That said, however, we find that surface structure provides us with no clue as to why two arguments are analogues of one another.

X and Y, the deep structure was very abstract, nothing more than the truth functional structure of *modus ponens*. The structure of *modus ponens* deserves the name of logical form because it is an abstraction that determines the logical appraisal for any argument exhibiting it. True, the abstraction (1) to (7) is nothing like a truth functional skeleton, but it too determines the terms of logical appraisal for any argument exhibiting it; and it is precisely in this sense that it too qualifies as logical form.

It is interesting to note, in passing, that even if Jarvis Thomson's comparison argument is correct, its logical form almost certainly is not exhibited in the facts about the abortion case. The foetus does not consent to the pregnancy; the pregnancy is not temporally indeterminate; and, in the general case: it is not a grievous impediment to locomotion and mobility; it is not an invasion of privacy; it does not threaten economic self-sufficiency; it does not constitute a social calamity. Thus the analogical argument fails. No doubt revisions could be cast about for, and found. But what is not clear is that any revision of the comparison argument — which would show a logical form exhibited also by the facts of the pregnancy case — would continue to justify the verdict of correctness either for the comparison argument or for the original.[14]

4. Counterexample

It is possible that we have succumbed to an over-regimented account of analogical argumentation. Consider:

D: 1. Bobby and Billy are both your young sons.
2. You're giving Bobby some pocket money.
3. So you should give Billy some pocket money, too.

Though not a meta-argument, why isn't this an analogical argument? Our answer can only turn on the exposure of some further structure of D, something along the following lines.

[14]This possibility is explored in some detail in [Woods, 1978, pp. 77–82]. There is also reason to think that the comparison argument may not be correct. This is suggested by a further comparison argument concerning Siamese twins *a* and b, who are now fifteen years of age.
1. Their attachment is such as to constitute for *a* a vital dependency upon b, but not vice-versa.
2. The attachment is such as to constitute a severe impediment to locomotion and mobility.
3. The dependency is temporally indeterminate, for though *a* could die at any time, it is possible that he will have a full life-span.
4. Honourable employment is all but out of the question.
5. *a* and b are subject to morbid and demeaning curiosity.
6. Their attachment is a perpetual and reciprocal invasion of privacy.
7. Both are innocent victims of circumstance. Would b be morally justified in severing the attachment?

D*: 1. Other things being equal, people should be treated equally.
2. Other things are equal in this present case.
3. A and B are people.
4. A and B alike bear some relation R to F of such a general type as to make it appropriate for F to do X for A and B alike.
5. F does do X for A.
6. Therefore, F should do X for B.

Putting A=Bobby, B=Billy, F=Father, X=providing pocket money, then the particular force of the supposed analogical character of D*, and thus of D, is indicated by conditions that would refute it. Thus

(a) Bobby is good.
(b) Billy is naughty.
(c) So other things aren't equal.
(d) Father needn't give Billy some pocket money.

The refutation of D* stands or falls on whether premiss 2 of D* is true. It isn't, and so one might say that Billy is disanalogous to Bobby in relevant respects.

Notice, however, that the analogy between Bobby and Billy (if there is one) is that they both bear R to F and that R makes appropriate a certain action. D* fails not because the analogy fails, but because premiss 2 does. That is, D*'s failure is independent of its analogy-making factors (if such there be), and so it is infelicitous to call this a failure of analogical argument. True, you could say that the failure of premiss 2 suffices for disanalogy and thus that 2 itself suffices for analogy. But that would mean that disanalogy just is dissimilarity and analogy just is similarity; and that, we say, would be false.

That dissimilarity does not suffice for disanalogy is a point which should be flagged. Not just any dissimilarity will do for a disanalogy. What is required is a relevant dissimilarity. In particular, the dissimilarity in question must be shown to affect the validity of the argument. It may be, for example, that there is a missing premiss essential to the validity of the Violinist argument which has no counter-part in the case of the rape-induced pregnancy. That is, identification of a dissimilarity as a disanalogy is a function of the logical form of the argument.

Still, it is possible that we haven't quite managed to understand the force of the complaint. Perhaps the example reproduced in D is better formulated in the following way. We might imagine that someone, noticing the pocket-money disparities of the Smith household, reasons that since Mr. Smith endows his young son Bobby with spending money, he should do the same for Billy, his other young son. The neighbor's inference turns on an expectation unfulfilled, and thus owes some of its features to other inferential commonplaces. Noticing that lobsters have been red when boiled, I expect the next boiled lobster to be red (but not that all boiled lobsters are red). "analogy thus pictured is an inferential leap, whereof the top of the trajectory is a slurred-over induction" [Quine and Ullian, 1970]. Now, if this captures the sense of the objection, we yield to it. But, note, this is a matter of inference, and inference isn't argument.[15] So we are not persuaded that the complaint succeeds against our account of analogical argument.

Even so, isn't there something generically analogical about D, for its refutation — a counter-example — turns on a disanalogy between Bobby and Billy? The question is important. This calls forth the suggestion that the concept of the negative analogy receives paradigmatic expression in the concept of the counter-example, the classical manoeuvre, surely, of philosophy and science alike.[16] Since counter-examples are not in general meta-arguments, isn't the meta-argumentational account of (negative) analogy somehow wrong? Let us be clear about what the present objection is. In giving a counter-example, C, you can be understood as giving an argument. Thus,

E: 1. It can't be so that all A's are B.

2. Because C — there is an A which isn't B.

But argument E is not about arguments; it's about generalizations and instances. So, again, isn't our own account too restrictive?

The notion that counter-examples turn on, or are, disanalogies has its temptations. Consider a further exchange between Bill and Sue.

E*: (Bill) Ravens are black.

(Sue) Albino ravens aren't!

Perhaps we can say that the ravens to which Sue alludes are disanalogous to the others, from the point of view of albinohood, so to speak. Up to a point, we can say what we like. But apart from its standing in for the idea of a falsifying dissimilarity or a negative instance, we persist in the difficulty of seeing the concept of analogy doing any real work in the theoretical elaboration of E*. We are reinforced in this reservation by a second example, F.

F: (Bill) Ravens are green.

---

[15]Thus, a system such as that found in Ilkka Niiniluoto, may well help to explicate the role of similarity in analogical inference, but it is not clear that this also suffices for analogical argument [Niiniluoto, 1988].

[16]Cf. "This mode of argument uses what the writers of antiquity called comparison by opposites" [Perelman and Olbrechts-Tyteca, 1971, p. 387].

(Sue) Every blessed raven that has ever been known has been black.

With F we have a momentous counter-example. Every instance of ravenhood so far observed falsifies Bill's generalization. Are we also to say that, likewise, we have a huge negative analogy? If so, it would bind us to the view that every single known raven in the history of the world is somehow a disanalogy, but, if disanalogous, disanalogous with what? Thus, if the counter-example of F doesn't capture in any theoretically illuminating and deep way the factor of disanalogy, we doubt that the counter-example of E does so anymore convincingly. For the counter-example of E is just a generalization of the counter-example of F. Since it is a generalization in which any theoretically useful notion of disanalogy has been leached out, we are prepared to hazard the same reservation for the counter-example of E*.[17]

There is nothing to be gained by dogmatism. We say that it is a distinctive feature of analogical arguments that they are in the general case meta-arguments. Perhaps this is mistaken; we would not be astonished if it were so. In the event, the account offered here would have to be down-graded; it would be true of only a certain class of analogical arguments. It would be a large and important class, however. Loss of generality would not be ruinous.

5. Relevance

Relevance enters our account in at least two important ways. It enters, first, in the specification of deep disagreements, which is what analogical arguments are intended to break out of. Fair-minded dialecticians will seek to settle a deepening disagreement about a disputed proposition by throwing up would-be premisses which not only might the disputants by prepared to accept, but which also might have some resolving effect upon the dispute. It would seem that the greater the depth of the disagreement, the greater the likelihood that proffered premisses, even when accepted by both parties, will be dismissed by one or other of them as not affecting the outcome of the dispute; that is to say, as irrelevant. As such, irrelevant would-be premisses, even when accepted by both, fail to "change the mind" of at least one. Thus, the idea that a bit of information is relevant for an arguer when, or by the extent to which, it changes his mind with respect to some fixed issue, calls for theoretical elaboration. This is attempted elsewhere by one of the present authors and will not be pursued here [Gabbay and Woods, 2003].

[17]Hence our reservations about an approach such as Mary Hesse's. On such an approach it is not clear how one should deal with a case where "differences outweigh similarities". Do we merely have two different cases, or a disanalogy? See for example, [Hesse, 1988; Hesse, 1966].

Relevance occupies an even more central place in the present theory of analogical argument. In its barest essentials, the idea of relevant similarities is a commandingly important dialectical notion. In the process of analogical argumentation, the comparison argument, however impeccable, will get us nowhere unless it is agreed (a) that its logical form is argumentatively definitive for it; (b) that the self-same logical form is exhibited by the facts of the original dispute; and (c) that this logical form is definitive for it.

In so saying, a theoretical grip of sorts, is available for this use of relevance. It is a matter of whether the comparison argument suffices for the settlement of the original dispute; that is, whether the accompanying analogical argument is correct. Thus the idea of the relevance of the comparison argument is cashed out in the idiom of "the correctness of the analogical argument". If, as we hope, the latter now enjoys some degree of theoretical clarity, the same may be said for the kindred notion of relevance.

We have been proposing that an analogical argument is an argument to the effect (schematically represented) that since

Argument A

1. p
2. q

⋮

n. Therefore, w

and another argument

Argument B

1. s
2. t

⋮

n. Therefore, u

both instantiate (or are cases of) an argument

Argument Q

1. S_1
2. S_2

⋮

n. S_n

and, furthermore, since B draws an assessment-verdict, V, by virtue of its relationship to Q, so too should A draw down the same verdict.

Now it might be said that since the relationship that the sentences of A and B bear to one another is that of a shared subject matter, evidence d by their instantial connections to just the same constituents of Q, this shows the existence of a prior conception of analogy; analogy as shared subject matter. Thus A. 1 and B. 1 might be said to be analogues of one another by virtue of the subject matter that they share on account of their instantial connections to Q.

True, one could speak in this way. One could say that the notion of analogical argument draws down a different and prior notion of analogy. Even so, this would not disturb our claim that analogical arguments are meta-arguments. Nor would it make our account circular, since the notion of analogy as the sharing of subject matter is not the concept of analogy which our account seeks to capture.

A more interesting and fruitful thing to say is that the shared subject matter of the constituent arguments in an analogical (meta-)argument discloses not analogousness but relevance. Thus analogues in our sense answer to and fulfill a relevance condition, and so relevance enters our account in a third way. Analogues are required to be topically relevant to one another, and they are. So the point at issue, far from damaging the meta-argument account, seems to lend it further support.

If this is right, then we have it that two arguments said to be analogues by virtue of sharing a valid truth functional form, say *modus ponens*, are so without fulfilling the relevance condition. This is actually a welcome consequence, since it reinforces the notion that in general there has to be something more to analogousness than the sharing of logical form under conditions of pure topic neutrality. So we will say that instantiations of *modus ponens* are analogues of one another at the limit.

Analogies are dialectically interesting in those cases in which the logical form of a given argument is obscured by its surface structure. By examination of an analogue of a second argument in which the logical form is more or less transparent in its surface structure, it is proposed that the first argument can now be read more perspicuously, that is, in ways that allow its form to show forth.

So it is quite true that even analogousness at the limit can be dialectically interesting, but usually it is not. The more interesting cases involve the presentation of pairs of arguments the sentences of which admit of an abstract re-description; the same abstract re-description for each sentence-pair. When it is conceded that this abstracted argument deserves verdict V, it becomes clear that its instantiations do too.[18]

[18]This is not to say that the explicit recognition of logical form is epistemologically prior. Most often its recognition is a byproduct of argument-pair comparisons. The priority of logical form is of a different order. It is meta-argumentation priority. It is the pivot on which analogical (meta-)arguments all turn.

Thus, in the general case, an argument from analogy won't go through unless the premiss-pairs and conclusion-pairs of the original and the comparative argument yield to the same abstract re-description. But this is just to say that in the general case the original argument and the comparative argument can't be analogues of one another unless they fulfill a condition of topical relevance. Relevance, then, enters our account in yet another central way. For relevance is a condition of the specification of logical form in the dialectically interesting cases.[19]

6. Analogical Predication

We said, some pages ago, that analogical argument and analogical predication call upon no tightly unified single concept of analogy. Can this be justified? What do we mean? In an analogical argument, two (or more) arguments are said to be one another's analogues. They are analogues when they share a logical form which is argumentatively definitive for them both; that is, when they are both good or bad arguments of a given type by virtue of their possession of that same form.

Analogical predication works on quite different structural principles, as the following example will show. Let us allow that "Philip is the First Lady of Britain" is a correct analogical predication. If so, Philip Mountbatten is an analogue of Hilary Clinton, but not, as we will see, she of him. Philip's First Ladyship of Britain turns on an abstracted description which both he and Mrs. Bush share, namely being spouse of his of her respective head of state. Let us now say that a's satisfying a description D is predicationally definitive for a with respect to F just in case 'Fa' is true. Let us also say that a's satisfying D is analogically definitive with respect to F just in case 'Fa' is a correct analogical predication. It is now easy to see that where D='is spouse of his/her head of state' and F='is First Lady of his/her country', that Philip's satisfaction of D is not predicationally definitive with respect to his First Ladyship, but is analogically definitive. However, in contrast, Mrs. Clinton's satisfaction of D is neither predicationally nor analogically definitive with respect to her First Ladyship. Being spouse of the president doesn't suffice; the First Lady must be his wife. So it is that, although Philip is Mrs. Clinton's analogue with respect to First Ladyship, she is not his. Argument analogues are symmetrical. Predicational analogues are not.

It also seems clear that argument analogousness is reflexive and that predicational analogousness is not. Every argument has the same logical form as itself, and that seems to be that for reflexivity. Admittedly,

[19]It is worth noting that topical relevance as here described is not captured by theories in which topical relevance is variable sharing. We think the exclusion salutary.

it sounds odd to say that every argument is an analogue of itself, but reflexivity frequently manifests itself in the limiting case (e.g., every proposition implies itself; every two-dimensional plane figure is congruent with itself), so we will let the oddity pass. If nothing else, reflexivity is secured by the over-all theoretical yield of the account which begets it. On the other hand, there is no intuition, no factor of theoretical fruitfulness that would sanction our saying, for example, that the wing is the-wing-of-the-bat of the bat, i.e., that the bat has a wing by analogy with the bat's having a wing [Woods and Hudak, 1992].

Transitivity is tricky. On the one hand, we have it quite straightforwardly that predicational analogousness is not transitive. Although the rudder is the tail of the boat and the tail is the rudder of the fish, there is no rudder that is the rudder of the fish. But argument analogousness is another matter. If, for arguments, M, N and O, M is analogous to N and N to O, then M and N share a logical form and N and O share a logical form. But we haven't ruled it out, especially for complex arguments, that the form shared by M and N is not the same as the form shared by N and O. In particular, we haven't ruled it out that N is multi-structural, and so possessed of multiple forms. So we remain uncertain about transitivity in this case.

The fundamental difference between argument analogues and predicational analogues is this:

I When arguments are analogous, there is some shared structural description by virtue of which they are both good or bad arguments. So, there is a form by virtue of which they are both some same thing.

II When things are predicational analogues of one another, there is some shared structural description by virtue of which they are not some same thing.

7. Deep Structure

We have been saying that two arguments are analogous when they share a deep structure by virtue of which they stand or fall as arguments. deep structure deserves the name of logical form when it binds logical appraisal in such ways. Of course, not every appraisal of an argument is determined by its deep structure; in simple cases, validity is settled thus, but not soundness.

Not everyone is happy with logical form as an instrument of theoretical elucidation.[20] Logical form, even for the purposes it is meant to serve in

[20]We certainly cannot pretend to have at our disposal a theory of logical form that meets the conditions set forth in [Harman, 1975].

the present account of analogy is both (i) obscure, and (ii) susceptible to explanatory collapse.

The charge of obscurity can arise in the following way. True, the idea that arguments X and Y share a logical form that is decisive for their validity is a reasonably clear theoretical device, for it is simply that of the truth functional structure of *modus ponens*. But in claiming that the Violinist has a logical form which it may or may not share with the original abortion argument, one resorts to nothing as clear as truth functional structure.[21] Pending some further clarification, imputations of logical form are here too obscure to bear much theoretical weight.

Clarification is, however, possible. In each of the lines of the abstract structural re-description of the Violinist, there occur key descriptions on which the argument turns, and to which corresponding descriptions in the Violinist stand in the relation "is a case of". This we can see as follows:

Line 1: kidney-sharing is a case of vital dependency.

Line 2: enduring something for 9 months or even much longer is a case of temporal indeterminacy.

Line 3: not being able to walk the dog or tie one's shoes is a case of impediments to locomotion and mobility.

Line 4: being overheard in everything one says is a case of the invasion of one's privacy.

Line 5: not being able to date, dance, shop is a case of serious social disruption.

Line 6: not being able to play for the Rams any more is a case of economic privation.

Line 7: severing the renal attachment is a case of termination of a vital dependency.

Thus the abstract structural re-description, which is the deep structure in which the appraisal of the Violinist is anchored, is clarified by the notion of casehood, therewith illuminating the appellation 'abstract'. 'Structural' is secured by the fact (if it is a fact) that any argument whose critical descriptions are in this sense cases of counterparts in the abstract re-description will call down the same logical appraisal as the abstract argument itself.

Clearly, deep structure gives a less tight notion of logical form in the case of the Violinist than it does in simple truth functional environments. In so saying, it is appropriate to call upon a notion of *soft analogy*.[22]

[21] Thus, "... logical analogy illustrates the fact that connections may be general without being, in the standard logician's sense, formal" [Govier, 1985, pp. 25–26].

[22] Agassi distinguishes between soft analogies and hard analogies [Agassi, 1988]. Ours are not soft analogies in his sense.

analogy can be said to be soft when (i) not every critical description in the deep structure of the comparison argument is instantiated in the original argument or (ii) when the relation of casehood between instantiating critical descriptions and their instantiations admits of degrees (as some believe to be the case with 'foetus' as a case of 'human being'). Counter-examples can also be more or less soft. Some people think that albino ravens constitute at best a soft counter-example to generalizations of blackness, since albinism can be thought of as much as a 'teleological failure' as a counter-example.

We can now appreciate the wisdom of the remark that analogy is an unstable form of argument.[23] In complex cases, there is often the question of whether deep structure has been adequately specified, and there is also the question of whether the analogy is too soft to be bothered with.[24]

Logical form also lies open to the charge of explanatory collapse. Consider, for example,

G: 1. This is red.
2. Therefore, this is coloured.

Many people will have no hesitation in seeing that G is semantically valid, that is, there is no valuation v such that v(This is red)=T and v(This is coloured)=F. Is it formally valid as well; that is, valid by virtue of its logical form? If so, how does one state its logical form? One possibility is to claim that G is enthymematic for

G*: 1. Anything red is coloured.
2. This is red.
3. Therefore, this is coloured.

which, if true, now nicely exhibits a logical form.

It is notorious that every invalid argument can be salvaged by just this manoeuvre, and this faces us with the quite general question of when it is legitimate to use it at all, especially in those cases in which it is agreed, as in the Violinist, that the premisses of a premissorily unaugmented argument do not semantically entail its conclusion. We will not pursue the matter of when premiss-augmentation (by way of the corresponding conditional) is all right and when not;[25] for we cannot bring ourselves to see that this would help us with the logical form of G. In brief, we think

[23]As Govier says, the "technique of logical analogy... depend[s] on the human skill [of recognizing] the essential point of an argument" [Govier, 1988, p. 31].
[24]Dialectically a common use of the analogical argument is to shift the burden of proof. "You agree that this is a good argument. Why, pray, is this other not a good argument?"
[25]Concerning which see [Woods, 1993c].

that if re-expression by way of G* is necessary to give sense to the idea of the logical form of G itself, it is better to face the fact that it is far from obvious that G owes its validity to anything deserving of the name. So we find it necessary to acknowledge that judgements of validity are not always anchored by postulations of logical form. In those contexts, logical form suffers from explanatory collapse.

Further trouble for logical form lurks in celebrated criticisms by Massey in [Massey, 1981]. First, every argument, good or bad, instantiates invalid argument forms. This puts pressure on our claim, repeated throughout, that the validity or invalidity of an argument turns on the validity or invalidity of its analogues — those arguments having the same form as it. If every argument instantiates an invalid form, we cannot set out to characterize an invalid argument as one possessing an invalid form. True, we might say that an argument is invalid just in case its every argument form is invalid. But this is tricky on two counts. First, it may be (as we have seen), that some semantically valid arguments (red, therefore coloured), will come out invalid on this test. Second, if we try to reinstantiate these arguments by premissory augmentation, nothing is an invalid argument.

Trouble though this is, it would be quite wrong to hold Massey or ourselves to a radical skepticism in which there can be no theory of invalidity-making structures. Massey's point is that though quantificational paraphrase is not the definitive test of either validity or invalidity, it does a much better job for validity and nothing else does as good a job for invalidity. In fact, Massey has "reason to hope that the gap, or the chasm, that separates our ability to show validity from our ability to show invalidity can be narrowed or even closed through a successful unification of logic and grammar" [Massey, 1975a, p. 62]. Perhaps unification can be found in something like Lakoff's natural logic, the logical theory of deep structure developed by generative semantics (though this is not something to inculcate optimism in the present authors), [Lakoff, 1970]. A more fruitful theoretical milieu for our longer term purposes would seem to be that set forth by William Lycan in [Lycan, 1984]. But by whatever means unification is tried, it seems to us a reasonable adequacy condition that it extend and develop a notion of logical form via deep structure which enriches the theoretical employment that we have made of it here. For all its imperfections, we are not ready to abandon logical form as a powerful concept, central to good theories of analogical argument.

Chapter 17

VERDI IS THE PUCCINI OF MUSIC[1]

In the preceding chapter, analogical arguments were construed as meta-arguments; as arguments about arguments in which it is claimed that they instantiate some common deep structure. Analogical predications aren't arguments at all; they are statements. They raise the hard question of what their truth conditions might be. This question is the business of the present chapter. It is appropriate to sound a warning right at the beginning. Some people are of the view that the secrets of analogical predication cannot be captured by truth conditions [Lichter, 1995]. In this they may be correct. Perhaps analogical predication is better investigated pragmatically, that is, in terms of how they are characteristically used in speech [Thagard et al., 1990]. This is not something to be dogmatic about. The best course is an (energetic) open mind. I am not ready to give up on the truth conditional approach. It is all grist for the mill of the larger question posed by the fallacy of false analogy: "What *are* analogies anyhow? What good are analogies? Why bother with them? What is the purpose of trying to establish a mapping between two things that do not map onto one another in reality? The answer surely is very complex, but the heart of it must be that it is good for our survival (or our genes' survival), because we do it all the time" [Hofstadter, 1986, p. 556].

1. Analogies

Yes, we do it all the time. Analogical characterizations come trippingly from the tongue. The knee is the Achilles' heel of the leg. Achilles'

[1] Co-authored with Brent Hudak.

Achilles' heel is not Achilles' heel (after the title of [Woods, 1965]). Philip is the First Lady of Britain. The subspaces of a Hilbert space are the lines and planes through the origin of ordinary Euclidean 3-space. To say nothing of the theological doctrine of Analogical Predication.[2]

Then, too, we routinely resort to analogical arguments, of which some of the most celebrated are also among the most controversial. We have the Design Argument for God and the Analogical Argument for Other Minds and numerous others of lesser standing and *cachet.*

There is also talk of analogies in induction, scientific theory construction and problem solving. Science employs models and perhaps they are a kind of analogy. In our inductions, especially our primitive inductions — in which, for example, the toddler learns in a simple encounter with the hot element on Mamma's stove "never to do that again" — there may be something like analogical pattern recognition.

We have space in this chapter only for *analogical characterization.* We deal with *analogical argument* in the chapter to follow.

Theories of analogy have been given a side-pew in cathedrals of respectable contemporary intellectual practice. The once mighty doctrine of Analogical Predication is, as we remarked in footnote 2, now seldom studied and is considered something of a theological relic.

As recently as the early part of the present century, no treatise on logic or scientific method could be considered finished if it did not have the mandatory chapters on analogical reasoning. Today our logic texts give little or no attention to analogy, and what they do manage to say about it is, by and large, pretty bad.[3]

Mind you, people these days quite often say that analogy is important, but one looks in vain for its serious *theoretical* working up. Thus we have it from Polya:

> We can learn the use of such fundamental operations as generalization, specialization, and the perception of analogies. There is perhaps no discovery either in elementary or advanced mathematics or, for that matter, in any other subject that could do without these operations, *especially without analogy* ([Polya, 1954, p. 17], emphasis added).

The title of his book notwithstanding, it contains no account of analogy. Still, it is pleasant to be able to recognize a recent and important exception to this theoretical dearth in [Holyoak and Thagard, 1995].

---

[2]The contemporary unimportance of which is convincingly regretted by James F. Ross in [Ross, 1981]. The direction in which Ross takes the theory of analogy in this book does not bear directly on our own proposals here, and so we shall reserve comment.

[3]A quick look at [Carney and Scheer, 1980; Woods and Walton, 1982; Copi, 1986; Giere, 1984] discloses that analogy receives respectively five pages worth of consideration, none at all, eight pages worth and two. A welcome exception is [Weddle, 1978, pp. 138–160].

With the exception noted, it is appropriate to ask why the theory of analogy should have gone into its shabby and genteel retirement. Perhaps the answer is altogether straightforward, namely, that since analogies are very reluctant to yield up their secrets, analogy theory is in fact very difficult. Possible, of course; but we also think that it possible that we have been led, contrary to fact and to the rich cognitive lives that lie under (or should we say 'behind') our very noses, to think that analogies are not terribly important, not terribly central to the operations of intelligence. We could be thinking, for example, that metaphor always has an analogical structure, and that, since metaphor is not central to language, and certainly not indispensable (after all, literal speech is all we ever really need), the same is true of analogy. There are two things wrong with this line of thought: (1) metaphors are not always analogical (see below), and (2) metaphors are absolutely pervasive in and central to human language and thought.[4]

As Hofstadter has said, the use of analogies "is good for our survival". That alone would justify attempts to revive a theoretical interest in them.

1.1 The Method of Cases

How does one begin to construct a theory of analogy? Two methods might be considered. We might accept theoretical guidance from a settled consensus in the intuitive pre-theory of analogy. Or we might use the method of cases.

As for the intuitive consensus approach, we do not here avail ourselves of it, for there is no settled semantic consensus. Not enough is agreed upon about the meaning of 'analogy' to have generated a corpus of intuitive knowledge which any good analogy theory must honour and preserve. Thus, there "is no word... which is used more loosely, or in a greater variety of senses, than analogy" [Mill, 1959, p. 364]. This situation can be likened to the endless wrangles between classical and relevant logicians about whether the paradoxes of strict implication do or do not do violence to the meaning of "entailment". The fact is that here too there is insufficient semantic consensus to constrain *á priori* any ensuing theory of entailment. So, in either case, we believe that the resort to semantic consensus will be unavailing.

[4]See, for example, [Cormac, 1983]. Cf. "We have found... that metaphor is pervasive in everyday life, not just in language but in thought and action" [Lakoff and Johnson, 1980, p. 3]. And "[b]eyond serving us at the growing edge of science and beyond, metaphor figures even in our first learning of language..." [Quine, 1981, p. 188].

The second approach the method of cases is somewhat more promising. It involves the scrutiny of accepted and rejected cases of analogy and, in particular, a careful consideration of the differences between them. These supply the basic data for a theoretical working up; they are the test cases for a theory of analogy. One looks for logical and semantical structure which is reflected in the accepted cases and not in the rejected cases; and then one draws inferences to the best explanation. The method of cases is, of course, maddeningly imperfect. There is plenty of room to quarrel about what the data are, but this is unavoidable and we shall simply have to put up with it.

1.2 Analogical Prediction

Consider the following candidates for recognition as analogical predictions:

1 Achilles' Achilles' heel was not Achilles' heel.

2 The knee is the Achilles' heel of the leg.

3 Philip is the First Lady of Britain.

4 Philip is the Laura Bush of Britain.

5 Ella Fitzgerald is the First Lady of jazz. (She holds that place even in death.)

6 The subspaces of a Hilbert space are the lines and the planes through the origin of ordinary Euclidean 3-space.

7 The eyes are the window of the soul.

8 Cherie Booth is the First Lady of Britain.

9 Theaetetus is the Frank Ramsey of Plato's Academy [Hintikka, 1987, p. 227].

10 "Try speaking English for a while without using the letters 'e' or 't' at all. Then you'll have an inkling of how Japan felt, having two of its major cities wiped out" [Hofstadter, 1986, p. 566]. (Cases (11) to (14) are also from Hofstadter).

11 Consider the sequences:
A: 1234554321
B: 12344321
What is the 4 of B, that is, what is to B as 4 is to A?

12 Compare A with
C: 12345666654321
What is the 4 of C? Is there a uniquely correct answer?

13 Compare A with
D: 11223344544332211
What are the possibilities for what is the 4 of D?

14 Compare A with
H: 123465564321
Could there be three possibilities here?

15 The process of division is the dynamic system with a structurally stable duplication of the cell.

16 The chromosomes and molecules are the dynamic systemic spectral elements or singularities of nucleic acid.

Case 1: (Achilles' Achilles' heel was not Achilles' heel.) The analogical characterization is apposite. In Lewis Carroll's logical parable, the Achilles' heel of Achilles was not his heel, but his mind; and the Achilles' heel of the classical story is a point of unique vulnerability. So we have all the difference between a mind and a heel; and we have this similarity — unique vulnerability — conveyed by the locution "Achilles' heel". "Achilles' heel" plays, if not the same function, certainly a same role in the different cases.

Case 2: (The knee is the Achilles' heel of the leg.) A similar story can be told here, which in the interests of space, we will leave undeveloped.

Case 3: (Philip is the First Lady of Britain.) In the United States, the First Lady (when there is one) is wife of the President. The President is both head of state and head of government. If it is true that the office or role (or is it just a title?) of First Lady of the United States is as wife of the head of state, then it would appear that Philip is at least a contender for the analogical role in Britain; but if it is as the wife of the head of government that a person is First Lady of the United States, then Cherie Booth would appear to have been in the running in Britain. If, as we understand, the First Ladyship is derived from the headship of state, then we shall stay with Philip.

If, in American terms, 'First Lady' means
(i) spouse of the head of state
then Philip is indeed the First Lady of Britain. But if it means
(ii) *wife* of the head of state then the post is currently vacant in Britain.

The limits of analogizing have something to do with the legitimacy of the abstractions one makes. In Carroll's Achilles' heel case, we abstracted to the idea of unique vulnerability (whether of mind or body) and that gave us the needed similarity amidst all that difference. In the present case, if we abstract from the idea of wife of the head of state to the idea of spouse of the head of state, we exploit that similarity between Laura Bush and Philip

Mountbatten to make the analogy work. Notice, then, that out of this point comes a (small) part of the logic of analogy. If Philip is the First Lady of Britain by virtue of his sharing with the First Lady of the United States the similarity-abstraction, being spouse of the head of state, then it is clear that the analogical First Lady need not be a lady, whereas the First Lady of the United States, that is, the real First Lady, must be. Philip is specified as the analogical First Lady of Britain by sharing with Laura Bush an abstracted relational property, fulfilment of which is not sufficient to make somebody First Lady of the U.S.

Case 4: (Philip is the Laura Bush of Britain.) Well, who is the Laura Bush of Britain? We may notice that the *role-specification* marker of "Who is the First Lady of Britain?" is muted in the present example. This means that it is now possible, in fact easy, to get two tenable answers, reflecting together the role-ambiguity of the American presidency, headship of state and headship of government, each of which offices admits of a kind of consorthood. So (8) is all right; i.e., it could be Cherie Booth, as spouse of the British head of government. And (4) is all right, too; it could be Philip, as the spouse of the British head of state.

Case 5: (Ella Fitzgerald is the First Lady of jazz.) Here we meet with a difficult case. So far, analogical characterization seems a matter of the ascription of appropriately abstracted similarity amidst significant difference, where the similarity is (perhaps) somehow functional. On this view, the woman who is the third cousin once removed of the widowed Governor of Kentucky is not (just so) First Lady of Kentucky. This is because the abstracted similarity between that lady and Mrs. Bush, namely, *being a relation of*, has gone too far. It is not that they do not genuinely share that similarity; it is the wrong similarity for the analogy. Of course, the Governor could have arranged for his third cousin once removed, the most vivacious hostess and most influential *éminence grise* in the entire state, to serve as consort. Then she would indeed be the First Lady of Kentucky, by analogy, sharing with Mrs. Bush the abstracted-similarity 'consort of'. We admit to some uncertainty about this case. We plead ignorance of State-level protocol in the U.S. So far as we know, a relative of the Governor might achieve the status of First Lady more directly; that is, by gubernatorial appointment. It might be better to waffle in favour of 'the unofficial First Lady of Kentucky.'

The claim of Case 5 comes to something like this: that Ella Fitzgerald is the most influential and accomplished senior woman performer in jazz over a lengthy and successful career. One should notice the expression 'First Lady' doing strikingly different work in Case 5 from what it does in Case 3. In Case 3, 'First' implies "first in rank after the head of state", although it is clear that this is an honorific. Not only is Ella Fitzgerald first in a very much less honorific sense in Case 5, there is also no notion of consortship or close (legal) relationship with a head of state, or head of anything. Ella did it all by herself. The proposed analogy-making similarity is *too* abstract: even though Ella Fitzgerald and Laura Bush are both *very prominent* in their respective roles, this won't support the analogy. So, if 5 is apposite it is not by analogy with the First Lady of the United States. That would be a bad analogy.

Case 7: (The eyes are the window of the soul.) (Again in interests of time, we will simply declare the appositeness of cases (6), (9) and (10).) A beautiful and affecting metaphor, but one must struggle to find any analogy here. It might be true that by looking into a person's eyes you can tell something about what kind of chap he is *inside*; it is also true that sometimes by looking through a window you can tell something about what's on the other side. The similarity, though there, is terribly strained. The looking is not the same ('into' vs. 'through'); the somethings-about are not the same (that he is a compassionate person vs. that there is a bittern on the mantle) and where the glimpsed goings on are very different ('in the soul' vs. 'on the mantle'). There is some kind of similarity that takes hold here, something like being-able-to-tell-by-looking, but it is too general to sustain the analogy. Something important emerges from this case, and perhaps also from the Ella Fitzgerald case; it is that a metaphor need not be an analogy, and a strong metaphor need not be even a weak analogy. And so we find it necessary to resist the opinion of Earl R. Mac Cormac:

> an analogy serves as a necessary but not sufficient condition for the existence of metaphor.... How can we draw the line between an analogy and a metaphor? One might argue that although the line of demarcation is not a sharp one, the difference between an analogy and a metaphor depends on the degree of difference between the two referents. Referents that differ substantially can be called metaphors, whereas those that possess more similarities are analogies [Mac Cormac, 1985, pp. 23–24].

I am puzzled by these remarks. It seems that the more likely conclusion to draw from the premiss (which we think is true) that metaphors sometimes involve greater degrees of difference even than analogies, is that sometimes metaphors *are not* analogies. That is, sometimes metaphors tolerate differences that are too great for analogy. I do not mean by this that some differences are so great that no similarity can be discerned by which to overwhelm them in ways suitable for the creation of analogies. I say only that in some cases, a given similarity can be forced which suffices for metaphor but not for analog. For one thing, over-radicalness of difference may damage the (quasi-)functionality requirement as in "the eyes are the window of the soul" case. (I discuss functionality in the next subsection.)

Quine has got this exactly right. "We generalize our application of the expression by degree of subjective resemblance of occasions until we discover from other people's behaviour that *we have pushed analogy too far*, exceeding the established usage... the crux of metaphor is creative extension *through* analogy..." [Quine, 1981, p. 188], emphasis added.

1.3 What the Cases Suggest

At this juncture, we break away from our review of cases in an effort to derive some tentative theoretical guidance from them.

It would seem reasonable to assume that a theory of analogy should be expected to fulfill at least the following three *adequacy conditions.*

AC1: A theory of analogy should give appropriately detailed structural descriptions of the similarities amidst differences that are the stuff of analogy. In particular, it should characterize the relevant similarities and differences and explain what their relevance consists in.

AC2: A theory of analogy should 'predict' the goodness of clearly good cases, the badness of clearly bad cases, and the unclarity of clearly unclear cases.

AC3: A theory of analogy should have something to say about the metamathematical, logical and semantic character of analogy. For example, it should try to determine whether the strategy of analogy employment is a programmable feature of intelligence, whether analogy is symmetrical, whether analogy obeys the laws of proportion, and whether correct or apposite analogies are true.

I think that the data furnished by our cases support the following principal claims.

I *The similarity view of analogy is badly mistaken.* John Maynard Keynes supplies us with a succinct expression of this notion, an account which Susan Stebbing approvingly describes as "[M]uch the clearest treatment of the nature and importance of analogy..." [Stebbing, 1961, p. 250]. It is the view that "[i]f some one thing is true about both of two objects, if, that is to say, they both satisfy the same propositional function, then *to this extent* there is an *analogy* between them." [Keynes, 1963, p. 223]. (Our emphasis in the first instance, Keynes' in the second.) See also Bunge, passion, in: a 'substantial analogy' arises from the sharing of properties; a 'formal analogy' is a homomorphic image [Bunge, 1974, p. 101]. (More of this below.) Consider also Joseph: "analogy meant originally identity of relation. Four terms, when the first stands to the second as the third stands to the fourth, were said to be analogies, or to exhibit an analogy" [Joseph, 1916, p. 532]. This Keynesian account is unsatisfactory on two counts. It endows *any* similarity with analogical significance, and thus provides that any and all red things are analogous to each other. It further provides that things satisfying exactly the same propositional functions are analogous, and so, in particular, that *to the extent* that identical things satisfy the same propositional functions, *identity* is the perfect analogy. Clearly such a view is not supported by our cases. They dispense as well with Stebbing's "two peas in a pod view"; that is, that two peas are made analogous to one another by virtue of having come from the same pod [Stebbing, 1961, p. 250].

II *The similarity-amidst-difference account, though better, is unsatisfactory.* This is perhaps the most common modern view about analogy. We have it from Mill in these words: "an argument from analogy is an inference that what is true in a certain case is true in a case known to be somewhat similar, but not known to be exactly parallel, that is to be similar in all the material circumstances" [Mill, 1959, p. 520]; and also from Perrin and Ebbitt: "[a]nalogy is a variety of comparison that points out resemblances, real or imagined, between two things that are not normally thought of as alike" [Perrin and Ebbitt, 1972, p. 479]. It is also the view of Ronald Giere, Mary Hesse and Nicholas Rescher ([Giere, 1984, pp. 79–80]; [Hesse, 1966, ch. 1]; [Hesse, 1988, p. 319]; and [Rescher, 1985, pp. 98–104]). Giere describes the role of "analog models" in the process of scientific discovery and the filling out of a fledgling scientific theory. Thus Rutherford and Bohr came to think of "the atom as being analogous to the solar system.... So Rutherford and Bohr were taking the solar system as the analog model for the atom" [Giere, 1984, p. 79]. Hesse takes a similar position, noticing that a theory of sound waves can be constructed by analogy with the wave theory of water. (See here [Stebbing, 1961, 11 ff]. Stebbing also refers to Labplace's resort to analogy in the formulation of his Nebular Hypothesis [Stebbing, 1961, p. 255].)

We should attend to the Rutherford-Bohr 'analogy' a little more closely. An atom is a tiny solar system, with electrons orbiting the nucleus in conformity with the laws of classical mechanics and electromagnetism. Electrons are fixed in those orbits because they are able to emit entire quanta of energy (photons), not the continuous radiation that is required by the classical theory. The orbits are *stable.* That is, they correspond to certain fixed amounts of energy, each a multiple of the basic quantum. Since fractional quantities of energy are impossible, there can be no intervening orbits. This is like saying that the Earth's orbit around the sun is stable and so is Mars', and there is no such thing as a stable orbit anywhere in between. *But* electrons are allowed to absorb or radiate a photon and then jump discontinuously from one energy level (orbit) to another. Instantaneously. It is as if Mars disappeared from its orbit and reappeared in the earth's orbit instantaneously. These are *quantum* leaps. It is obvious at once that the solar system is a terrible *model* of the atom. What are its prospects as an analogy?

There is in this, the similarity-amidst-difference view, an important improvement over the previous one. It is altogether necessary to emphasize that analogies are matters of similarities *amidst difference*, and this is done on the present approach. But it fails in other ways. It does not *characterize* the similarity and difference that is needed for analogies and it does not attend to the dynamics of the similarity acting on the difference.

Not any old similarity amidst any old difference will turn the analogical trick. Our cases suggest that the analogy-making similarity should be abstract enough to provide for a genuine similarity but not so abstract as to downplay the difference. One thing anyhow seems quite clear; if you rise to too abstract a similarity, the analogy will fail. Philip is the First Lady of Britain because he is spouse of the head of state, not because he is a relation of the head of state. Ella Fitzgerald is not the First Lady of jazz because of the similarity she bears to Laura Bush, namely, distinguished, prominent and leading accomplishment in what they respectively 'do'. This abstracts too much from their substantial differences — one being a jazz singer, the other the consort of the U.S. head of state. In this particular abstraction one loses sight of the point that First Ladyships have something essential to do with role-relationships of a very particular nature. So we think that "Ella is the First Lady of jazz" is not an analogy. It borrows from, but hardly extends the operatic approbation 'prima donna'.

III *The theory of similarity-amidst-difference needs to be supplemented by an account of similarities of role or function between things that are analogous to one another.* So, for example, it is common in biology to distinguish between homology and analogy. One organ may be said to be *homologous* to another if and only if they are basically similar in structure, stand in similar positional relationships to other parts of the organism, are similar vascularly, in nerve-supply and in embryonic development. One organ is said to be *analogous* to another if and only if it is not homologous to the other but has a *similar function.* The wings of a bird are homologous to the wings of a bat, but a fly's wing, though functionally similar, has a quite different structure. It is analogous, not homologous, to the wings of the bird and bat. Biologists also claim that homologous features suggest a common evolutionary background. This is instructive. It suggests that with analogies, *radical* or *extreme* difference are significant.

The attractiveness of the present view, namely, that analogies arise from functional non-homologous similarities (as is plainly suggested by the word 'analogous' in its biological use) speaks for itself. It answers the question of how much similarity analogies can tolerate (not too much); it answers the question of how much difference is required (quite a bit); and it leaves open the question of whether there is an *a priori* upper bound on degrees of difference required for analogies. This is desirable. It helps emphasize the point that analogies are *drawn* and not found; they are created by language users and are forced upon objective difference.

Possibly it was something like this that Polya had in mind: "analogy is a sort of similarity; It is... similarity on a more definite and more conceptual level.... The essential similarity lies... in the intentions of the thinker. Similar objects agree with each other in some respect. If you intend to reduce the aspect in which they agree to definite concepts, you regard those similar objects as analogous" [Polya, 1954, p. 14]. Analogies succeed when they manage to overwhelm this difference. It is not prudent to insist in advance that *here* is a difference that will never be overwhelmed. Proof of the pudding is in the eating.

That said, we make so bold as to say that it does not seem to us that the Rutherford-Bohr case presents us with a credible analogy. True, we have, between the atom and the solar system, some similarities; but we also have a great deal of difference. Further, that difference does not seem overwhelmable. After all, the similarities involved in this comparison are not helpful until great liberties are taken with the solar system. If there is an analogy here, it is between the atom and a solar system which owes substantial allegiance to radically nonclassical laws.

If the present account has a serious drawback, it is probably this. It overstates the functionality condition on (analogous) similarity. It is perfectly true that the appositeness of the analogy between Philip and Laura Bush has a lot to do with the fact that they perform similar functions, that is, that they are both consorts of a head of state. But one needs to strain oneself to discern any identity or close similarity of *function* between Achilles' mind in Lewis Carroll's logical parable and Achilles' heel in Homer's tale. True, they are both points of unique vulnerability, but it is over-reaching to think of this as functional similarity. It seems to me better to think of the similarity as role-like, and we propose to modify the functional similarity condition so that it requires the relevant similarities to be similar in function *or role.*

Perhaps we have had enough theoretical canterlevering for the present. Let us return to our cases.

1.4 Numerical Analogical Prediction

Consider now the simple numerical cases furnished by the Clossman-Hofstadter sequences (11) to (13). (So named, in part, after Gray Clossman, Hofstadter's graduate student in 1981. See [Hofstadter, 1986, p. 549]. We here closely follow Hofstadter's treatment of these cases.)

The importance of these sequence-problems lies mainly in their simplicity. They expose us to pattern recognition possibilities undistracted by the *Sturm und Drang* of heavy metaphor or the dialectical drag of full-scale argumentation. They give us analogical characterization in its stripped down versions. They deal with the basics.

Case 11: (Where A=1234554321 and B=12344321, what is the 4 of B, that is, what is to B as 4 is to A?) It seems that the 4 of B is 3; both 4 in A and 3 in B are numerically and positionally closest to the central pair. The abstracted similarity is, thus, *being numerically and positionally closest to the central pair.*

Case 12: (Compare A with C=12345666654321. What is the 4 of C?) If you stay at the same level of abstraction as in the previous case, then 66 is the central pair, and 6, being closest numerically and positionally to the central pair, is the 4 of C. But if you advance the abstraction from 'central pair' to 'central plateau', in this case 6666, then by the more abstracted similarity, *being numerically and positionally closest to the central plateau*, 5 is the 4 of C.

Case 13: (Compare A with D=11223344544332211. What are the possibilities of the 4 of D?) Notice that here we have no central pair, but rather a central singleton; a limiting case, surely, of a central plateau. So if, in general, for any sequence S, the 4 of S is the number closest numerically and positionally to the central whatever, then, in particular, the 4 of D is 4. But we might also notice that in D the sole singleton replaces the only pair in A and that the singletons of A are now replaced by their respective pairs in D. This is easier to see if we rewrite D as follows:

1-1 2-2 3-3 4-4 5 4-4 3-3 2-2 1-1,

with the elements being the singleton, 5, and all the other pairs. Suppose now that the rule is: for any S the 4 of S is the element closest in position and magnitude to the central whatever. So the 4 of D is 4-4.

Case 14: (Compare A with H=123465564321.) Here there would seem to be three possibilities:

(α) the 4 of H is 6 (via the abstraction "the element closest numerically and positionally to the central pair "[i.e., 55]).

(*beta*) the 4 of H is 4 (via the abstraction "the element closest numerically and positionlly to the central crater "; [i.e., 6556]).

(λ) the 4 of H is 5 (via the abstraction "the next largest number in S").

Now (α) and (β) seem to us to be admissible analogical characterizations, with (β) the stronger, and (α) a somewhat weaker instance. Concerning (λ), it is certainly the weakest of the three, perhaps to the point of inadmissibility. (β) is strongest because its similarity abstraction principle involves more structure and more pattern recognition then the much thinner abstraction principle in (λ). This verdict is reinforced, it seems, by considering with Hofstadter the obviously weak (perhaps inadmissible) answer that 4 is the 4 of H because 4 is the fourth element of H just as it is the fourth element of A. The weakness of this answer reflects the *triviality of its abstraction*: "the 4 of S is the nth element of

S, where n is the position-place of 4 in A". Once we notice that there is no S-admissible element whatever which couldn't be the 4 of S, we see the sense in which the abstraction is trivial.

1.5 More Theory

At this juncture, it is possible to reveal a little more of the logic or the formal semantics of analogousness. For example, we think it appropriate to try to discourage

IV *the widespread mathematical habit of thinking of analogies as isomorphisms, or some such thing.* Isomorphisms are not in general good analogies. Often enough the defining principle of the isomorphism is a trivial abstracted-similarity in precisely the sense mentioned only lines above. For example, consider a standard isomorphism between the positive numbers, P, and the odd numbers, O, which preserves precedence. It will look something like this:

P=1, 2, 3, 4, ..., k, ...
O=1, 3, 5, 7, ..., k+2, ...

Now we ask, under this isomorphism what is the 3 of set O, that is, what in O plays the role of 3 in P. The similarity abstraction that would (correctly) give 5 as the 3 of O just is our old friend: being the nth element of S, where n gives the place of 3 in P. And this is trivial.

It is clear again that there is a tension between what is wanted by abstracted similarities. On the one hand, the abstraction has to yield a relevant similarity amidst all that difference; on the other, it should not be so generalized as to leave unreflected such richness of structure as is compatible with the level of abstraction required by the similarity requirement.

It also appears that

V *the goodness of a theory of analogy will turn in part on the ability to notice analogies between analogies.* A similar point was raised by the Polish mathematician Banach: "Good mathematicians see analogies between theorems or theories, the very best ones see analogies between analogies" [Ulam, 1983, p. 203].

Banach would, no doubt, have been fascinated by Hofstadter's concept of meta-analogy, an analogy between analogies. "I was unconsciously developing a parallel in my mind between the First Lady example and the numerical example", writes Hofstadter. "In this meta-analogy I see structure A as corresponding to the United States, structure B to Britain, 4 to Nancy Reagan and the unknown number to the unknown person" [Hofstadter, 1986, p. 550]. Recall that Hofstadter is here writing in the Reagan era. We eliminate the circularity of [A] and [B] in the next section. And B can be allowed to stand, for now, as it is, for ease of exposition.

We think that our "Achilles' heel was not the Achilles' heel of Achilles" is a fairly convincing example of a meta-analogy. In Carroll's parable, Achilles is the logical opponent of the Tortoise (thus they are the *racers* of the logical dialogue); and this (analogical) Achilles has (analogically) the Achilles' heel of Achilles (the Greek hero). The analogue of the Greek hero's heel (a point of unique vulnerability) is, for our logical Achilles, his logical acumen, which does him in in the logical joust of Lewis Carroll.

We also have at hand the resources to defeat a fairly common misconception. It is

VI *the idea that 'x is analogous with y' is symmetrical.* It is not.

Philip is the First Lady of Britain by virtue of his sharing with Laura Bush the relational property of being spouse of the (respectively *'right'*) *head of state.* Though Laura Bush is spouse of (her) head of state, it is not on that account that she is the First Lady of the United States. *She* needs to be *wife* of (her) head of state (and is), which is precisely what Philip can't be with respect to (his) head of state. So even if we granted, without analogical qualification, that Philip is the First Lady of Britain, Laura Bush could not have that role in the United States by analogy with Philip's having it in Britain. There is no relational property that she and Philip share that confers the role upon her in the United States. She is precluded from having the job analogically by virtue of her having it *really.*

But why not say that Laura Bush is the First Lady of the U.S. not only 'really' but also by analogy with any First Lady of any country having *that* status analogically by virtue of sharing an appropriate relational property with The First Lady of the United States? That is, why not say that, in addition to her actually being the First Lady in the United States, Laura Bush is also First Lady of the United States by analogy with Philip's being analogically the First Lady of Britain; that she is First Lady by analogy with everyone else's being First Lady of their countries by analogy with her really being First Lady of hers?

It won't work, thank goodness. The claim is preempted by the nature of its first element, the real First-Ladyship of Mrs. Bush. It rules out her analogical incumbency in that same office; for "x is (analogically) First Lady of Y" implies "x isn't really First Lady of Y", and this is false of Mrs. Bush.

At this point, we wish to make a further brief remark about functionality. In the simple numerical cases of the previous section, the structure-forming concepts (central whatever; relative position; numerical weighting) are *relational.* The descriptions of their interactions need not be functional, or so it would appear. Of course, one *can* say such things as "6 in H does the same job as the 4 of A", but this is a metaphor. Being positionally and numerically closest to the central pair isn't a job; it isn't *doing* anything. This is a useful reminder of the wisdom of our having weakened the functional similarity requirement in favour of similarity of function or role. For it seems quite all right to say, for example, that 6 plays in H the *same role* as 4 does in A. This is functional enough for our tastes, and it preserves a nice analytical point; for a while at least.

Our first adequacy condition, AC1, obliges us to give some appropriately detailed structural description of the by-play between similarity and difference that is the very stuff of analogy. This we have attempted to provide, in approximately the following way.

[A] if x is analogous to y in respect of M, then there is a role-similarity abstraction S shared by x and y such that

(i) x's satisfaction of S entails x's possession of M analogically.

(ii) y's satisfaction of S does not entail y's possession of M analogically or non-analogically (really).

(iii) the level of the abstraction of S must be minimal; that is, just sufficient to fulfil (i) and (ii) above, but otherwise constrained by the differences between x and y.

[B] if x is analogous to y, there must exist between x and y significant difference D, where the idea of significance may be expressed somewhat as follows:

(i) if x is analogous to y in respect of M, then difference D between x and y must falsify the ascription of M to x non-analogously; and

(ii) the requirement of significant difference is best secured when D reflects a radical difference, that is, a difference of category, natural kind, or semantic kind between x and y. For discussion of kinds, see, for example, [Quine, 1977] and several others in [Schwartz, 1977]; as well as [Woods, 1974b]. We think, by the way, that Bosanquet may have been attempting a similar point, though the passage is obscure, in Bosanquet: "The essence of this judgement [of analogy] is that it is neither purely subsumptive; as expressing a de facto conjunction of attributes in a simple object, nor purely constructive as expressing a de jure connection of attributes independent of this immediate subject in which they may exist, but is something intermediate.... [Thus, the] analogical judgement, like the aesthetic judgement, is essentially outside relativity and necessity.... " [Bosanquet, 1911, pp. 214–215]. Moreover, it would seem that we also have a generalized syntactic description of analogical characterization, the canonical form of which is:

Subject-marker + Role-definition-marker + Preposition + Object-marker.

It's latitude is striking, as indicated by a partial list of its admissible instance: Philip is the First Lady of Britain; Kim Philby was the Mata Hari of MI6; The tail is the rudder of the fish; Ezra Pound is the T.S. Eliot of Idaho; Oxford is the Cambridge of Oxford; The hyperreals are the infinite magnitudes of the small numbers; The subspaces of a Hilbert space are the lines and planes through the origin of 3-space; Cherie Booth is the Laura Bush of Britian; Britain was the mother of Rhodesia; The process of division is the dynamical systemic spectral elements or singularities of nucleic acid.

We think, therefore, that we have been able to offer some degree of compliance with AC1.

As for AC2, this obliges us to predict good, bad and unclear cases. It is difficult to determine compliance in the absence of the judgements of others. Certainly our own intuitions are satisfied by the verdicts we have given in the case reviews above, and these judgements have been by and large confirmed by others whose responses we have sought out. But our sample is nevertheless very small, and so we count on and solicit the responses of our readers.

Condition AC3 requires us to say something of the formal and semantical properties of analogousness. We can meet this condition, at least in part, by saying why we disagree with Mary Hesse's characterization of certain logical properties of analogy. Hesse attempts to reveal something of the logical structure of analogues by embedding analogousness in the general theory of relations.

Her attempt is welcome, but the results must be disputed. Our reservation is that it systematically confuses proportionality, which does indeed satisfy her logic, with analogousness, which does not. Her confusion may well have etymological origins since 'analogy' derives from the Greek '*analogia*' for proportion.

Hesse notices that the proportionality

$$\frac{a}{b} :: \frac{a}{b}$$

holds, that is, that proportionality is "trivially reflexive" as it surely is (Hesse [Hesse, 1966, 65 ff]). She infers from this that analogousness is likewise *reflexive*. We think that this is mistaken. There can be no analogousness in a proportionality that admits no difference. If Hesse were right, then just because the wing of a bat stands to a bat as the wing of a bat stands to bat, we should have it that the wing of a bat is by *analogy* the wing of the bat, which is nonsense. We acknowledge by the way that Hesse's more considered views pertain to what she calls "material analogues" as opposed to "formal" ones. Still, Hesse's earlier work continues to win adherents (see the passages on Giere, above), and requires comment. Furthermore, even though it is not clear whether, on her present account, analogousness is reflexive or transitive, it is at least clear that symmetry and inversion hold. (See, for example, [Hesse, 1988].) Since these are the primary notions which we wish to lay to rest, our criticisms remain. We must grant, however, that Hesse's more recent work, does not, to our knowledge, deal explicitly with the formal characteristics of analogy.

It follows that analogousness cannot be anything but weakly symmetrical; but it is not even that, as we have already attempted to show above.

It follows, too, that analogousness cannot be transitive, as Hesse rightly acknowledges. But she also asserts that analogousness obeys the inversion property. We agree that proportionality obeys it, that is, that if

$$\frac{a}{b} :: \frac{c}{d}$$

then

$$\frac{b}{a} :: \frac{d}{c}$$

But we doubt that it holds for analogousness. Proportionality considerations invade analogy in interesting ways. We wish that we could be clearer about them. On the one hand, it seems quite clear that analogical set-ups imply proportionalities: Since Philip is the first lady of Britain, we would seem to have it that Philip is to Britain as Mrs. Bush is to the U.S.A. Of course, the implication isn't formal; it turns rather on semantic and/or contextual features of the role-marker "is the First Lady of". (For more of this, see the following section.) Moreover, since proportionality obeys the formal equivalence of the inversion law, then our analogical sentence may also imply the inverse "Britain is to Philip as the U.S.A. is to Mrs. Bush". Even this is not clear, since it equally may not be true that non-formal implications are preserved under formal equivalences. That said, we don't mind letting the implications of the inverse proportionality stand, since the concession will not help the cause of the closure of analogousness under inversion.

We do not, however, have it the other way. While it is true that many a proportionality statement suggests an analogical connection between its terms, no proportionality statement expresses the analogy. Some proportionalities, for example, the identity-proportionality, leave no room at all for analogies. That a is to a as b is to b holds just on account of the self-identity of everything. But nothing is analogous to anything simply by way of their respective self-identities.

We conclude that, although analogical sentences may imply the inverse of proportionality sentences that they imply, this is no reason to say that analogousness itself is inversely proportional. This conclusion is reinforced by a similarity with the symmetry relation, as previously noted. It may seem alright to say that "x is an analogue of y" is equivalent to "y is an analogue of x". In fact, such sentences are always misleading; they are ellipses for sentences in the form "x is an analogue of y in respect of property or circumstance or factor P", and the interchange of 'x' and 'y' in sentences of this kind is routinely non-truth-preserving. It is the same way with standard proportionalities. They understate the analogy-making facts. When made to call down those facts, as in "x is to y as z is to w with regard to property P", inversion is likewise routinely non-truth-preserving.

The logic of analogy is strangely elusive. We have certainly not captured much of it here. Our logical lessons are mainly negative, but not entirely uninformative. Analogousness is (i) irreflexive. (ii) asymmetrical. (iii) non-transitive. (iv) non-inversive.

Bosanquet, recall (see note 33), was of the view that analogical assertions were something less than "*de facto* judgements". This suggests, what in any event we believe to be true, namely, that analogies are not *objets trouvrés*, that analogies are made, not found. However, in private correspondence, Douglas Hofstadter holds this view to be deeply wrong, unless that is, we could show that all predication is creative. We quote the point in full:

> Among your conclusions was that analogies are not objects *trouvrés*, but are invented or created. With this I would strongly disagree. I believe analogy to be no different from the assignment of categories to objects, i.e., no different from the perception of a scene, which involves segmenting it into subunits, choosing appropriate labels for them, and also for their interrelationships, and perhaps more such labelling on variable levels of abstraction. The only way I would accept your thesis is if I took it as meaning that all perception has the same arbitrary or pseudo-arbitrary characteristics as analogy-making does" [Hofstadter, 1988].

However, we do not as yet have the details of Hofstadter's opposition, we do not at the moment have a specific case to answer. When that time comes, we can expect that Hofstadter's reservations will excite intense and important debates of significance, not only for analogy theory, but also for the theory of predication itself.

Suffice it to say for now that in making the point that analogies are created, not found, we took ourselves to be agreeing with Hofstadter in the lines quoted under the title of this paper. "What are analogies?" asks Hofstadter. "What is the purpose of trying to *establish* a mapping between two things that *do not* map onto one another in *reality*?" ([Hofstadter, 1986, p. 556], emphasis added in the first and third instances). We return to this issue in the next section.

2. Analogical Semantics

We have said that it is not on to regard analogical characterization as (when apposite) true or true *tout court*. Of course, we think that it is true that Philip is the First Lady of Britain, but also that it is not true. (How could he be, not being a lady?) On the face of it, we seem doomed to give apposite analogical characterizations incompatible semantic assessments. What are we to do about this?

A common strategy is to dissolve a contradiction with a plea of ambiguity. The soundness of such a strategy here quickly becomes apparent. To see that this is so we need to invoke a distinction between Meanings$_I$ and Meanings$_{II}$;

I Meanings$_I$ of sentences are functional components of the lexical meanings of the expressions within.

II Meanings$_{II}$ of sentences are their truth conditions.

It is sometimes supposed — though not without contention — that by and large the Meanings$_I$ and the Meanings$_{II}$ of a sentence tend to converge. However, whatever might be the case 'by and large' it cannot be allowed that for

3. Philip is the First Lady of Britain.

Meanings$_I$ and Meanings$_{II}$ do converge. On the contrary, it is necessary that they diverge radically.

So we ask, "In virtue of what is (3) true?" and, as we have seen, the answer is that it is Philip's satisfaction of the condition that he be the husband of the head of state of Britain. Thus we may say that (3) means$_{II}$ that Philip is the husband of that head of state. This way of looking at the matter is reinforced by the fact that the contained expression in (3), "is the First Lady of", resists standard closure moves, most notably the implication of "is a lady". Strawson had a nice way of flagging expressions subject in this way to such niggardly closure conditions. These expressions, he said, had grown capital letters; and so "The Holy Roman Empire" was neither holy nor Roman. It is also worth noting *en passant* that it is quite wrong to hold that analogical assertions are semantically inert, that is, that they are insusceptible of literal paraphrase. Giving the meaning$_{II}$ of (3) is to do precisely that.

All the same, it would also be wrong to think that the meaning of (3) is exhausted by its meaning$_{II}$. It is necessary that (3) also retain the meaning$_{II}$ of its contained expressions. That is, the expression "is First Lady of" must also bear the meaning got by semantic composition of its lexical items. If it didn't, (3) couldn't be striking in the way that it is. If all that (3) meant were what it means$_{II}$, it would be impossible to account for its analogousness. For if (3) meant nothing more than what it means$_{II}$, that would stick us with the consequence that the whole meaning of (3) is that Philip is husband of Elizabeth. But this, though true, is no analogy.

There is a more vivid way of capturing the notion of analogy, more vivid but not better. It requires us to recognize that although (3) is true, there is a sense in which it isn't true. Saying so gives us occasion to think that (3) also retains its lexical meaning, its Meaning$_I$. And so we can propose a further thing — that the analogousness of (3) consists in its simultaneous possession of a Meaning$_{II}$, for which it is true, and a Meaning$_I$, for which it is false. Though a more vivid characterization of analogy, to be sure, it is too vivid by half. For now the fact that (3) is an analogy straightforwardly entails that it is both true and false. It is no good to plead that the ambiguity of 'meaning' as between Meaning$_I$ and Meaning$_{II}$ dissolves the contradiction. (3) possesses both Meaning$_I$ and Meaning$_{II}$ and thus possesses properties in virtue of which (3) is both true and false. Admittedly, one might insist that the ambiguity of 'meaning' leeches into our truth-predicates — such is the close kinship between meaning and truth — giving us in the end that (3) is 'true$_{II}$' and "false$_I$', or some such thing. But this is too much ambiguity for our tastes, that is, too much ambiguity for too little explained.

We might wish to say that by and large, Meaning$_I$ is a route to truth. If that held true here, then we could say that the Meaning$_I$ of (3) is a route to the truth of

3.* Philip is the lady who is highest in civil rank in the British population minus the Monarch.

But (3*) is no truth, and we will want no route that takes us there. One way to keep ourselves clear of (3*) is to put a hermetic seal on the Meaning$_I$ of (3), to make (3)'s Meaning$_I$ truth-conditionally inert. This is something that nature cannot do. It is up to us, and we do it all the time. Here is perhaps the best reason for saying that analogies are made and not found. Let us say again, in passing, that there is also a sense in which it is true that analogies *cannot* be paraphrased literally. They cannot be paraphrased, as with (3) via (3*), in such way as breaks the truth conditional seal on Meaning$_I$.

And so we have it. To say that (3) is both analogical and true is to say that (3) possesses Meaning$_I$ and Meaning$_{II}$ such that

(i) Meaning$_{II}$ is truth conditionally active and (3) is true by its truth conditions; and

(ii) Meaning$_I$ is truth conditionally inert and that, if it were not, (3) would be false for its Meaning$_I$.

"Placing the Meaning$_I$ of (3) under truth conditional seal". What is this all about? How is it done? Are we saying that the implications aren't there, that (3) doesn't imply (3*) for the Meaning$_I$ of "is First Lady of"? No, we acknowledge that they are 'there', but we also claim that they are not drawn. But how can you not draw an implication that you recognize as being 'there'? Here is one way. Some philosophers of science regard physical theories as set-theoretic predicates. Sometimes a physical theory can be adequately reconstructed using the resources of naive set theory. But naive set theory contains unwanted consequences, thanks to the Russell Paradox. What is 'not drawing' here? It is restricting the set theoretic sentences used in describing the physical theory T to a selected proper subset of naive set theory's inconsistent set of theorems. What is the principle of selection? It is this: Select what you need for T.

It is much the same way with the making of analogies, only more so. For here we draw none of the consequences of (3) under its Meaning$_I$ (self-implication excepted). What is our principle of selection? It is this: Select what you need for (3). We see, then, that the situation with regard to (3) is a limiting case of procedures for T. T helps itself to an edited selection of naive set theory's consequences; (3) helps itself to none of the consequences of its own meaning$_I$.

It may seem that our closure-crimping semantics is too eccentric, too ad hoc to bear serious scrutiny. We demur from such a view. The semantics sketched here for analogical sentences is an extension of a well-established way of handling fictional discourse. Among the going accounts of literary semantics, David Lewis is perhaps preeminent in the severity with which he constrains elementary closure operations. "We should not" says Lewis

> close under the most obvious and uncontroversial sort of implication: the inference from conjuncts to conjunctions [Lewis, 1978, p. 278 of preprint].

Of course, the trouble with fictional discourse is its alarming propensity for contradiction. The fictional sentence

(α) Sherlock Holmes lived in Baker Street

is true of the Conan Doyle stories yet false of the world, speaking loosely both times. Numbers of authors (for example, [Lewis, 1978] and [Woods, 1974a, ch. 5]), have pleaded the case for postulation of an unexpressed bit of syntactic structure with which to handle such cases — thus the fictional operator 'f' which is assumed to attach to the deep structure of fictional sentences. By these lights (α) would be ambiguous between

(α') Sherlock Holmes lived in Baker Street, which is false, and
(α'') f(Sherlock Holmes lived in Baker Street),

which is true. The plea then, is for recognition of something like syntactic ambiguity in sentences like (α).

Much the same treatment of fictional discourse can be provided without express resort to a fictive operator, and with very similar results. We may say that (α) possesses Meaning$_{II}$ and Meaning$_{II}$. By virtue of Meaning$_{II}$,

(α*) "Sherlock Holmes lived in Baker Street" is a member of maximal account[5] of the Conan Doyle stories

is true and (α), too, since this is what it means$_{II}$. But given its Meaning$_{I}$, and under the gentle guidance of modest semantic ascent, we have problems. For we have it by implication that

(α**) $\exists$x(x=Sherlock Holmes and "lived in Baker Street" is true of x),

which, if true at all, is true only for existence-neutral quantifiers. We also have it that

(α***) $\exists$x(=Baker Street and "Holmes live in" is true of x),

which is nothing but false.

Here, too, it is possible for the theorist to insist in view of the trouble caused by the Meaning$_{I}$ of (α) that (α) should be denied all meaning except its Meaning$_{II}$ — the truth condition peculiar to sentences of this sort. But if this were *all* that (α) and its ilk meant, the verisimilitude of fiction would be destroyed utterly. By *reductio ad absurdum*, (α) must also possess Meaning$_{I}$; and though 'there', its implications are not drawn, lest contradiction ensue. So here, too, the best counsel is to let (α)'s Meaning$_{I}$ be, but to crimp closure and not to draw the consequences that land us in the mess of (α**) and (α***).

The close semantic similarity between fictional discourse and analogical discourse is, let us note, but a reflection of the close similarity between the making of stories and the making of analogies, and further reinforcement of our claim that analogies are not *objets trouvés*.

However, there is a further peculiarity of the analysis of analogousness itself which calls this contention into serious question. Consider the sentence

(β) Philip is the Guiseppe Verdi of Britain

concerning which there is strong intuitive reason for saying that it nicely exemplifies the classical notion of *false analogy*. It is easy to see that (β) fails our condition [A] on analogousness; for there is no role-similarity abstraction S shared by Philip and Verdi sufficient to overwhelm their

[5]The term is Parsons'. See [Parsons, 1980]. A story's maximal account is the text of the story supplemented by what a competent reader would furnish by way of specification of unexpressed content.

large differences in a way to make of them one another's analogues. (This we say with suitable tentativeness, anticipating hopefully no biographical surprises from either camp.) If this if right, two interesting consequences drop out.

(γ) A 'false analogy' is not an analogical sentence which happens to be false, but is no analogy at all (by the failure to satisfy condition [A]). So false analogies, so-called, are here represented as failed analogies.

(δ) Falsity gets a grip nevertheless. For we also have it that (β) lacks Meaning$_{II}$. There is no truth condition for (β), hence no condition that (β) satisfies. Thus from the perspective of Meaning$_{II}$, (β) is false (or anyhow, let's not quibble) not true.

Jointly, then, (γ) and (δ) provide as follows for false analogies. They are sentences proclaiming analogies that don't exist and they are false (or not true). There is no analogical sentence that is not true, that is, true by way of objectively determinate features of how the world is. It is in their possession of Meaning$_{II}$ that analogies are 'found'. Point to Hofstadter. But there is a further condition on the analogousness of sentences. They must be expressed not in the language of their truth conditions — after all, "Philip is the husband of the British head of state" is no analogy — but rather in language bearing a Meaning$_I$ which, if it were implicatively cashed, would make them false. The selection of language appropriate to the expression of analogical sentences is constrained by the necessity to find Meanings$_I$ which would be objective falsifiers of their bearers if honoured in the ordinary way. The creativeness of analogy resides not in its indifference to what is objectively so, but in its selective hostility to it. The analogy-maker imposes his will against the semantic grain of Meanings$_I$. It is in his refusal to draw the consequences of Meanings$_I$ that analogies are made.

The dispute between Hofstadter and us has turned, as we can now see, on a false contrast. What is interesting about analogies is that they have a distinctive way of being both 'found' and 'made'.

Finally, let us now redeem our pledge to remove the circularity of conditions [A] and [B] on analogousness. Both conditions employ the notion of something's possessing a property analogously. The idiom can be expunged as follows:

C x has M analogically iff "x has M" has Meaning$_{II}$, by which it is true, and meaning$_I$, by which it would be false if its implications were drawn. Thus, x has M analogically iff "x has M" has Meaning$_I$ and Meaning$_{II}$, its meanings diverge radically, it is true for one meaning, and for the other no valuation is made.

3. Concluding Reflections: Fregean Sense and a Counterexample

We think that it is implicit in the foregoing pages that a theory of analogical characterization will need to acknowledge as a significant theoretical determinant something like a Fregean notion of sense. In short, analogies are functions of something like Fregean senses. What we are trying to get at may be made clearer by some further cases. (The qualification "something like", we shortly return to.)

15. Amsterdam is the Venice of the North.
16. Venice is the Amsterdam of the South.

We are sure that (16) is no good. The reason that it's no good seems to be that in (16), the expression 'the Amsterdam' operates on the full power of the Fregean sense of its embedded proper name, as does 'the Venice' in (15). It is not enough that Venice and Amsterdam alike have canals that make their cities lovely (and inconvenient and smelly). It is rather that the sense of 'Venice' contains dominantly the idea of 'city of canals'; and although the sense of 'Amsterdam' may contain that idea too, it does not do so dominantly. Other features of that city are sense-dominant, its being the financial capital of Holland, for example. So we have it that the features of Venice that Amsterdam shares with it for it to be the Venice of the North are not sufficiently sense-dominant for those very similarities to make of Venice the Amsterdam of the South; that is, that city of the South distinguished by having the canal-significance of Amsterdam. There is no such city.

All of this is unapologetically subjective. We can suppose that an Amsterdam parochialist, on arriving in Venice wholly innocent of its character, might in his first rapture exclaim, "Why, Venice is the Amsterdam of the South". Worldly fellow-passengers would smile indulgently, no doubt. The poor chap's equally parochial travelling companions might think him a poet, or anyhow a clever fellow with a way with words. These differential reactions would be explained if it should be true that the worldly travellers are operating with different Fregean senses, to which analogical characterizations are of course relativized. The relativization to X's Fregean sense [= the (one of the) attributes of X most known; known as 'common knowledge'] provides that many people will not know how to evaluate

9. Theaetetus is the Frank Ramsey of Plato's Academy.

The easiest explanation of this fact is that many people don't know who (for example) Frank Ramsey was. And this is a fact tailor-made for any view according to which analogies are functions of something like Fregean senses.

'Something like' requires a brief comment. Frege's *Sinn*, like Locke's 'idea' as common acceptation, reflects a common store of thoughts and propositions that is transmitted from one generation to another. It is not clear that *Sinn* will brook the suggestion of relativization to different linguistic communities. So (9) may turn out to resist analysis by way of unqualified Fregean *Sinn*. Then, too, it can hardly be said that the example of Case 14 turns on *Sinn*. This is the case in which, when A=1 2 3 4 5 5 4 3 2 1 is compared with H=1 2 3 4 6 5 5 6 4 3 2 1, it is proposed that six is the four of H via the abstraction "the element closest numerically and positionally to the central pair". The analogy here is not constructible out of any semantic properties of the expression "is the four of H" without contextual specification of H, and A too. Fregean senses are free of such fine-grained contextual dependencies. The range of cases from "is First Lady of Britain" to "is the Frank Ramsey of Plato's Academy", and thence to "is the four of H", stretch the notion of Sinn from its free-standing place at the one end a goodly way in the direction of the Fregean 'associated idea' or Vorstellung. One person's Vorstellung need never be another's, and the stretch from *Sinn* to Vorstellung is the stretch from common acceptation to uncommon. Our own idea of sense need not extend quite that far but, if it is to accommodate a full range of cases, it must be a more relaxed notion than *Sinn*, which our own will include. I don't know how to draw the line precisely, but our sense will be the interval from Sinn to some point well along the way to Vorstellung — a conception common enough to deal with First Lady analogues and uncommon enough to handle Hofstadter's numerical analogies.

Then, too, some analogies are opaque because of their reliance on technical information. For example, cases (15) and (16) arise in the following context. It can be asked in what sense the DNA of chromosomes convey a message. We might notice that in the geometry of different systems a structurally stable dynamic system (think of a gradient field), is nearly determined completely by the spectral elements of singularities of the field (these are the equilibrium points of the system, given that the field is zero). If we consider a cell in division as a dynamic system with structurally stable duplication, then the singularities in this duplication (that is, the set of points where spatial duplication begins), correspond to the chromosomes and the molecules of nucleic acid. We draw this example from [Thom, 1975, p. 158].

We said in section 6 that the relation of analogy is not symmetrical, but Gerard Verhoeven has challenged this claim in conversation with the author. It seems analogically apposite that (17) Hitler is the Nero of modern Germany, and also that (18) Nero is the Hitler of the ancient world.

I shall try to say in a moment why we don't find ourselves convinced on this question of symmetry. For now, Verhoeven's examples may seem to cause us trouble in other ways more central to our account. For it may appear that (17) and (18), good analogical characterizations each, resist the structural analysis imposed on them by condition [A], in the latter part of section 6. Condition [A] asserts in part that if x is analogous to y in respect of M, then there is a role-similarity abstraction S shared by x and y such that

(i) x's satisfaction of S entails x's possession of M analogically.

(ii) y's satisfaction of S does not entail y's possession of M analogically or non-analogically (really).

Now it is easy to see that several of our cases manage to fulfil condition [A] quite nicely. In case (3), we test for:

Philip is analogous to Laura (in respect of M),

with M=First Ladyship of Britain, and S=spouse of head of his/her head of state. Philip's satisfaction of S entails his possession of M analogically, whereas Mrs. Bush's satisfaction of S does not entail her possession of M analogically or non-analogically.

Similarly for case (11), in which the 3 of series B is analogous to the 4 of series A. Here S=closeness to its series' central pair. Clearly both the 3 of B and the 4 of A satisfy S, and 3's satisfaction of it endows it with M, being the 4 of B. But 4's satisfaction of S does not endow it with M, for 4 is not the 4 of B.

But consider now Hitler and Nero. If, as we have it in (17): Hitler is analogous to Nero (in respect of modern Germany), then M=the Nero of modern Germany, and S=the most criminal politician of his day. Hitler's satisfaction of S endows him with M. But Nero's satisfaction of S endows him not with M but with M*=the Hitler of the ancient world. So it would seem, after all, that condition [A] is not breached. It is anyhow not breached *unless* Nero's satisfaction of S implies that Nero is the Nero of modern Germany. And if he is *that*, we may also have it that Nero is *the* Nero of modern Germany of the ancient world. While I admit to feeling some pressure to yield to such iterations, I am not in the end ready to admit them whole-heartedly. More settled decisions will require the disentanglement of referential and role-attributive constructions involving 'Nero'. For the present I think that condition [A] resists the counterexample.

PART VIII

INDUCTION

On the present-day understanding of these things, our final three chapters all deal with inductive difficulties. On this understanding, hasty generalization and *secundum quid* are the same fallacy, each having to do with inductive misperformance. Aside from some brief and puzzling remarks in the *Rhetoric*, this is decidedly not Aristotle's view of *secundum quid.* But, even if it is not an inductive fallacy, what Aristotle has to say about *secundum quid* is genuinely interesting and is the basis for the first of the next three chapters. Those that follow are centrally about induction, especially the last.

Chapter 18

SECUNDUM QUID

Readers will have noticed my habit of calling upon Aristotle to assist in the task of saying something sensible about fallacies. In so doing, I cut across the grain of what appears to be the received view of Aristotle as logician and fallacy theorist. On this view, Aristotle has been wholly displaced by subsequent developments, notably the Fregean revolution in logic (discussed in the Prologue of this book), and present-day developments in fallacy theory. My view is different [Woods, 2001]. The account that Aristotle gives in *On Sophistical Refutations* of syllogisms and of arguments having syllogisms as their core is extremely rich, notwithstanding the fragmentariness of the discussion. But if read with patience and sympathy, Aristotle's contributions are ingenious, and have a resonance even today. There is, in my view, no better discussion of *ad hominem* arguments that the passages in the *Metaphysics* (and discussed in chapter 7). In that same chapter I laid out the structure of Aristotle's theory of refutation. Here, too, there is no better account.

This is the spirit in which I approach the subject of the present chapter. The received view, if I am not mistaken, is that Aristotle's characterization of *secundum quid* is thin and superficial. My view is that on a thin and superficial reading of Aristotle the characterization of *secundum quid* is indeed thin and superficial, but that this is not the best way of reading Aristotle.

1. Omitting a Qualification

First mention of *secundum quid* occurs in Aristotle's treatment of the fallacies in *On Sophistical Refutations.* It is the mistake of omitting a qualification on a term or generalization. For example, if one is going to predicate the term 'white' of Geronimo or Mr. Mandela, one had better apply the qualification 'teeth' or 'hair', or some such thing, lest the predication turn out to be false. Similarly, if one intends to generalize on the correlation between day-in-Amsterdam and rainy-day, one had better add the qualification 'that I have known'. Thus

1. Every day I have known in Amsterdam has been a rainy day.

rather than

2. Every day in Amsterdam has been a rainy day.

A further example, namely,

1. Sherlock Holmes exists in fiction.
2. So, Sherlock Holmes exists.

looks like something else. The qualification, whose omission makes this a bad argument, is a qualification neither of a term nor of a generalization. But it seems to be the same fallacy. Can we extend Aristotle's characterization? Can we do it in a non-*ad hoc* manner? The answer is that we don't have to. The Holmes case is already covered in modern logic. An open sentence is a term. A closed sentence is a limiting case of a term. So the qualification 'in fiction', in qualifying the sentence, "Sherlock Holmes exists", also qualifies a term.

It is important to emphasize that, for Aristotle, the fallacy of *secundum quid* is inherently the fallacy of mismanaging a qualification (i.e., omitting it when it is needed for truth). It is possible that some inductive arguments imbibe this fault and that some general inferences do so as well. But inductiveness and generalization do not inhere in the fallacy. For consider

1. This black man is white-toothed.
2. Therefore, all black men are white.

and

3. This wet match didn't light when struck.
4. Therefore, the next match won't light when struck.

Our two cases show that some inductions and some generalizations commit the fallacy of *secundum quid.* But there are examples of *secundum quid* that are neither inductions nor generalizations, as again our Holmes example makes clear.

Hamblin reminds us that in 1847 DeMorgan published his influential *Formal Logic.* In this work, DeMorgan represents the *secundum quid* as a misgeneralization from a single case, concerning which see chapter 19 above. So,

5. This man, Socrates, is Greek.

6. Therefore, all men are Greek.

Hamblin seems to think that this marks a significant discontinuity in the tradition that comes down to us from Aristotle [Hamblin, 1970, p. 28]. If so, it is a discontinuity so radical as to draw the charge that

(i) DeMorgan got the fallacy wrong

and

(ii) Since the DeMorgan type of example has become something of a paradigm, everybody else has got the fallacy wrong in the years following 1847.

In fact, and perhaps a little more charitably, we can take Hamblin as suggesting that with DeMorgan the concept of *secundum quid* changed (and without prior notice); that it came to be understood as *hasty generalization* or as a *sampling error*; and that this view has come to be the dominant contemporary position. If this is understood as the claim that people now believe that sampling errors are not construable as errors of inappropriate qualification and that they are right in thinking so, then I agree with the first part of the claim and demur from the second part.

Recall that, for Aristotle, *secundum quid* is the fallacy of mismanaging a qualification. In the examples given by Aristotle, the mismanagement is always by way of omission. On reflection, there is little reason to think that it couldn't go the other way as well. That is, sometimes we make things go wrong by attaching a qualification inappropriately. Thus,

7. Everyone has a right to the liberty of his person.

8. Therefore, everyone who is a convicted murderer has a right to the liberty of his person.

or

9. E.M. Barth exists.

10. Therefore, E.M. Barth exists in fiction.[1]

or

11. No black person is white.

12. Therefore, no black person is white-haired.

So it seems to cohere with Aristotle's fundamental insight to define the *secundum quid* as follows.

Definition One: In claiming that K, one commits the *secundum quid* if and only if there is a claim P and an expression Q which admissibly (i.e., grammatically) applies to P or to a term of P such that either (i) $K = P$, P is false and $Q(P)$ is true; or (ii) $K = Q(P)$, P is true and $Q(P)$ is false.

[1]By the way, here is an example that allows us to correct the all too common mistake of confusing the fictional with the legendary. Else Barth is a legendary figure in logical theory and analytical philosophy; the last thing that she is fictional.

If this is an acceptable definition, it certainly cannot be thought to end the matter. For notice, Aristotle gives us no general theory of the conditions under which the omission or application of Q to P constitutes a mismanagement of qualification. Aristotle gave us the basic vocabulary ('inappropriate qualification') and a basic pathology (a list of examples). He didn't, however, specify sets of conditions which map the technical vocabulary onto sets of generalizations from and beyond his pathological cases. Aristotle's account is, therefore, substantially incomplete.

2. DeMorgan's Approach

With regard to DeMorgan's paradigm, it is said that (6) is a bad induction from (5), and so it is. It is also said that, for DeMorgan, this is the fallacy of *secundum quid* only in some new sense of that term. This is a mistake. Notice that had we inferred

6.* Some men are Greeks.

the inference would have gone through. Comparing (6) and (6*), we see that this difference is all a matter of which quantifier is used. If we understand quantification in this way, 'all' and 'some' are predicates of terms and so they are qualifications — the right qualification in (6*) and the wrong qualification in (6). DeMorgan's paradigm thus commits Aristotle's *secundum quid*, if we understand quantification in Aristotle's way.

Of course, DeMorgan was no Aristotle, and 1847 is a long way from circa 344 BC. There is much that is modern about DeMorgan's treatment of logic, and there is reason to think that his understanding of the quantifiers was, if not Aristotle's own, very like the one conveyed by the following definition.

Definition Two: A quantifier is a second order predicate of sentences, open or closed.

Modern logic, as we saw, understands open sentences as general terms, and closed sentences as limiting cases of them. Jointly, then, quantifiers are qualifications of terms.

In the case of the inference of (6) from (5), 'all' is the wrong qualification, and was applied, and 'some' is the (or a) right qualification, and was omitted. So whatever else we may want to say of this argument,

(a) it does commit Aristotle's *secundum quid*.

(b) DeMorgan says that it commits *secundum quid*.

(c) DeMorgan does not say that it commits *secundum quid* in some new or different sense.

So, it is reasonable to conclude that

(d) DeMorgan, too, thought that it committed *secundum quid* in Aristotle's sense.

(e) It is wrong to 'blame' the hasty generalization departure on DeMorgan.

Even so, things may strike us as fishy. Aristotle gave many examples of the *secundum quid* (or more carefully, Def. 1 provides for many different kinds of ways of omitting it). One way is by inferring (6) from (5). But it is a hardly a paradigm or the best example didactically speaking. Black men with white teeth is a much more accessible example, as is Sherlock Holmes existing in fiction. So what accounts for DeMorgan's choice?

We will take it as given that if we cannot satisfactorily answer this question, we will have significantly weakened the claim that DeMorgan was not breaking with tradition and was not pioneering the hasty generalization conception.

The solution of this puzzle is conjectural or, as I would prefer to say, circumstantial. Recall our earlier observation that Aristotle's account is seriously incomplete, that it offers no theory of inappropriate qualification. Recall, too, that DeMorgan is writing in 1847, four years after the publication of Mill's *A System of Logic* (1843). This was an ambitious and influential work. Among other things, Mill was strongly positioned in a logical tradition which sought to rehabilitate a form of argument which was disallowed in Aristotle's syllogistic: Arguments all of whose premisses are singular. Mill agrees that such arguments can be deductively invalid, but he also recognized that there are conditions under which they could be inductively strong. These conditions are approximately what constitutes Mill's conception of scientific method. So, when singular premisses are collected, and general conclusions derived from them in conformity with the principles of scientific method, such arguments are inductively strong.

DeMorgan knows this. He now — in 1847 — has at his disposal Mill's theory with which partially to repair a gap in Aristotle's theory. DeMorgan can now explain what inappropriate generalization 'really is' in the case of the inference from (5) to (6). It is an inference which involves a qualification made inappropriate by Mill's theory of induction.

Such a theory DeMorgan did not have for other examples of inappropriate qualification. For example, the black man/white-toothed kind of case requires us to have a good semantic theory of adjectives, and similar cases require a good theory of adverbs. We in the 21st century don't have these theories, much less DeMorgan, and it may well be that this is why he didn't deal with other cases. He was just following the principle of "Playing to your strength" or "Playing your best cards."

3. Extending the Reach of the Fallacy

It may be thought that construal of DeMorgan's paradigm as a case of Aristotle's *secundum quid* is open to question. Aristotle nowhere endorses *Definition 1*, though nothing he says in *On Sophistical Refutations* vitiates it. Even so, clause (ii) is problematic. It provides a sufficient condition on the *secundum quid*: Any claim falsified by addition of a qualification to a true claim counts as Aristotle's fallacy. Seen this way, very large classes of rather different kinds of claims will turn out to be cases of *secundum quid*. For example, many examples of the fallacy of composition will turn out to commit the 'further' fallacy of *secundum quid*, as witness

13. All the members of the mathematics department are male and first-rate.
14. Therefore, the mathematics department is all-male and first-rate.

Notoriously, (13) could be true and (14) false, as when all the members of the department work and teach exclusively in topology. No such department would realistically qualify as a first-rate mathematics department. Notice, however, that omission of the qualification 'first-rate' in (14) would produce a truth and would make of the passage from (13) to (14) a valid argument. So it can be said that here is a case in which the addition of the qualification 'first-rate' converts a truth into a falsehood and, by clause (ii) of Definition One, the argument commits not only a composition fallacy, but *secundum quid* as well. The same is true for a large class of modal fallacies. For example,

15. All the coins in my pocket are centimes.
16. Therefore, necessarily all the coins in my pocket are centimes.

is a modal fallacy. It is also true that in (16) the qualification 'necessarily' converts a truth into a falsehood and invalidates an argument which, without it, would be trivially valid.

It may strike one as unlikely that Aristotle either intended or would have endorsed a conception of *secundum quid* as broad and superficial as this one. To be sure, there is an indication to the contrary, namely, that Aristotle had in mind a general theory of inappropriate qualification, with regard to which a rather broad notion of *secundum quid* would not have been problematic. The theory that Aristotle may have had in mind is that of the *Categories*, in which every *secundum quid* commits a category mistake (i.e., the mistake of applying a predicate appropriate to a certain kind of object to an object of a different (an inappropriate), kind). Whatever may have been in Aristotle's mind, it would seem that his hunch is misconceived. For whatever we say of the error of the argument from (13) to (14), it is certainly not a category mistake to predicate 'first-rate' of a department of mathematicians. Yet, if Def. 1 is allowed to stand, this argument commits the *secundum quid.*

In any case, short of having a general theory of inappropriate qualification, it is not very illuminating to characterize a claim as fallacious on this count; and the illucidity is intensified, not diminished, by having very many and very different types of structures characterized this way. This may be true, but it leaves room for a conjecture. There is textual evidence that Aristotle did not think of the *secundum quid* as a determinate and compact fallacy on a par with the other twelve. If this is so, it is an interesting question as to how Aristotle may have thought the *secundum quid* to relate to the others. It can be conjectured that the *secundum quid* loosely identifies a claim as committing an error in one or other of many different structural misarrangements. So one way of being an inappropriate qualification is having the structure of a modal fallacy, that is, a structure of which a decent modal theory would reveal, in fine, particularities of the qualification's inappropriateness. Another way for something to count as having an inappropriate qualification is by having a structure which a good theory of the part-whole relationship would reveal, in fine, as a composition fallacy.

It is of some possible importance that precisely the same has already been said of DeMorgan's paradigm. The argument from (5) to (6) is a *secundum quid.* If Definition One is allowed to stand. But a good theory of scientific method will tell us what the inappropriateness of the qualification comes to in this particular kind of case. What it comes to, in fine, is what has come to be known as the fallacy of hasty generalization. These things being said, a further possibility announces itself. Whether a given, determinate, fallacy was a member of Aristotle's list of thirteen or arose later, there is a good chance that it can be constituted as a case of *secundum quid* in Aristotle's generous sense of the term. The history of fallacy theory thus has an interesting structure. As various theories developed that gave good accounts of inappropriateness of qualification for particular ranges of cases, the loose proto-fallacy of *secundum quid* was displaced by peculiarities of these particular structures. Thus we don't any longer emphasize the fact that the argument from (5) to (6) is a *secundum quid* because there is something more detailed and precise to say about it: It is a sampling error, an offense to inductive reasoning, and so is (in particular) the fallacy of hasty generalization, which was not on Aristotle's original list. So, too, with the argument from (13) to (14). If Definition One is all right, it is a *secundum quid.* But it is also such that a good theory of the part-whole relation would find fault with it on more determinate grounds. So it is the fallacy of composition, also on Aristotle's list, but a more determinate type of fallacy than *se-*

cundum quid, and some explanation of why Aristotle may have thought that *secundum quid* was not on a par with the others.

We are suggesting that Aristotle's *secundum quid* is a master fallacy, a general and somewhat indeterminate indication of the way in which arguments go wrong. We are also suggesting that certain of the more determinate fallacies are particular kinds of instantiations of the general mistake. Modal fallacies, composition and division, and hasty generalization seem to fit this picture. It is fair to ask whether other fallacies do so, as well. It would appear that they do. Ambiguity is a case in point. The fallacy of ambiguity is committed when an ambiguous term converts an obviously bad argument into an apparently good argument which is, in fact, just as bad.

If it is built into Aristotle's notion of the *secundum quid* that an omitted qualification makes bad an argument that would otherwise be good, then ambiguity is an error of misqualification (omitting the qualification, "but not in the same sense"), and it seems also to be an argument which is good, except for the omission of the qualification. But this is wrong. What the qualification 'does' is to remove any temptation to complete the argument. For then it would be a transparently bad argument. But transparently bad arguments cannot be apparently good ones. So the transparently bad argument, though bad, can't be a fallacy in Aristotle's sense of a bad argument which seems good. No one would ever accede to it.

Perhaps we need to rethink the generalization

> SQ: An invalid argument, A, commits the *secundum quid* fallacy, if there is a qualification Q which A does not contain and which, were it made, the result, $Q[A]$, would be a valid argument.

One way in which we might reformulate SQ is this:

> FSQ: An argument, A, commits the *secundum quid* fallacy if A is an invalid argument which seems valid (cf. the definition of fallacy); and there is a qualification Q which A does not contain and which, if it were made, the resulting argument, $Q[A]$ would be either
>
> (i) valid
>
> or
>
> (ii) invalid, but transparently so.

FSQ now provides that in either case, the omitted qualification if not omitted would have converted a fallacy into a non-fallacy. So we may conclude that arguments guilty of Aristotle's fallacy of ambiguity are also guilty of Aristotle's fallacy of *secundum quid*.

4. *Petitio Principii*

Let us now consider *petitio principii.* Once again, a fallacy for Aristotle is an argument which seems to be a syllogism, but is not in fact. Aristotle here intends 'syllogism' in a rather broad sense; as any argument whose conclusion is necessitated by its premisses (and by no proper subset of them) and which is not identical to any premiss. Let P and Q be propositions such that the sentence expressing P and the sentence expressing Q are synonymous. If we allow that synonymous sentences express the same proposition, then the argument

17. P
18. Therefore, Q

is a *petitio* fallacy. It is a valid argument which is not a syllogism, since premiss and conclusion are the same proposition. (It may also fail to be a syllogism because it has only a single premiss.) And since it is a transparently valid argument both in our modern sense, and for Aristotle, too, it has a propensity to seem to be a syllogism in Aristotle's sense. In any event, it will seem a syllogism for anyone who appreciates the synonymy between 'P' and 'Q' but who does not appreciate that this synonymy implies that the proposition expressed by 'P' and 'Q' are one and the same.

Our question is whether the argument, (17)-(18), also commits the fallacy of *secundum quid.* It would seem that it does. For consider the argument

17. P
19. Therefore, synonymously, Q.

This argument is just like its predecessor except that its conclusion, (19) contains a qualification, "synonymously", which the previous conclusion, (18), omits. Moreover, it is a qualification whose presence in (17)-(19) guarantees that it commits no *fallacy* of begging the question. The qualification does not save (17)-(19) from being a non-syllogism, but it does save it from being an a non-syllogism that is a fallacy. Anyone who knows the meaning of "synonymously" in this context is also disposed to know that 'P' and 'Q' express the same proposition. Knowing that, one knows that (17)-(19) is transparently not a syllogism. Knowing that, it cannot be a non-syllogism that seems to be a syllogism. So it cannot be a fallacy. We have it, then, that the prior argument (17)-(18) omits a qualification whose presence would spare it the embarrassment of being a *petitio* fallacy. Hence (17)-(18) commits the *petitio* and it commits *secundum quid.*

Here, too, with ambiguity fallacies and question-begging fallacies, each is construable as the *secundum quid* fallacy. In the case of the ambiguity fallacy, it is a particular type of inappropriate qualification which a good theory of ambiguity allows us to specify for appropriate classes of argumentative structures. It is the same way with question-begging. The *petitio* is a special kind of inappropriate qualification which a good theory of propositional identity would specify for appropriate classes of structures. In each case the inappropriateness of qualification is the omission of a qualification whose presence would have been appropriate — so appropriate, in fact, as to save the arguments in question from lapsing into fallacy.

If this line of reasoning is right, it can be expected that, as human knowledge enlarges and deepens, there will be many types of *secundum quid* that qualify for detailed and specific theoretical analysis. It can also be expected that these various 'new' instances of *secundum quid* will significantly outrun Aristotle's list of thirteen. This is precisely what has happened throughout the centuries. 'New' fallacies have cropped up, depending on the development of our theoretical capacities to recognize them and to take their measure. It is also said that these are 'Aristotle's' fallacies, and this has drawn the scorn of those who think that they are no such thing. They are wrong. These are Aristotle's *secundum quid*, not in every case, but in a truly striking range of cases. And in every case, their *secundum quid* identities have been obscured by the more interesting and theoretically robust details of particular types of cases, as witness again the fallacy of hasty generalization.

It remains true that Definition One may well overstate Aristotle's conception of *secundum quid*. Perhaps clause (ii) will have to go, and we will need to make do with the idea that Aristotle's fallacy is always the fallacy of omitting a qualification. If this is so, it will do damage to our conjecture of late; but it is possible all the same to go some way in rehabilitating our claim that DeMorgan's paradigm of hasty generalization is an Aristotelian *secundum quid*. Consider the argument

5. This man, Socrates, is Greek.

6.** Men are Greek.

in which (6**) is a generic sentence carrying the sense that it is characteristically true of men that they are Greek. The argument is bad, a violation of Mill's theory of scientific method. But it also omits a qualification which would save it. It is the qualification 'some', as in

6.* Some men are Greek.

The argument from (5) to (6**) commits a *secundum quid* fallacy in Aristotle's sense. The conditions under which it is also a hasty generalization constitute an analysis of Aristotle's proto-notion of inappropriate qualification for a broad range of structures known as inductive.

Chapter 19

HASTY GENERALIZATION

I said in chapter 1 that most of the fallacies in the gang of eighteen are research programmes in their own right. In the present chapter I shall try to make good on this claim. Hasty generalization is usually classified as an inductive fallacy. The research programme that inheres in a correct understanding of it is little short of the project in inductive logic. Inductive logic is too much to handle in a single chapter, even if I knew how. But headway can be made with the problem of induction and a well-known attack upon the confirmation relation.

1. Hasty Generalization: The Basic Idea

Also called *secundum quid* by modern writers, hasty generalization was not much discussed in antiquity. Aristotle discusses *secundum quid* in two places. In *On Sophistical Refutations* (e.g., $166^b38, 167^a20$: Barnes [Barnes, 1984, p. 282]), it is the fallacy of omitting a qualification, as in an argument which concludes that Mr. Mandela is white from the premiss that he is white-haired. An apparently different conception is proposed in the *Rhetoric* ($1401^b35 - 1402^a28$: [Barnes, 1984, p. 2237]), suggestive of the fallacy of hasty generalization in the sense of a sampling error or a generalization drawn without adequate inductive support. It may be that the two Aristotelian conceptions can be reconciled. Testing for rancidity in butter, if an examiner confined himself to samples of butter uncooled for at least seventy hours, his generalization could be said to be hasty if it were the assertion that all butter is rancid. Aristotle might say that, unacceptable as it stands, the generalization is nevertheless correct 'in a certain respect' (which is the meaning of *secundum quid*). It is correct if we add the qualification 'uncooled for at least seventy hours'.[1]

In later writings, *secundum quid* and *hasty generalization* are usually distinguished. In [Arnauld and Nicole, 1996], *secundum quid* is approximately the *secundum quid* of *On Sophistical Refutations*, and hasty generalization is approximately the *secundum quid* of *Rhetoric.* However, hasty generalization is not discussed by that name in the 9th edition of Copi's *Introduction of Logic* [Copi and Cohen, 1994]. It is analyzed there as *converse accident*, the "fallacy we commit when we move carelessly or too quickly *to* a generalization" [Copi and Cohen, 1994, p. 125].

The modern hasty generalization takes on a greater theoretical complexity with the development of work in inductive logic, confirmation theory, probability theory and statistics, and theories of causality. It has proved to be a rather bumpy theoretical ride.

Induction is met with the difficulty of securing a non-circular justification that inductive practices are sound [Hume, 1964, p. 370], though some contemporary writers are unconvinced that the circularity, which is unavoidable, disables the justification [Goodman, 1965, p. 61]. The probability calculus is thought by several writers (e.g., [Skyrms, 1986, p. 20], to give an inadequate or underdetermined analysis of the concept of probability. Others allow that it does a good job in capturing what can be called *aleatory* probability or a concept of probability appropriate to the analysis of games of chance, but that it misdescribes a probability concept — sometimes called *Baconian* — involved in the description of inductive support [Cohen, 1989, ch. 1]. One contentious issue is whether experimental evidence that human beings have a propensity to err in the compounding of probabilities makes for a convincing case (see, for example, [Cherniak, 1981; Stich, 1980]). Critics of these and other psychological studies argue that subjects do miscalculate conjunctive probabilities only if they are working with the aleatory concept of probability, rather than the Baconian, and moreover that there is reason to think that in the reported experiments it was the Baconian concept that they were manipulating ([Cohen, 1989, pp. 19–20] and chapter 1 above).

Confirmation theory, in turn, runs into a number of snags, the most celebrated of which is the Raven Paradox [Hempel, 1943; Hempel, 1945]. Causal reasoning has since Hume [Hume, 1964, p. 373] been vulnerable to the charge that all attributions of objective casual significance to

[1]For more of the connections, or lack of them, between Aristotle's and modern treatments of *secundum quid*, see chapter 18.

positive correlations commit the fallacy of false cause or of *post hoc, ergo propter hoc.* I will come back to these matters.

It seems intuitively clear that many of the generalizations comprehended by the term 'hasty' are bad and that judgements of their badness are rather obscured and called into question by theoretical difficulties in the various domains of non-demonstrative reasoning. Of 'hasty', however, it needs saying at the outset that it is hardly *le mot juste.* If, for example, from the true observation that

1. Cynthia is a bad driver

we inferred that

2. All women are bad drivers

this would be an unacceptable — and insulting — generalization. It would be an error irrespective of whether it were made after a second's reflection or a week's so haste seems not to be the heart of the matter. Rather the problem with the inference is a function of the narrowness of the sample, as in (1), and the width of the generalization, as in (2). So evidently part of the inference's badness turns on a sampling error. It is also a fact that (2) is false. But the falsehood of the generalization is not an essential part of the story. One commits the misinference equally convincingly in moving from

3. Here is the first observed emerald and it is green

to

4. All emeralds are green.

The truth of (4) gives the *inference* no comfort, and this tends to reinforce the idea that such mistakes are fundamentally sampling errors (see, for example, [Carney and Scheer, 1980, pp. 63–64]).

It is also true that inductive inferences can be achieved without resorting to universally quantified sentences. Generic sentences will also do and they are more resistant to counter-example. The generic counterpart of (4), for example, would be

4.* Emeralds are green.

Generic statements stand to general, or universally quantified, statements as "Horses are four-legged" stands to "All horses are four-legged". Knowledge of generic truths resembles what Minsky has called a 'frame-system', an idea we met with in the previous two chapters [Minsky, 1977]. If we possess a frame-system for horses, and if the horse we are now inspecting is three-legged, then the system's variable for *number of legs* takes the value *three.* But in the absence of specific information about a horse's leg-multiplicity, the value of the *number of legs* variable is the characteristic value *four.* Four, then, is the *default value* for this variable. Default values specify attributes that are characteristic of a thing or state of affairs. If we are competent default-reasoners, a characteristic feature of something is the default value of a variable supplied by its frame system (the logic of default reasoning was pioneered in [Reiter, 1980]).

Projection is the fundamental inductive skill of human beings. Projection is that device by which we are enabled to learn from experience. Projections, thus, run the gamut from primitive to generic to fully general. Primitive inductions need to be explained. There are two current usages of the term. In one, a primitive induction (or projection) is an induction of any width from a single case. In the other usage ([Quine and Ullian, 1970, pp. 59–61], and [Quine, 1981, p. 56]), primitive induction is induction from single cases to single cases, sometimes as few as one each. Thus on the first usage, all three of our examples are primitive inductions and on the second usage, only the third is. We will say that primitive inductions are projections of any width from single cases.

2. Primitive Induction

Consider the following case, which we briefly considered in chapter 1. A toddler is in his mother's kitchen. Mother, distracted by her ringing phone, interrupts her vigilance and the youngster places his hand on the stove's front element (only recently turned off). A hand is burned. It may seem that the toddler learns in one encounter *never to do that again*, a primitive induction. The first sense is the more general, and we shall opt for it here. There seems little doubt that the received view is that *no* induction from a single case (whether to a single case, a generic conclusion, or a full-bore generalization) is any good, and that any such induction commits the fallacy of hasty generalization in a paradigmatic way.

This is not to say that the name of hasty generalization needs to be reserved for the one-shot case. The fallacy will also be loosely specifiable for multi-shot cases, that is, for samples of greater than one, where features of sample size can be shown to be vitiating or at least problematic. For example, even large samples can be problematic, as when they are unrepresentative. Inferences from unrepresentative samples are said to commit the fallacy of biased statistics and are quite naturally thought of hasty generalizations [Carney and Scheer, 1980]. A still more general notion of hasty generalization is discernible in the literature. We might, for example, follow Johnson and Blair, and define the more encompassing fallacy of 'hasty conclusion' as any argument whose conclusion is not given 'sufficient support' by its premisses [Johnson and Blair, 1983, pp.

41–46]. Hasty generalizations would then be a special case of the former — hasty conclusions whose conclusions happen to be generalizations. It is worth noticing that hasty conclusions would not be restricted to statistical or inductive (projective) contexts. Certain abductive inferences (or 'inferences to the best explanation'), and certain mathematical proofs could qualify without strain.

The toddler's nasty adventure with the stove would also count as a primitive induction if the toddler projected his new wisdom to the next observation of a kitchen stove. It is empirically underdetermined whether the projection is local. Is it to the next observation of his mother's stove or to any next-observed stove? It is unlikely that the toddler has made a full-width projection to all. At sixteen months of age he lacks the cognitive resources for universal quantification. Despite these uncertainties, the case of the toddler is fascinating and important in a number of ways.

It seems quite clear that in some sense the toddler's inference is a good one. Whatever the finer details of the case, the toddler has learned a certain risk-aversion strategy for dealing with stoves (whether his mother's own or any next-observed or all–we leave these details unsettled). His projection resembles a case described by Stephen Stich. Rats will reject any food similar in smell to any food ingested prior to a toxic illness [Stich, 1985, p. 124]. Assuming now, simply for expository convenience, that rats possess cognitive resources for belief and reflection,[2] it is apparent that the strategy, good though it is, encumbers them with a multitude of false beliefs, beliefs in the form "This food is toxic". If we allow nevertheless that such a rat pursues a rationally competent strategy, it is necessary to mark a distinction between strategic and cognitive success, and a kindred distinction between strategic and cognitive manifestations of rationality. What is striking about the rat's case is that strategic success seems to require cognitive failure; or, in any case, is untroubled by it. The rat pursues an extremely conservative risk-aversion strategy. He is a minimaxer about food. His security reposes in a tangle of falsehoods, and the soundness of the strategy may well turn on his not having cognitive resources sufficient to fine-tune his projections. Given his cognitive limitations, it may be supposed that this is the best (cognitively) that the rat can do.

[2]It should be noted that according to David Perkins, in conversation with the author, the toddler's 'inference' is subcortical. Still, the *example* is instructive.

3. Strategic Rationality

It emerges from the rat's case that the concept of rationality is competed for under different criteria. The rat is strategically rational even when he is cognitively subrational.[3] Thus his hasty generalizations can be good and bad simultaneously. Stich, avers that truth has little evolutionary significance for rats, that strategic rationality carries a dominant priority over cognitive rationality [Stich, 1985, p. 124].

An interesting question now presents itself. Do we assimilate the toddler case to the Cynthia case (as in the inference from (1) to (2)) or to the rats case? Or should we try to find for it an analysis *sui generis*? It may be granted that the toddler's projection is more like the rat's than the projection from Cynthia's difficulties as chauffeuse. But there are differences. The rat gets things wrong and the toddler gets things approximately right. That is, the rat projects to falsehoods whereas the youngster projects to (approximate) truths. This may be thought to leave room for a *sui generis* analysis of the latter case. The rat's performance is a strategic success and a cognitive failure. The child's performance is a strategic success and something of a cognitive success. If true, this suggests that unlike the rat, the young human is hard-wired for survival and for truth alike, and that getting at the truth of things primitively is a loose pre-condition of certain ranges of human strategic success. The suggestion possesses a certain anthropocentric appeal in the idea that humans are distinctively endowed with the capacity for primitive knowledge, and in the kindred idea that such knowledge is distinctively a precondition of human survival. There is, however, reason to resist this anthropocentrism.

The possible limitations of the anthropocentric view can be approached in the following way. As we have seen, it is Stich's view that, for the rat, truth is of doubtful evolutionary significance. The rat's hasty (and false), generalizations are his best strategies, given his lack of supplemental and corrective cognitive wherewithall. It should be noted, however that Stich could be right without the rat's thinking him right. Having, for expository ease, agreed to posit for the rat a certain reflectiveness about his beliefs, it can hardly be denied that rats *may* think that they have got at the truth of things, and that proof of this is that they haven't perished.

It may be the same way with us. We may say that we differ from rats and other creatures precisely in the fact that truth for us is evolu-

[3]The distinction between cognitive and strategic rationality is fruitfully discussed in [de Sousa, 1987].

tionarily significant. For one thing, it takes time for a human's cognitive engineering to mature. With maturity comes re-consideration and self-correction. As our toddler grows up, he comes to understand the minimaxingcharacter of his previous primitive induction from Mama's stove. He now has resources for selective adjustment of prior conviction, for he now shows early proficiency with Mill's Methods, as we will say for short. In correcting his early stance, he does something salient. He implies that it was cognitively suspect. He now exchanges the conservative and extreme protection of minimaxation for the comforts of more complex inductive reflection — and in so saying, we see that the maturing youngster is prepared to replace his former costly security with equal security of lesser cost. For he now knows that Mama's stove is at worst specifically tricky in contextually sensitive ways. It is nothing but tempting to say that our maturing human is replacing merely successful strategies with successful strategies that also tell the more precise truth about things.

This is what the rat would say, too. But he is wrong, and is wrong without being able to know or surmise that he is wrong. The skeptic says that it is the same way with us. We are all rats in the same boat, so far as we really know.

4. Practical Reasoning

Peirce says that truth is the limit at which enquiry converges [Peirce, 1955, pp. 5–22]. This may be right. If so, then truth structures human life. We all aim at the ideal and may have been naturally selected for it. The ideal of truth pays off for us in extraordinary strategic success. But the skeptic will say that it cannot be inferred that successful strategies are true ones. The skeptic might even insist that we cannot *know* that our strategies are successful ones. He can cheerfully concede that they may be the best that we have thought of as yet and still be unimpressed. As Lord Acton observed, in the long run we are dead.

There is a reply to this: It is bad *strategy* to bet the skeptic's option with respect to our prior judgements of strategic correctness. Of course, the skeptic can recur at this point and attack our meta-strategic animadversions. It seems that the skeptic can always catch his opponent at the next step. Should the structure of this pattern of attack turn out to be appropriately recursive, the skeptic can generalize to a guaranteed standoff with his opponent. No answer will adequately contain the skeptic.

It is not perhaps for the fallacy theorist to agitate himself over skepticism about objective truth. But he should lend every effort to rehabilitate objective access to judgements of strategic correctness. The skeptic argues that we cannot know that it is bad strategy not to trust in strategies we believe have worked well for us.

How do we block this move? It is possible that invocation of what has been called "one of Aristotle's best discoveries" will turn the trick against the skeptic [Anscombe, 1957, p. 58]. The discovery commended to us by Anscombe is the so-called *practical syllogism.* Aristotle is not very clear about the practical syllogism. For one thing, although Aristotle gives many examples of the premisses of a practical syllogism, he never states the conclusion of one, and at times says that they are not propositions, but actions. Anscombe gives an amusing contemporary example of what Aristotle had in mind [Anscombe, 1957, p. 60].

Vitamin X is good for all men over 60
Pig's tripes are full of vitamin X
I'm a man over 60
Here are some pig's tripes

With these as premisses, a practical syllogism is concluded by way of the speaker's choice of an appropriate action. And so — in conclusion, as we might say — he takes some tripes.

Aristotle doesn't say very much about the conditions whose fulfilment makes a practical syllogism a good piece of reasoning. But he does disallow that practical syllogisms are apodeictic or necessitating. If, in modern terms, this means that their premisses do not entail their conclusions, then Aristotle's demurrer is quite right. But if he means to deny that sometimes the premisses of practical syllogisms rationally compel the actions that are their conclusion, then we may join with Anscombe in dissenting from this claim [Anscombe, 1957]. So we will say that a practical syllogism is apodeictic or valid just in case its premisses rationally compel its conclusion.

Our skeptic is determined to uphold the position that so far as we really know our position is like the rat's, that strategic success in both cases may be rooted in false belief and may even require it. Thus, our skeptic invites us to assimilate the toddler case to the rats case. The core of the skeptic's position is that for all we know it is not true that the best meta-strategy is continuing fidelity to strategies that work.

Those who are of a mind to resist the skeptic may avail themselves of a single-premissed practical syllogism, as follows:

our various strategies have worked well so far,

the conclusion of which just is continued behavioural fidelity to those strategies. They may also think that the syllogism is apodeictic. If they are right, that counts as a strong reply to the skeptic. Are they right?

At this juncture it is helpful to re-invoke our distinction between 'painful' and 'painless' theories (see the preceding chapter). A theory is counted painful to the extent that, although it can be disbelieved, acting on the disbelief would be ruinous for survival. Painless theories are the obverses of these: Though they can be believed, acting on the belief would be ruinous for survival. A painful theory, paradigmatically, would be the ordinary causal theory of middle-sized, middle-distance physical objects of our common experience and routine affection. It is the theory that tells us to stay out of traffic and to feed the kids. A paradigm of a painless theory would be the Buddhist doctrine of the illusoriness of material reality and of the falsity of all our first-order beliefs about the material world. Though different, painful and painless theories can come together in mixed theories. Consider, for example, the surprising empirical fruitfulness of category theory in organizing the methodology of mathematical physics. Recall, too, that the permanent stoppage of the heart is explanatorily opaque without a theorem from topology.

5. Notional and Behavioural Belief

An equivocation lurks in the foundations of our distinction. A painful theory can be disbelieved but in some significant sense also cannot be disbelieved. A painless theory can be believed but in some significant sense also cannot be believed. Twice-over a distinction is evoked between *notional belief* and *behavioural belief*. Something is believed notionally when it is sincerely assented to and, behaviourally, if it is given dominant and routine strategic import, that is, if it is acted on in circumstances characterizing ordinary work-a-day life.

When we consider the skeptic's resistance to our claim that it is good strategy to bet the skeptic's option with regard to our prior estimates of strategic correctness, we can now notice that the skeptic is inviting us to relax our belief in our painful theories. The proposed relaxation is by way of an invitation to consider as possible a rival painful theory. But, as we have seen, a cognitive agent cannot take the invitation up, since he cannot behaviourally disbelieve his painless theories, or cannot anyhow short of grave pathological defect. Thus they are rationally compelling for him. It may be that our behavioural certainties with respect to our painful theories are functions of hard-wiring. Whether or not this is so, it is profoundly important that painful theories are rationally insusceptible of behavioural disbelief; and that one of our painful theories is that our painful theories are not just strategically correct, but true or approximately true.

Thus the skeptic's invitation to acknowledge the possibility that our correct strategies do not get at the truth of things is an empty possibility or, as William James has said, a dead option. What is offered as a possibility is a painless theory, one which a rational agent cannot believe behaviourally when it contradicts a painful theory. We now have a reason for saying that our practical syllogism is apodeictic. Its conclusion clearly implies that the skeptic's possibility incorporates a painless theory and thus a theory which a rational agent cannot act on.

It is worth noticing, as an aside, that the cognitive agent who, like the orthodox Buddhist, believes rival and mutually inconsistent theories jointly is in something of the same situation of the smoker who defects from his own anti-smoking policy, only worse. His situation is a standing invitation to accusations *ad hominem*. How well he might parry such charges will depend on what we make of notional belief conjoined with behavioural disbelief with regard to some or other common proposition. If the assent of notional belief is detached from any possible behavioural expression of it, that would seem to leave too little of assent to count as belief in any credible sense. There also arise obvious questions about self-deception.

It now seems possible to deal with a larger question. A skeptic about induction will grant the impressive strategic success of our inductive practices and may even grant the strategic advisability of persisting with those strategies. But, as before, he has a possibility to dangle before us. It is the possibility that our meta-strategy will not reflect the world as it truly is. This is true, but empty. The possibility incorporates a painless theory, a theory which a rational agent cannot act on and to which he cannot give his behavioural belief. The practical syllogism which concludes with our persistence with inductive practice is nothing short of rationally compelling — rationally compelling in the strategic sense. Thus is solved the problem of induction (and, of course, it is in all essentials Hume's resolution).

6. The Raven Paradox

With the solution of the problem of induction now in hand, are we out of the inductive woods, so to speak? Possibly not.

1 Consider the generalization, G, "all ravens are black" or, in the notation of standard logic, "$\forall x$ (raven $x \supset$ black x)".

2 We now define a *positive instance* of a generalization as the conjunction of the antecedent and the consequent of its scope, in which individual variables are replaced by individual constants. (Hence "Tweety is a raven $\wedge$ Tweety is black" is a positive stance of "$\forall x$(raven $x \supset$ black x)".)

3 It is a fundamental principle of confirmation theory that in the absence of any negative instance, every true positive instance of a gen-

eralization gives it some non-zero degree of confirmation, no matter how slight.

4 It is further asserted that the confirmation relation is closed under equivalence, i.e., anything confirming a statement to a certain extent also confirms to that same extent any statement equivalent to it.

5 By the Contraposition (also called Transposition) law of logic, generalizations are always equivalent to their contrapositives. Hence "$\forall x(Rx \supset Bx)$" is equivalent to its contrapositive "$\forall x(\neg Bx \supset \neg Rx)$".

6 Consider now the huge set S of conjunctive statements in which a contrary of 'raven' and a contrary of 'black' are asserted of an individual. This set includes, for example, "That is red and a shirt", "This is white and snow", "2 is prime and even", and on and on. Every statement in S implies a positive instance "$\neg Ba \wedge \neg Ra$" of the generalization, K, "$\forall x(\neg Bx \supset \neg Rx)$". And every successive positive instance of K raises its confirmation. [Note: check the definition of 'positive instance'. You will see that there is nothing in it which rules out positive instances containing the negation connective.]

7 Because K is G's contrapositive, and hence equivalent to it, and because confirmation is closed under equivalence, then the set of S confirms the generalization that all ravens are black.

8 The set of S contains a multitude of true sentences none of which mention any raven or any black thing. In other words, there is in S absolutely no reference to black ravens.

9 Hence S does not confirm the generalization that all ravens are black.

10 Line (7) says the precise opposite of line (9). Our argument reaches an inconsistent conclusion.

This is Hempel's Paradox. It raises a number of critical questions ([Hempel, 1937; Hempel, 1945], reprinted with changes in [Hempel, 1965, pp. 3–46]). One is whether confirmation is closed under equivalence. Another is whether confirmation theory can safely be transacted under the careful syntactic constraints which Hempel fashioned for it. A third is whether anything entailing a positive instance of a generalization is itself a positive instance of it. A fourth is whether anything entailing a positive instance of a generalization confers on the generalization the same degree of confirmation conferred on it by the implied positive instance of itself. I want here to take up the last three of these questions.

In the classic formulation, the Raven Paradox presents a huge relevance problem. Anything not black and also not a raven confers positive confirmation on the general proposition that all ravens are black. Since none of the statements reporting any of these nonblack and nonraven states of affairs so much as mentions a raven or a black thing, it can hardly be supposed that these propositions bear any relevance to the law that ravens are black. Yet without relevance, how can they be considered as confirmatory?

Whatever we decide to make of the irrelevance complaint, I want to press two different and, we may come to think, more powerful objections. One is a cardinality objection, which says that the sheer size of the confirmation class for "All ravens are black" constitutes the more wrenching difficulty. The other is an inconsistency objection, which alleges that the confirmation class for "All ravens are black" confirms pairs of mutually incompatible statements.

Problem 1. Let K be the class of true sentences of the form "$Fx \wedge Gx$" in which 'F' is a contrary of 'Y' and 'G' is a contrary of 'X'. Then each such sentence confers positive confirmation on any generalization (let's call it *gen*) of the form "For all x, if not-Y then not-X" and thence by contraposition on "For all x, if Xx then Yx". K is a confirmation class of *gen*. So is K^*, the class of *gen*'s observed positive instances. So we shall say that $K''=\cup(K, K^*)$ (i.e., the sum of K and K^*) is *the* confirmation class for *gen*. It is clear that K contains every true sentence about, for example, the natural numbers (we ignore the issue of indenumerable expressibility posed by the real numbers). Thus the confirmation class of *gen* is (infinitely) bigger than the class of its true positive instances K^*. Even if we decided to restrict the membership of K to sentences whose predicates are empirical, the problem remains. What *makes gen* true is vastly larger than what *gen* itself if true makes true.[4]

Problem 2. It is not true, in general the case, that 'F' is the sole contrary of 'Y', nor 'G' of 'X'. This is because contrariety is not equivalent to predicate negation. This being so, then for a considerable number of cases in which a member of K confers positive confirmation on "For all x, if not-Y then not-W", it will also confer positive confirmation on "For

[4]This might be a good place to touch on what I call Philosophy's Most Difficult Problem. Consider any valid argument whose conclusion is highly counterintuitive. There are two ways of interpreting this argument. One is to see it as a *reductio ad absurdum* condemnation of at least one of the implying premisses. Or it can be seen as a sound demonstration of a surprising — or even shocking — truth. The problem is that there seems to be no generally efficacious way to determine which interpretation is correct, case by case. Perhaps it is not wholly surprising therefore that whereas most philosophers of science tend to see the Raven as paradoxical, some see it as a demonstration of a surprising truth.

all x, if not-Wx then not-Vx", in which 'W' and 'V' are contraries of 'F' and 'G' respectively, and of at least one of 'Y' and 'X' respectively. Thus K confirms to the same degree inconsistent pairs of generalizations. Not only will "Red $x \wedge$ Shirt x" confirm "$\forall x$(not black x $\supset$ not raven x)" but also and to the same degree "$\forall x$(not green $\supset$ not raven)". Hence by contraposition we have confirmation for the blackness of all ravens and for the greenness of all ravens.

Problem 3. Of course, some people won't like restricting K to sentences whose predicates are empirical. Some will insist that 'empirical' is not a load-bearing qualification. Others will insist that even if it is, mathematical predicates hardly warrant banishment from one of the centre-pieces of scientific methodology. Suppose that we let these objections stand. Then $K''-K^*$ confers the same degree of confirmation upon *gen* as K'' itself does. That is, even if we remove from the confirmation class of *gen* every observed truth of the form "$Fx \wedge Gx$", the weight of confirmation for *gen* will not be one jot less than it was before. The *weight* of *gen's* confirmation owes nothing to what is made true by *gen* if *gen* is true.

Problem 4. We also run into some methodological heavy weather. The problem of multiple contrariness requires a theory of predicate-contrariety. Presumably it will have to include an account of genus-species contrarieties ('square' and 'circular'), determinate-determinable contrarieties ('black', 'green'), and other semantic kind-contrarieties ('raven', 'logician'). No such theory currently exists, and future prospects are not immediately encouraging. (Shades of the Logical Inertia problem of chapter 3.)

As long as K is allowed as a confirmation class for *gen*, problem 1 and 2 are insoluble; problem 3 may have a solution but this is of no help with 1 and 2; and problem 4 threatens to be at least a long-term difficulty. On the other hand, not permitting K to be a confirmation class for *gen* takes care of all four problems in one fell swoop. It is a course to be recommended. It gives us principled reason to say *either* or *both* of the following:

(a) A statement implying a positive instance of a generalization is not thereby also a positive instance of it.

(b) A statement implying a statement conferring positive confirmation on a generalization does not confer that same degree of confirmation on that same generalization. The converse of degree-of-positive-confirmation is not closed under deducibility.

The definition of positive instancehood, in any event, *requires* that we deny (a). The ugliness of our problems *argues* that we should also deny (b). In denying (a) and (b), our problems lapse entirely, including (4).

For if (a) and (b) are false, there isn't the slightest occasion to demand for confirmation theory a logic of species and determinates, determinates and determinables, or any other semantic kinds (notwithstanding attempts such as [Searle, 1959; Woods, 1974b]).

7. The Problem of Induction Again

Induction, or something like it, is a basic human skill. We cannot do without it. The skills of induction answer, as we have said, to an important distinction — a distinction between our cognitive and strategic endowments. Cognitive skills are used in getting at the truth of things and, in particular, encompass the talent for making good alethic arguments (Recall that 'alethic' derives from the Greek word for truth. An alethic argument is an argument for the truth of its conclusion). Strategic skills are employed in figuring out what is the best (or a better) thing to do and so, in particular, encompass the talent for making good prudential (or deontic) arguments ('Deontic' derives from the Greek sentence "it is right". A deontic argument is one whose conclusion is about what should or shouldn't be done). Alethic and strategic skills are frequently exercised jointly. Having got hives following a first encounter with strawberries and cream, a human agent may wish to know what is rationally best to do when offered a subsequent bowl. Whether to accept or decline will turn on what projections are true from the prior correlation between strawberries and hives.

Projections are the heart of induction. We make projections from observed cases to (as yet) unobserved cases. All ravens heretofore observed are black, so all the unobserved ones are likewise black. As a special case, we also make projections from past and present correlations to future correlations. All ravens presently and in the past have been black, so ravens in future will be black. Ravens are not, perhaps, an especially interesting case. Greater poignancy attaches to Bertrand Russell's chicken who, noticing that he has not yet died, predicts a good day ahead. But today an inhospitable farmer wrings the chicken's neck.

One of our most abiding inductions is that, since induction has been, so far, a good guide to truth and a good guide of action, it will continue thus to serve. One way of expressing such confidence is to hold that the future will resemble the past or that nature is uniform. It is noticeable with little effort that our confidence in the uniformity of nature is itself the result of an induction. Nature has been uniform so far, so nature will be uniform in the future. A problem is posed for inductive serenity. If one undertakes to justify inductive practices by appeal to abiding uniformities in nature, we embrace an inductive argument and the putative justification is circular.

Not everyone worries about the circularity. Goodman, for example, holds that the rules of induction are justified by their conformability to successful inductive practice, and that inductive practice is made successful by its conformity to correct inductive rules [Goodman, 1965, p. 64]. And that is that, the circle is 'virtuous'. Most writers are unconvinced by the 'that is that' approach to the justification of induction. But Goodman has pointed out another difficulty with induction, and most writers are convinced that here he is on to something that cannot so easily be dismissed. This further difficulty first saw the light of day in [Goodman, 1965, ch. 3], and it has come to be known as the *Grue Paradox*, to which we now turn.

Green emeralds confirm to some non-zero degree that all emeralds are green. This insight is reformulable as a central tenet of confirmation theory, already discussed in the previous section:

1. A generalization g is confirmed to some non-zero degree by its true positive instances.

A positive instance of the generalization:

2. For all x, if x is an emerald then x is green

is any true sentence of the form

3. a is an emerald and a is green

where, in (3), 'a' names the value of the variable 'x' in (2), and 'emerald' and 'green' are occurrences of the two respective predicates contained in (2).

It seems clear that the more positive instances there are for our generalization, the better-confirmed it is — provided, of course, that there is no true negative instance of it, that is, no true sentence in the form

4. b is an emerald and b is not green.

It is also clear that a given stock of positive instances can serve as confirmation for more than one sentence. It would seem that if some instances confirm a generalization g to some non-zero degree, they likewise confirm to that same degree any sentence either equivalent to g or entailed by it. What we may not wish to have, however, is a stock of instances confirming to some non-zero degree two sentences g and h where g and h are inconsistent with one another. The Grue Paradox suggests that whenever some instances confirm any sentence g to some non-zero degree there will always be another sentence equally confirmed that is incompatible with g. Some writers say that if this is so, no sentence whatever admits of any non-zero degree of confirmation, neither the sentence that all emeralds are green nor the sentence that all ravens are black, nor any other.

To see how this grim result is got, let us introduce the predicate 'grue'. We introduce it as a technical term and by stipulation, but there is no harm in that. Science bristles with artful neologisms. So we will say that

5. x is grue iff x is green and has been inspected or x is blue and has not been inspected.

The class of grue things is, then, simply the class of green inspected things and blue uninspected things. Straightaway we have it that

6. All inspected emeralds are grue since they are green.

By the central tenet of confirmation theory — that is, by proposition (1) — since all examined emeralds are grue, it will be confirmed to a non-zero degree that

7. All emeralds are grue.

Equally, since all inspected emeralds are green, it will be confirmed to the same non-zero degree that

8. All emeralds are green.

It is true that, when it comes to generalizations about the colour of emeralds, we favour (8) over (7). And it might be thought that this is explicable by the fact that 'green' is a familiar predicate and 'grue' is not. Furthermore, since a given body of evidence can confirm more than one statement, surely there is no harm in taking it that all the emeralds that we know about so far establish equally well the two claims that they are all green and that they are all grue. After all, the vital statistics of the New York Registry Office confirm that $n\%$ of adult New Yorkers are bachelors in 2000 *and* that $n\%$ of adult New Yorkers in 2000 are unmarried men who are not divorcés and not widowers.

It won't wash. The generalizations (7) and (8), though different generalizations, are also logically inconsistent with one another. To see this, let us define the 'next' emerald as some particular emerald not included in our emerald data-base. Generalization (8) implies that the 'next' emerald will be green and generalization (7) implies that the 'next' emerald will be blue. But nothing can be both green and blue, that is, green only and all over and blue only and all over.

Worse still, for any generalization

9. For all x, if Fx then Gx

which is confirmed by some data, d, there will always be another generalization, h, which is just like g except that its predicate 'Gx' is replaced by a predicate 'Hx' having the structural features of 'grue' with respect to 'green'. That is, 'Hx' will be defined thus

10. x is H iff x is G and inspected or x is I and uninspected, where 'I' is incompatible with 'G'.

Consequently, for all such g and h, anything, such as d, confirming g will also confirm h just as well; but since g and h are logically incompatible with one another, no data d will confirm either. No generalization whatever is confirmable, and this is bad news for science and practical affairs alike.

There has been a good deal of controversy about what exactly the Grue Paradox actually shows.[5]

The idea of the lines just above, that it shows that no generalization has a non-zero confirmation, has been resisted. This is worth pausing over. For why, it might be asked, must it be the case that if a body of evidence equally confirms incompatible propositions that neither has *any* confirmation? Does the detective, in acknowledging that his evidence to date equally supports the suspicion that Jones alone committed the crime and that Smith alone committed the crime, place himself in the position of having to say that, on that evidence, Jones and Smith are now completely in the clear? That Jones and Smith must now be dropped from the list of suspects?

It is worth noticing that confirmation is said to be closed under conjunction. That is, if

11. d confirms g

and

12. d confirms h

then

13. d confirms g-and-h.

If g is incompatible with h, then d will confirm a logically inconsistent sentence, namely, the conjunction of g and h. And surely, it might be insisted, logical inconsistencies are made distinctive by the fact that nothing whatever confirms them. Whether it would be profitable to persist with this line of thought will turn on whether confirmation *is* closed under conjunction. Though some writers think not, it is not essential that this controversy be settled. There is a prior, and perhaps more interesting, issue which commands attention.

[5]Goodman himself draws out a different moral. What is anomalous about the grue-case — and nothing less than an "intolerable result" — is "that anything confirms anything" [Goodman, 1965, p. 75]. For example, on the basis of our observations about emeralds to date we will be able to confirm that roses subsequently inspected will be blue [Goodman, 1965, p. 74 n. 10].

The prior issue is that

14. Competent inducers predict not (7), but (8)

and

15. It seems that they are right to do so.

If so, then

16. The central tenet of confirmation theory, that is, proposition (1), is defective because it does not discriminate between green-cases and grue-cases, and so does not explain why it is we treat (and are right to treat) the cases differently.

By these lights, the principal challenge of the Grue Paradox is the rehabilitation of principle (1) or, more carefully, its replacement by a more explanatory and discriminating successor.

It should be emphasized that an acceptable solution of the Grue Paradox must succeed in doing two things:

(a) it must replace (1) with a principle that acknowledges our inductive favoritism for 'green' over 'grue';

and

(b) it must establish that the new principle is objectively preferable to the old one.

Goodman's own solution appears to fulfill condition (a) only. This is the doctrine of entrenchment. Induction, says Goodman, is the manipulation of entrenched predicates [Goodman, 1965]. 'Green' is entrenched, 'grue' is not.

What makes for a predicate's entrenchment? Some writers, though not Goodman himself, thinks that it has something to do with the complexity of its extension, that is, of the class of things of which the predicate is true. The class of green things is fairly homogenous, notwithstanding that it comprehends many different kinds of things — from lizards to greenbacks and beyond. On the other hand, the class of non-green things is sprawlingly more complex. It comprehends everything whatever minus just the green things.

Goodman sometimes speaks of the projectibility of predicates. 'Green' is 'projectible'; 'grue' is not. If by 'projectible' it is meant that a projectible predicate is one proof against 'grue'-some exploitation, this is true but unhelpful. For then the claim that induction should be confined to projectible predicates is no more than the claim that our inductions shouldn't be made by way of predicates that land us in the Grue Paradox. Why they shouldn't is left unexplained.

If we take 'non-green' as a paradigm of an unentrenched predicate, it may be thought that owing to the vast heterogeneity of their extensions, unentrenched predicates cannot be efficiently reasoned with, that their extensions are too complex for human beings to 'get their minds around', so to speak. This cannot be right. We have no difficulty in reasoning that the complement of the class of green things is precisely the class of non-green things.

It may be that a mediaeval distinction will here do us some good. The mediaevals distinguished between *ordo essendi* and *ordo cognescendi.* By *ordo essendi* was intended a ranking of all the things there are in the order of their ontological priority, with God occupying first place. *Ordo cognescendi* ranked items of knowledge by way of their epistemic (or in some versions their learnable) priority. Given this distinction, it could be ventured that entrenched predicates possess learnable priority over unentrenched predicates, that they dominate in *ordo cognescendi.* Thus we have an empirical thesis: Entrenched inductive vocabularies are more learnable than unentrenched inductive vocabularies. True or not, the hypothesis leaves untouched the task of showing that unentrenched inductions are somehow objectively bad inductions.

Granting, still, paradigmatic status to 'non-green' as an unentrenched predicate, it is clear that 'grue' makes no serious claim upon that status. The grue things are only slightly less homogeneous than the green things, and such heterogeneity as the class of grue things may possess is no obvious impediment to the mind's encompassing stretch. So it would seem that we haven't yet got right the concept of entrenchment.

What, then, is wrong with 'grue'? Let us press the question of what exactly it is in virtue of which green things also count as grue. It is the fact that they have been inspected. It becomes noticeable at once that the property of being inspected is in some sense less a property of emeralds than the property of being green is. Green emeralds are green independently of whether anyone sees it so. Emeralds that are inspected are not so short of someone inspecting them. It is true, by the way, that colour-properties may require perceptual inspection in order to be fully instantiated, but differences persist nonetheless. Perception-independent colour properties can be said to be dispositions of the things having them, whereas the property of being inspected is certainly not the disposition to be inspected or any such thing.

Still, we haven't yet got to the heart of the matter, for it cannot be said that, different though they are, being green and being inspected are not alike objective facts about the properties of the things having them. What can be said, however, is that given that all inspected emeralds are green, it is obvious that the uninspected ones lack a property thanks to which all the inspected ones are grue; and that this is the property of being inspected. At this juncture, the idiom of projectibility presses for reinstatement. The predicate 'grue' is now seen to be unprojectible in a clear-eyed way. Consider just the green things that are also grue. If we seek to generalize upon what we know to be grue, we will generalize upon the conditions in virtue of which the known things *are* grue. Let us say that inductions are projective just when they are generalizations in this way. And let us also say that predicates are projective just when they enter into inductions of just this character. One of those conditions in virtue of which the emeralds of our data are grue is that of having been inspected. Having been inspected is thus a necessary condition of a green thing's being grue. But it is precisely this condition that is *not* generalized in the induction from our data to the statement that all emeralds are grue. The induction is, therefore, not projective. Not only is the predicate 'grue' not projective, it cannot be. Any projection of the property of being inspected beyond our data will run afoul of the very meaning of 'grue'. Thus 'grue' is unprojectible.

We may grant that sense has now been attached to the notion of unprojectibility, and that we now have the means of rewriting the principal tenet, (1), of confirmative theory. Instead of saying that generalizations are confirmed by their positive instances, we can say

1.* Generalizations are confirmed by their *projectible* positive instances,

where a projectible positive instance is a positive instance all of whose predicates are projectible.

That said, it may not yet be clear as to why (1*) is an objectively preferable induction rule than (1). This can be shown as follows. We must ask ourselves "What would falsify the generalization that all emeralds are grue?" Suppose that the next-observed emerald were blue. This would falsify the generalization that all emeralds are green and, so too, that all emeralds are grue, since our "new" emerald has been inspected and found to be blue. What we could not do is falsify that all emeralds are grue by finding a blue emerald that hasn't been inspected. Finding such an emerald just is inspecting it. So the generalization that all emeralds are grue is unfalsifiable by way of the failure of one of its defining conditions. That all emeralds are grue is a phoney induction. That said, we have ready to hand a further and equally central tenet of confirmation theory:

17. No generalization g is confirmed by any data unless g is falsifiable.

The objective preferability of (1*) over (1) as a fundamental principle of confirmation is now straightforwardly established. Principle (1) violates (17) and principle (1*) fulfills it. The Grue Paradox is disarmed.[6]

This returns us at once to the alleged virtuosity of the circular justification of induction itself. Induction rules are justified by their compliance with inductive practice; and inductive practice is justified by its fulfillment of the induction rules. If we take Goodman's solution to be an answer to the question of why 'grue'-some generalizations are less good than the others, we must interpret that solution in the following way: Projectible inductions are better than the others just because they are the inductions that constitute established inductive practice.

But Goodman himself has undermined his own solution, so construed, and his justification of induction, as well. "Why", he asks, "worry about... 'grue'? ... since we are unlikely to use them in making predictions...? We get along well enough without them in daily life and in scientific research" [Goodman, 1965, p. 80]. True enough, but

> if we seek a theory at all, we cannot excuse gross anomalies resulting from a proposed theory by pleading that we can avoid them in practice [Goodman, 1965, p. 80].

If we count Goodman's solution of the Grue Paradox as part of his theory of confirmation, and if we count his justification of induction as part of this theory of induction, then Goodman is hoist on his own pétard. In each case, the justification of what we do is rooted in the commonplace that it is, after all, what we do. In the particular case of 'grue', 'grue'-some generalizations are less good because the grue things are less well entrenched.

And the grue things are less well entrenched because we make 'grue'-some generalizations with strikingly less frequency than non-'grue'-some ones. And so, in the end, is that all there is to it?

8. Strategic Rationality Again

It wouldn't surprise me if some readers were unconvinced by our purported solutions of these two paradoxes. Certainly the case that I have made against Hempel is not a strict proof that Hempel's argument is flawed. I rather doubt that a strict proof is possible. The same can

[6] An historical note. The solution here proposed is not Goodman's own. For one thing, although it is argued that 'grue' is unprojectible, Goodman doesn't want to rule out 'grue' completely: 'grue', he thinks, is sometimes projected. Goodman's own solution is a matter of degree. 'Grue' is *less* projectible than 'green' and this is because green things are more entrenched than grue things. But saying this is only to say that the frequency of projections from 'green' is greater than those from 'grue'.

be said for my solution of Grue. But I also think that it is possible to strengthen the case against Hempel's and Goodman's arguments in ways that satisfy appropriate standards of plausibility.

We say that our treatment of the Toddler Problem and relatedly, of Hume's Problem of Induction, required the deployment of a concept of stragetic rationality. Strategic rationality in turn is an *economic* concept, a concept of rationality involving estimates (often tacit) of cost and benefits. I think that we can put this notion to good use in the matters before us.

Suppose that Hempel's argument were sound. Then there would be two equally good ways to determine the confirmation of the generalization that all ravens are black. We could examine some ravens and check for colour. Or we could examine some non-black things and check for non-ravenhood. If we went the first way we would be trying to sample a population of (let's say) 100,000,000 ravens. If we opted for the second procedure, we would be sampling a massively larger population of, say, a number whose numeral is '1' followed by forty zeros (i.e., 10^{40}). (I have borrowed these numbers from [Rescher, 1978, ch. 4].) Let us now say that an adequate statistical control of a population that is N, is $\sqrt{N}$. Then to exercise the first option requires a controlled sample of 50,000,000 ravens. But the second option requires a sample of 10^{20} non-blacks. Now these numbers are for illustrative purposes only. The only thing that really matters is the scale of their difference. Given the hugeness of the difference, it is simply too costly to assemble the sample required by option two. And that is a *strategic* reason to favour option one.

Against this it might be argued as follows.[7] If you take the line found in most statistics textbooks, for a fixed confidence level, or sampling margin of error and confidence level, population size becomes irrelevant beyond a high enough finite number — a million, say. If this is so, it couldn't matter whether the population is 100 million (as with our raven example) or 10^{40} (as with our non-black things example). But perhaps this is a textbook simplification. On the other hand, suppose that our claim about population size is indeed correct. Do we then have a further puzzle. Suppose we have a simple random sample of 1400 ravens and an equal number of non-black things. Suppose that all 1400 ravens are black and all 1400 non-black things are non-ravens. Suppose that we now inductively generalize to the respective populations. If what I've been suggesting here is true, then the sampling error for the non-black

[7]As did Leslie Burkholder in correspondence, for which my thanks.

population will be a great deal bigger than for the raven population. But given that the generalizations are equivalent, how can this be? Why would the sampling margin of errors at a fixed confidence level not be the same? The answer is the same as before, I suggest. Contrary to what is commonly supposed in statistics, the irrelevance of population sizes beyond a certain (not overly big) finite number is indeed a simplification.

Now for grue. We suggested above that a 'grue' colour-taxonomy is more difficult to learn than a 'green' colour-taxonomy. This is true and is made true by the fact that the latter type of vocabulary, but not the former, can be learned by ostension. One learns what green things are by attending to their phenomenal appearance. What things are grue *cannot* be learned in this way. In order to determine whether the *phenomenal* colour of a thing underwrites to a non-zero degree the proposition that all things of that kind have that colour, I must confine myself to the 'natural' colour taxonomy rather than a 'grue'-some taxonomy. So it is very much more costly in terms of time and investigative effort to make 'grue'-some inductions, since they can't be made on the basis of the phenomenal appearance of the things in the sample.

This gives us a simple and direct way of describing entrenchment. Observational vocabulary is entrenched if it is economical in the sense at hand. All of which take us full circle: Entrenched predicates are more congenial to hasty generalization, or what we have been calling primitive induction, than unentrenched predicates.

9. Conclusion

Our discussion of hasty generalization raises the question whether hasty generalization is always bad inductive practice. The paradoxes we have been examining, as well as Hume's celebrated problem, bear on this issue rather dramatically. If Hume's problem cannot be solved, and if Hempel's and Goodman's paradoxical arguments can't be dented, then the analysis of hasty generalization is trivial, and is so for the following reason:

1 All inductions are defective (or unjustified)

2 So, hasty inductions are defective (or unjustified). Full stop.

This has the feel of a wrong answer. It pins the blame of hasty generalization in the wrong place. It pins it not on haste but rather on induction. Whether we like it or not, hasty generalization takes us to the very heart of inductive logic. If the fallacy theorist wishes to say something deep about hasty generalization he had better be prepared to tackle these larger and more central issues.

Unless I am mistaken in what I have said about Hume's problem and the inductive paradoxes, we can now pin the blame of hasty generalization where it belongs, namely, on haste. But we must do so with appropriate circumspection. For some hasty generalizations are strategically rational, and not all are cognitively shaky, especially when one is claiming a plausible (rather than truth-preserving) connection between one's generalization and frame-considerations, such as the standard number of legs for tigers.

Chapter 20

THE PROBLEM OF ABDUCTION

I said in the previous chapter that the inductive logic project would be too much to try to take on in a single chapter. As anyone knows who has undertaken to move that project ahead, this is an understatement. At the conclusion of that chapter, I proposed that the badness of hasty generalization should not be attributed to the badness of all inductions, hasty or not. This clearly presupposes that some inductions are good. How this assumption can be made to square with Hume's infamous argument that induction as such lacks a justification is the business of the present chapter to determine. The reader may be puzzled by our decision to get this determination going by way of discussion of abduction, rather than induction. But, as will be seen, it is the right way to go.

1. Abduction

The term 'abduction' was introduced into logical theory by Charles Peirce late in the 19th century. The introduction was not wholly original, since 'abduction' is a passible translation of Aristotle's *apagogē*, which is also translated as 'reduction' and was given the Latin rendering *abductio* by Julius Pacius.

Peirce thought the following to be typical of an abductive inference:

> The surprising fact C is observed. But if A were true, C would be a matter of course. Hence there is reason to suspect that A is true [Peirce, 1958, pp. 5-189].

Critics of abduction are quick to point out that there exist interpretations of 'A' for which the second premiss is true, or at least plausible, but for which the conclusion, in spite of its tentativeness, is not at all plausible. This is so when all that is required of premiss two is that it exhibit a true conditional from A to C. But this is not what Peirce intended for premiss two. He required that it present a true conditional from A to the claim that C *would be a matter of course.* Given the evident importance of the qualification, it is regrettable that Peirce says little by way of direct clarification. However, there is reason to believe that what Peirce sought to capture with his paradigm was the idea of abductive reasoning as inference to the *best* explanation, in the manner of [Harman, 1965]. For my purposes in this chapter, such an interpretation of abduction will be sufficient.[1]

Some commentators regard abductive reasoning as a species of inductive reasoning, where inductive reasoning is understood in a generic sense as any kind of ampliative reasoning. In one way, it hardly matters how we distinguish abductive from other kinds of reasoning, provided that we attend to what makes them different. Others are of the view that such differences are sufficiently marked as to justify a taxonomy in which abduction and induction are different species of ampliative reasoning, where induction is now understood as non-demonstrative reasoning from samples (sample-induction for short).

I want to begin this note by examining briefly a controversy over the design argument for the existence of God, and, in particular, Hume's celebrated criticism of it. What I wish to point out is that if the design argument is taken sample-inductively, Hume's derision is justified; but if the argument is given an abductive interpretation, Hume's criticisms are unavailing. If I am right in so saying, we have a principled reason to favour a taxonomy in which induction and abduction are distinct species of ampliative reasoning; for it is a distinction whose relata produce utterly different verdicts for the design argument.

The design argument has been advanced since antiquity by thinkers of the first rank, although perhaps the best-known version of it is the argument of [Paley, 1822]. Hume's celebrated attack preceded Paley's book by twenty-six years [Hume, 1947]. So Hume was not attacking Paley. Even so, there is no reason to think that there is anything in Paley's actual argument which would have made Hume re-shape his criticisms. I shall use Paley's version, ignoring the anachronism.

Hume takes the design argument to have the following schematic form

1 Watches are the result of intelligent design

2 Watches and organisms are similar to each other

3 Therefore, organisms are the result of intelligent design.

---

[1] Although it can be claimed with confidence that there are also non-explanationist variations on abductive reasoning, we shall not here require this more capacious conception. See [Gabbay and Woods, 2004, ch. 4].

Hume sees the design apologist as holding that the argument at hand is inductively strong and that its degree of strength is the same or approximately the same as the degree of similarity between watches and organisms. One of Hume's criticisms is that the degree of similarity between watches and organisms is extremely small. Accordingly, the design argument is very weak at best.

Hume marshalls a second criticism against the design argument. Although he does not frame his objection in the terms that I shall use in formulating it, there is no serious doubt that the formulation is wholly faithful to Hume's intent. Hume's point is that the design argument commits a serious sampling error.

If we are to suppose that in the real world, in the world as we actually have it, organisms are the product of design, we would have had to examine other worlds to see whether the target correlation also obtained there. Since this is impossible, the argument's sample size is zero.

Hume has other objections to the design argument, which I shall not take time to review here. Suffice it to say that all of Hume's criticisms fail if the design argument is interpreted, not inductively, but *abductively*, as an inference to the best explanation. So construed, the argument displays the following form.

1 The watch is complex and well-adapted for telling the time

2 The watch is the result of intelligent design

3 The watch is the result of random processes.

Paley holds (in effect) that the likelihood of (1) given (2) is substantially greater than that of (1) given (3). And he concludes, therefore, that (2) is the more reasonable hypothesis. Likelihood is not probability. As here used, it is a measure of conjectural strength. Paley applies the same argument to organisms. Which, he asks, is the hypothesis that better explains the fact that organisms, too, are complex and well-adapted for survival and reproduction. Is it the hypothesis that organisms result from random processes? Or is it the hypothesis that organisms are the product of intelligent design? Paley's answer is well-known. Given that the Darwinian hypothesis, was not available for another fifty-four years, it can be argued that Paley was right to chose the design hypothesis over the sole alternative that was actually in contention [Darwin, 1964]. And, so construed, Hume's criticisms of the argument fail trivially.

This chapter has a three-part organization. In Part I, I present a justification of induction which, I conjecture, would have satisfied Hume had he thought to give to abductive reasoning a place in the argument for justification. Part II is given over to objections and replies. Part III concludes with some observations about skepticism.

2. Induction Has a Justification

If the argument of the past several paragraphs is sound, then it must be granted that Hume's rather spectacular failure to disarm the design argument derives from a blindspot about abduction. It is scarcely credible that Hume would have been able to persist with his criticism of the design argument had he chanced to think of the abductive option. What is more, since Hume evidently missed the abductive boat in connection with the design argument, it would be interesting to consider whether he also missed that same boat in other matters of critical moment. Did he, for example, display this same blindspot in the reasoning that constitutes Hume's Problem of Induction?

Hume wished to know whether there exists an epistemic justification of the reliance we place on our standard inductive practices. Collectively this is the problem of determining whether induction is provably rational. Of course, Hume's answer was in the negative. He asks his readers to consider the set of conditions under which his question could attract an affirmative answer.

I. There exists a sound deductive argument for the proposition that our inductive practices are rationally reliable.

II. There exists a tenable inductive argument that our inductive practices are rationally reliable.

Hume's attack on induction takes the form of denying that either of these conditions is satisfied and that jointly they exhaust the relevant justificatory options. In what must surely be one of the first things philosophy undergraduates come to learn about Hume, he claims that condition II is unsatisfiable, owing to the intrinsic circularity of an inductive argument that sought to establish the justification of, among other things, inductive arguments.

The trouble with condition I is not as well-known or as easy to fathom. It is a condition that sustains two readings. One is that no deductive argument could *possibly* turn the trick for induction. The other is that there is no *known* deductive justification of induction. It may be that Hume leans toward the first of these readings, and does so on the strength of a style of argument that has come to be known as *autoepistemic*, itself a benign variant of the modern conception of the *argumentum ad ignorantiam.*

1 If there were a sound non-question-begging defence of induction, we would know of it by now.

2 But we don't.

3 So there isn't.

Counting against the present conjecture is that, although the autoepistemic argument is deductively valid, and sound if premiss (1) is true, it can be argued with some plausibility that, in pleading (1), Hume would have committed the same offence that he so emphatically asserts would be committed by any defence of induction that satisfies condition II. For how else would premiss (1) of the autoepistemic argument be sustained, if not on some purportedly *inductive* principle that realities (of certain kinds) have a disposition to reveal themselves in a timely fashion?

My principal purpose here is not to assess Hume's argument against the rationality of induction (though I will return to this issue briefly toward the end of the chapter), but rather to raise the question whether Hume is right in supposing that there are just two *types* of argument that could be marshalled in favour of induction (or of anything else, for that matter); and whether should Hume's assumption of exhaustiveness prove incorrect, there might be some new hope for the justification of induction.

We say that in attacking the design argument, Hume took it to be an argument of a certain type, and in so doing quite overlooked the possibility (indeed the fact) that it is an argument of a different type, of a type known as abductive. This was Hume's blindspot towards the design argument. Is it possible that this same blindspot affected (and spoiled), his treatment of the Problem of Induction?

There is reason to think so. If we grant the distinction between sample-induction and abduction, we see that there is a third condition under which it may turn out that the problem of induction is solved, where induction *here* is generic induction or that form of ampliative reasoning exhausted its species, sample-induction and abduction.[2] The third condition turns out to be a pair of conditions that vary systematically depending on which species of (generic) induction is up for justificatory consideration.

Here is the defence of generic induction that Hume's blindspot for abduction caused him to miss. For ease of reference I shall call it "Hume's Lost Chance for Induction" or *LCI*, for short.

1 Induction is either sample-induction or abduction and nothing else.

2 Consider, first, sample-induction.

   a. It is not in doubt that on the whole our sample-inductive practices to date have been reliable.

[2]Perhaps sample-induction and abduction do not in fact exhaust the class of ampliative modes of reasoning. If not, the case that I am trying to make easily extends to this more capacious inductive universe.

b. The reliability-to-date in those practices is best explained by their general rationality.

c. Accordingly, sample-induction has a tenable *abductive* justification.

3 Consider now abduction.

a. It is uncontested that on the whole our abductive practices to date have been reliable.

b. Their continued reliability can be inferred from this sample.

c. Accordingly, abduction has a tenable *sample-inductive* justification.

4 Thus, sample-induction has a tenable, non-circular abductive justification, and abduction has a tenable, non-circular sample-inductive justification. Since this is all there is to generic induction, generic abduction has a tenable, non-circular justification.

It takes little effort to see that the form in which I have constructed *LCI* omits some important qualifications and nuances. So, for example, I haven't made it explicit that the proposed justification is not a strict proof, but is a proof in a looser non-deductive sense, and that the claim of inductive justification is a claim whose instances are sustainable only defeasibly. Nor have I sought to make explicit the role in these matters played by the so-called Principle of Uniformity. These omissions have not been made out of carelessness, but rather in the belief that, for purposes of appraisal, enough of the structure and detail of the defence of induction is present in our present schematic version of it, *LCI*. Perhaps I am wrong in so thinking. After all, there are different ways of composing samples and, correspondingly, different ways of producing inductions from them — projectively, by generalization, statistically, and so on. Similarly, a defeasibly rational practice is one whose precepts will occasionally let us down. But there are practices and practices, and correspondingly different thresholds of rationally tolerable setbacks. In saying nothing further of these things, do I not run the risk that an appraisal of the schematic structure of the present argument would be compromised by consideration of these other matters? After all, am I not here claiming that Hume's own approach to induction was too coarse-grained and that the argument he mounted against the rationality of induction *would* have been compromised, indeed overturned, had he paid attention to a distinction that even a slightly more detailed approach to induction would have made apparent?

Of course. But appraisal must start somewhere. I shall stay with the schematic version of the argument because even in this less than fine-grained version, it attracts objections. If those objections proved sound, the need for more detailed scrutiny of nuances would have lapsed. What, then, are the objections?

3. Objections

Objection 1. The argument assumes a "fact not in evidence" (as the lawyers say), namely, that inductive practices of the one kind are justified by an induction of the other kind.

But what hasn't been shown is that the type of induction that establishes the intentionality of the other type is itself well-justified.

Reply 1. On the contrary, sample-induction is justified abductively, and abduction is justified sample-inductively.

Objection 2. Hume's objection against an inductive justification of induction was, quite rightly, that it was achieved by an argument that is circular in one step. In the argument that (I say) Hume should have considered, the circularity is achieved in two steps, even if not in one.

Reply 2. If in the argument that (I say) Hume should have considered, there is no argument circular in one-step for the justification of either kind of induction, then there are embedded in that larger argument, *LCI* two non-circular and apparently tenable justifications. How can an argument made up of non-circular subarguments be made circular by them?

Objection 3. That's easy to answer. What *LCI* proposes is that sample-induction is justified if abduction also is, and that abduction is justified if sample-induction likewise is. So the schematic argument is a circular argument in the form,

1 If S1 then A and if A then SI
2 Hence S1 iff A.

Reply 3. That's not circularity. It's a proof of the justificatory equivalence of sample-induction and abduction. (A rather nice bonus!)

Objection 4. Be that as it may, it surely does not follow that if an argument has no circular subarguments, it itself is free of circularity. Consider the analogy with the consistency of sets. It may be that on a given partitioning of a given set S all proper subsets are consistent and yet S itself is inconsistent.

Reply 4. Yes, care needs to be taken not to rely wholly on the non-circularity of *LCI*'s proper subarguments in claiming that *LCI* itself is circularity-free. However, if *LCI is* circular, the onus is on its critics to demonstrate it. What has been shown so far is not *LCI*'s circularity but, to repeat the point, the justificatory equivalence of both forms of generic induction.

Objection 5. You're missing the point. If you justify a certain kind of epistemic practice by appeal to the justification of another kind of epistemic practice, then your appeal is *unfounded*, if it itself can't be justified, and is *circular* if *its* justification is by appeal to practices the rationality of which is secured by appeal to *it*.

Reply 5. Consider an analogy with geometry. Riemannian geometry is consistent relative to Euclidean geometry, and Euclidian geometry is consistent relative to Riemannian geometry. There are two points to take particular note of. One is that these proofs of relative consistency are not proofs of the consistency of the one geometry relative to *a proof* of the consistency of the other. (There are no such proofs.) The second is that the lack of a one-shot consistency proof for all of geometry or geometry as such is insufficient grounds for the skeptical assertion that our geometric practices cannot be supposed to have any kind of rational justification.

Taking these points in reverse order, the same would appear to be the case with induction. That there is no one-shot proof of the rationality of induction as such is insufficient reason for skepticism across-the-board. The two varieties of induction have relative justifications. If this sort of relativity is good enough to deter blanket skepticism about the consistency of geometry, it is hard to see why the same slack should not be cut to the rationality of induction.

The other point tries to make something of the distinction between practices that are rational and *proofs* of their rationality. The same distinction is evident in the difference between relying upon a justification furnished by a principle of such a practice and relying upon a *proof* of its justification. In general, inductive reasoning is a form of reliance in the first, not the second, of these two senses. When we infer from a stratified random sample of appropriate size in which 70% of the F are Gs, that something close to this percentage obtains in the population at large, we are employing a justified principle of statistical reasoning. We are not employing the proof (such as may be), that that principle is justified.

It is the same way with justification of induction. When we justify sample-induction, we employ rational principles of abduction. We do not employ the proof that abductive principles are rational. Similarly, when we justify abduction, we employ rational procedures of sample-induction. We do not employ proofs of their rationality.

If, in either of these situations, we *did* employ proofs of the rationality of the principles we have employed, our justifications *would* be circular. But as we have seen these respective justifications go through well short of any such over-arching reliance on such proofs. So, sample-induction is justified by abductive reasoning, concerning which there is a proof of its rationality; and abduction is justified by sample-inductive reasoning, concerning which, too, there is a proof of its rationality. But these proofs *aren't* involved in these respective justifications.

Thus is solved the Problem of Induction.

Objection 6. Not so fast. If the object of the present exercise is to construct an argument on Hume's behalf, which, Hume himself would certainly have produced had he not had a blindspot about abduction, it is not enough that the argument produced here, *LCI*, gives a plausible defence of induction. It is also necessary to show that it is an argument of a sort that Hume would have found appealing. Another way of saying this is that the mere recognition of abduction as a form of ampliative reasoning is insufficient to make of *LCI* an argument to which Hume would have been drawn. In fact, even if it had occurred to Hume to give an expressly abductive justification of sample-induction, this alone does not show that Hume would have thought this abductive manoeuvre to be *justified.*

Reply 6. Of course. The heart of *LCI* turns on the distinction between a mode of justification which draws on principles which are themselves rational and a mode of justification which draws on a proof of the rationality of those principles. In the defence of induction presented here, the argument exemplifies the first sense of reliance but not the second. The question is whether there is any reason to suppose that Hume himself would have sanctioned this *modus operandi.*

There is. To see that this is so, it sufficies to re-examine Hume's original argument. One way, he said, in which induction could be justified (if it could be justified at all), is by way of a sound deductive argument. This would be an argument in which the conclusion asserted or immediately implied the justification of induction by deductive consequence from other propositions that are not in doubt. Having considered this option, Hume gave up on it. Why? As we saw, there are two possible explanations. One is that neither Hume nor anyone else knows of such a proof. The other is that Hume believes (without showing) that there *is* no such proof. But what there is no evidence for is a third possibility in which Hume requires that a deductive justification of induction must rely directly on a proof of the justification of *deduction.* What is wholly evident is that Hume believes that deduction *is* justified, never mind what else he may have thought about the provability of its justification. Thus were a deductive proof of the rationality of induction (not of its own rationality) ready to hand, Hume would have grasped it, and that would have ended his skepticism about induction. What this tells us is that in our choice of the principles that do justify induction, it is unnecessary (and in some cases, question-begging) to include a proof of their own justifiability. Had the derived deductive proof been available to Hume, he could have said that there is a justification of induction relative to deduction. He would *not* have said that there is a justification of induction relative to a proof of the justification of deduction. Accordingly, Hume's reasoning about a deductive justification of induction has the same structure as our two justifications of the varieties of induction set out in *LCI.* It is a *modus operandi* of which Hume would have approved.

4. Skepticism

In all that has so far passed between my re-jigged Humean and his imaginary critic there has been an assumption that it has not occurred to anyone on either side to question. It is the assumption that a best-explanation inference is always a relevant consideration in the context of an assessment of the justification of induction. Thus, even if we grant that the best explanation of the reliability to date of our sample-inductive practices is that those practices are reliable in the general case, and even if we grant that there are lots of contexts in which it would be reasonable to infer the truth of this best explanation, it may not be an allowable inference *in the context of a skeptical attack* on sample-induction.

Our question now is whether explanatory attractiveness trumps skeptical force. There is reason to think that it does not, contrary to the claim advanced by *LCI.* Counting against inference to the best explanation in the context of an examination of the *bona fides* of induction is that the very idea of explanation is saturated with presumptions of inductive normalcy. For large ranges of cases, what best explains something does so only if it is inductively normal. What best explains that my pet raven Fred is black is that ravens are black. What best explains Harry's bleeding nose is that Hector punched it and punching noses cause them to bleed. And so on.

Sometimes an attack on induction is an argument on behalf of *skepticism* about induction. In its most damaging form, a skeptical attack on the justifiability of a practice is an argument to the effect that even if those practices seem to us to be perfectly all right, the proposition that they *are* all right does not meet some contextually indicated standard of proof.[3] In offering a best-explanation defence of such a practice the

[3]Of course, the central question for skepticism is what is taken as the standard of proof, and what it is that makes it an appropriate standard to invoke. In the case of skepticism about induction, the standard of proof is irrefutable proof of future reliability. There is a huge literature about this.

defender fails to take the measure of the critic's skepticism. He fails to see that he is begging the question against the skeptic.

A skeptic who has an adequate appreciation of his own argumentative designs should be willing and eager to press the following argument as a counter to the best-explanation move:

1 Suppose that, contrary to how things appear, our inductive practices are in fact untenable.

2 Then it cannot be the case that even if the best explanation of their tenability to date is their tenability as such, their tenability as such is a correct inference.

3 Given what we presently know, it can't be ruled out that the proposition assumed in (1) will turn out to be true.

4 That fact would be wholly compatible with the other fact that our inductive practices so far give no comfort to that supposition.

5 Consequently, the proffered best explanation cannot overturn skepticism with regard to induction.

The argument here sketched would also seem to generalize to skepticism about causality, the external world, memory, and so on. In so doing it furnishes us with a limitation theorem for abduction.

Limitation Theorem Abductive reasoning is insufficient to overcome skeptical claims (whether about induction, causality, the external world, memory, etc., etc.).

Still, it would be folly to make too much of the *Limitation Theorem.* It is a discouragement to those who would offer an abductive answer to skepticism about induction. It is no discouragement at all to those who would offer an abductive answer to the claim that induction lacks a justification. It is true that if skepticism about induction holds water then induction lacks a justification. It is also true that, if induction does indeed lack a justification, skepticism with regard to induction holds water. But what is *not* true is that any way of successfully *countering* the claim that induction lacks a justification must also be a way of successfully *countering* skepticism about induction. One response, not uncommon to skepticism among philosophers, is to grant it and forget about it. In the case before us, granting it would amount to a concession that the reliability of induction fails to meet a certain standard of proof. Forgetting about it would amount to pressing the abductive justification of induction *even so.*

It is textually ambiguous as to whether Hume was a skeptic about induction or some looser breed of non-justificationist. Given the weight of his skeptical leanings in other matters, perhaps we should also give the nod to inductive skepticism as well. But as we now see, if our own interest is merely in whether induction has a justification, then the answer is that it does. It is the justification furnished by our argument *LCI*.

EPILOGUE: THE WAY AHEAD?

We have seen that a recurring criticism of the Woods-Walton Approach to the investigation of fallacies is the general lack of success — indeed lack of interest — in producing a unified account of the fallacies. In a number of earlier chapters I have tried to handle this objection as best I can. But I must admit that I have never been completely "sold" on this complaint. By and large, a theory should minimize differences only with considerable care, just as it should postulate similarities also with requisite circumspection. Of course, as with all good policies, there are exceptions. One of the most interesting is a theoretician's opting for undisguised, outright *reconceptualization* of the theory's target concept. If for various reasons, it strikes the investigator that his target concept, K, will continue to resist solid theoretical articulation, then he may propose to drop K in favour of a successor-concept that he thinks he can make more of, theoretically speaking. It was in this spirit that eliminative materialists gave up on the ordinary notion of belief as a theoretically realizable target of scientific psychology, in favour of the ontology of the emerging brain sciences. Pragma-dialecticians have done much the same thing with the concept of fallacy. Rather than plugging away at the traditional notion of fallacy as an argument that seems to be good when in fact it is not good, they proposed a reconceptualization in which a fallacy is simply the breaking of a rule of critical discussion. It is a bold idea, not at all dismissable out of hand. One of the things that I have not been able to understand is why, after having achieved this liberation, pragma-dialecticians got cold feet[1] and tried to reconnect their liberated notion with the vexed and stressed old one.

[1]Needless to say, in saying this I intend no deficit of the moral virtues; the difficulty rather is a certain lack of strategic clarity.

Much the greater difficulty with the Woods-Walton Approach has been its relative indifference to the Fallacies Dilemma, briefly discussed in the Prologue. On the traditional view, fallacies are bad arguments that seem not to be bad. If one of the principal tasks of fallacy theory is *fallacy-avoidance*, then it would appear to be necessary that the would-be avoider be able to recognize a fallacy when it presents itself. But to do *this*, the would-be avoider would have to perform two tasks:

1 He would have to see the argument as bad (since it *is* bad)

2 He would have to fail to see the argument as bad (since it is not an argument that one *sees* as bad).

For me the real interest lies not in how to resolve the Dilemma. Like the attraction of anything fake, the true connoisseur, whether of paintings or of arguments is one who must be able to see the object in question as one that the ordinary observer would not see as bad, without himself yielding to the perception that indeed it is not bad. Thus the connoisseur is able to do what the ordinary observer cannot (or cannot without some helpful tutelage). The connoisseur is able to objectify two properties, namely, *being bad* and *looking bad*, whereas the ordinary observer mistakes the subjective property *looking good* for the objective property *good.*

It seems to me that interesting though the connoisseur's abilities are, there is an issue of more direct salience for fallacy theory, which is about one-half, so to speak, of the Fallacies Dilemma. It is, I would say, the half that matters. A fallacy is a bad argument that looks good. Consequently, a theory of fallacies must account for two things. It must subsume a subtheory of argumentative badness, and it must also subsume a subtheory of argumental *not looking bad.* If the Woods-Walton Approach is not much responsive to the Fallacies Dilemma, it is even less responsive to the Problem of Looking Good, as we might call it. It is well to note that the Problem of Looking Good is *not* the problem of recognizing the apparent goodness of an argument, but rather is a problem of determining what it *is* for a bad argument to look good, to have that appearance.

Some interesting work is being done by cognitive and experimental psychologists on the Problem of Looking Good. Various examples of it are mentioned in a number of the preceding chapters of this book. The common presupposition of this approach — and one that we dallied with in chapter 1 — is that a fallacy is an argument with *delusional force.* Thus, when I yield to a fallacy, there is an important respect in which I don't know what I am doing. A gentler way of saying this is that succumbing to a fallacy is always a matter of falling into a kind of phenomenologically inapparent cognitive error.

Considerable encouragement is given to this view by the plain (and now experimentally probed) fact that there are lots of cases of phenomenologically inapparent cognitive deficits. Perhaps the best known of such seductions is the (apparent) failure of intelligent, well-educated people to compute correctly conjunctive probabilities in everyday sorts of situations.[2] There is no doubting the attractiveness and tenacity of this sort of practice. Equally there is no doubting that it is an error that has attracted the name of a *fallacy*. Perhaps this is an oversight which we could easily put right with some slight stipulation, say, the conjunctive probability fallacy. But an explanation is also possible, or so I will now suggest.

On this view, a fallacy is not a matter of failing to see (or disposition thereunto) the badness of an argument or piece of reasoning because of its apparent goodness. Rather a fallacy is an argument that is good in some respects and is bad in certain other respects. Note that this fits the traditional notion of fallacy, as an argument or piece of reasoning that looks good but is bad. For on the present suggestion, a fallacy is an argument that *is* bad (since it is bad in a certain respect) and *looks* good (since it *is* good in certain other respects). Moreover, the present account gives a remarkably economical yet sure-footed account of looking good. An argument *looks* good when there are standards against which it *is* good.

This idea of fallacy gives the theorist only the most promissory and schematic guidance. What truly matters is whether the schematic sketch can be discerned in a more fleshed-out state in the actual practice of reasoning and arguing; or more briefly, whether it is manifest in actual practice.

It is. Or is, on what can be called a *resource-based approach* to cognitive agency. Such an approach is advanced in some detail in recent writings.[3] I shall restrict myself here to a fairly basic *précis*. On this account, a logic is a formalization of what is done by a cognitive agent. Logic thus is *agent based*. Agency, in turn, presents itself in various types, depending on the agent's access to the cognitive resources requisite to his (or its) task. Among the more important of these resources are *information, time* and *computational capacity*. This engenders a typology of cognitive agency, partially ordered by the relation of *commanding more [of these] resources than*. Thus is formed a hierarchy of agency-types. At the (idealized) top of the hierarchy are agents with unfettered

---

[2]A claim disputed by bounded-rationality theorists in the manner of, e.g., [Gigerenzer and Selten, 2001].

[3][Gabbay and Woods, 2001b; Woods et al., 2002; Gabbay and Woods, 2003].

command of the requisite cognitive resources. At the (concrete) bottom of the hierarchy are beings like us, beings who discharge their cognitive agendas under press of scarce resources: incomplete information, lack of time and limited computational capacity. The distinction here between low and high converges on the contrast between *individual* and *institutional* agency, which in turn also answers to one way of understanding the difference between *practical* and *theoretical* agency. Given these concurrences, a practical agent is an individual cognizer, who operates, as he (or it) must, on the cheap — with less information and time than it would be helpful to have had, and with real constraints on computational complexity. By these same lights, a theoretical agent is an institutional agent whose command of the requisite resources is greater in the degree to which it is positioned in the upper reaches of the hierarchy of types.

Seen in these ways, cognitive agency is, among other things, an intrinsically *economic* enterprise. An agent's agency-type is reflected in the cognitive targets that he (or it) can afford to set and the resources available to bring them off with appropriate finesse and timeliness. Thus in 1961 it was a coherent target for NASA to put a man on the moon before the end of the decade; but it would have been a wholly incoherent target for an individual agent. The best an individual of such interests could have done is join the army of individuals that constitute, in complex ways, the corporate or theoretical agent NASA.

In general, then, what NASA commands by way of information, time and computational capacity, a practical agent has less of at best. This constrains practical agency in two important ways.

1 It limits what can be coherent cognitive targets for the agent

2 It influences what counts as an adequate realization (or attempt at realization) of those targets that are coherent for it.

Cognitive tasks have inbuilt standards of successful or adequate performance. These are what lawyers call standards of proof. The standard of proof will vary with the nature of the cognitive target and the level of resource-adequacy available to agency-types for which that target sets a coherent task. In general, the greater the grade of agency-type, the higher the standard of proof. Correspondingly, the larger the typological marker of agency-type, the greater the agent's command of the requisite resources.

Most of the cognitive tasks that are coherent for human individuals end up being performed under press of scarce cognitive resources. They are tasks for which there is too little information, too little time, and too little computational fire-power. Of course, within any given agency-type, these cognitive economies play themselves out with contextually sensitive variability. It lay in the nature of that particular task that Descartes could have taken longer with, and with greater care, the *Meditations* than with the content of tomorrow's early morning chat with Queen Christina. Still, it would be a mistake to acquiesce in an over-literal use of the qualification "too". Individual agents nearly always do have too little in the way of information or time or fire-power to discharge any cognitive agenda that may be coherent for them in fulfilment of a certain standard of proof. The same is true of ranges of institutional or theoretical agents as well. But it may well be the case that resources inadequate to meet a certain standard of proof are perfectly adequate to meet a lesser standard that is nevertheless wholly appropriate to the nature of the cognitive task and the agent's type.

It flows from this that a cognitive performance lies open to different standards of proof, depending on adequacy-type and therefore on the degree of command of the requisite resources. This bears directly on the matter with which we opened this Epilogue. Cognitive performances can be both adequate and less than adequate, successful and less than successful or, in plain English, good and bad, relative to these various factors. If we see the standard of proof of a cognitive problem as set by the agency-type for which it is a coherent problem, then we attain a useful economy in saying that a cognitive performance is good or bad in relation to the degree of command of the requisite cognitive resources. Cognitive performance is a resource-bound enterprise, and must be judged accordingly.

Consider now an individual agent, plying his cognitive agendas minute-by-minute, hour-by-hour, and day-by-day. How likely is it that he is wholly unaware that his cognitive performances answer to standards that differ from these that apply higher up? I take it as obvious that there are individual agents galore whose generalizations from samples have involved samples that fail every test of representational safety imposed by inductive logic. Lots of individual agents have come to the realization that ravens are black on the basis of slender acquaintance with some ravens, each of which is black and none otherwise hued. But how likely is it that, in so generalizing, such agents have given the slightest notice to the question of sample-reliability? Of the numberless people who have learned from instantial experience that ravens are black, what percentage would have the slightest idea of what a reliable sample would actually be?

Even so, there is nothing in the idea of scarcely-resourced individual agency that requires the individual agent to be oblivious of the fact that there are higher standards that his present cognitive successes do not meet. Accordingly, if we see fallacies in such terms, there is no requirement that a fallacy not reveal its inadequacies even to the agent whose implementation of it counts as cognitively adequate given his agency-type. So the idea that fallacies are inherent by delusion is nowhere in evidence on the present account. Accordingly, the Fallacies Dilemma lapses, and The Problem of Looking Good likewise ceases to be a delusional difficulty.

To take just one example, individual agents — practical agents — are hasty-generlizers; for them, hasty generalization is as natural as breathing. What is striking about the individual's propensity to generalize on small and inductively dubious samples is not just that he negotiates the necessary economies (for it is much, much cheaper to generalize hastily than in inductively pure ways), but also that his track record is not noticeably worse than that of the inductive purist. Practical agents do not, of course, generalize on just any small sample or on most. Part of their getting things right, or approximately right, has to do with their talent for picking the "right" small samples to make haste with.

We may speak of hasty generalizations are "good" when they form a proper subclass of generalizations that we sound by the strict canons of confirmation theory or whatever else inductive logic may relevantly provide. What is striking about the practical or individual agent is his (or its) propensity to restrict generalizations made hastily to those that are good. Can we see in this concrete fulfilment of the idea of a fallacy as an argument or a piece of reasoning that is good and bad relative to different levels of access to the necessary cognitive resources? Yes. Even good hasty generalizations are bad in a relevant sense. Being good, generalizations made are true or true-like. Being hasty, they fail the standards of inductive proof. The question of central importance is whether this is a standard to which it is appropriate to hold the individual reasoner, even for those of his generalizations that while hasty, are also good. It bears repeating that the hasty generalizer's generalizations that fall into the category of the good are not any subset of generalizations he makes arbitrarily or randomly. Thus his good "hasties" are not a significantly small subset of his hasties, but rather are nearly them *all* ("nearly" indicating the defeasibility of most of an agent's cognitive performances.)

My own tentative answer to this question is that by and large an individual's hasty generalizations are cognitively adequate notwithstanding their failure to meet the standard of proof of inductive logic. Beyond, that is to say, their failure to meet the standard of proof which, if failed by NASA, would have made NASA's fidelity to that self-same generalization cognitively inadequate. If this is right, hasty generalization meets our new definition of fallacy, never mind that being a fallacy is not its cognitive most important property. It is a type of reasoning that is bad by the standards of inductive logic, but good by the lighter standards imposed on individual agency. It is a fallacy because it is an argument or piece of reasoning that is bad (by the standards of inductive logic), and that looks good (because it *is* good, by the standards that apply to individual agency.)

Hasty generalization is but one of several examples. *Ad populum* and *ad verecundiam* are also on our list, as are denying the antecedent, affirming the consequent, and others still from the gang of eighteen. But not each of the eighteen meets the current test of fallaciousness (indeed not all of the eighteen meet the *old* conception of fallaciousness). *Ad baculum* is a case in point. If we allow that *ad baculum* sometimes makes a prudentially rational claim on someone, it is not a lesser such claim if pressed by an institutional agent (by Revenue Canada or the IRS, say) rather than an individual agent. It is true that sometimes an institutional agent will have the time and other resources to make possible the making of a successful non-baculic case for compliance. But resource-scarcity is not by any means the driving consideration in determining whether an *ad baculum* argument is all right. The *ad baculum* is therefore not a resource-sensitive kind of argument. It is not a fallacy on our present conception.

The fate that awaited *ad baculum* also awaits *ad hominem* arguments on their modern interpretations. Whatever we think of abusive, circumstantial and *tu quoque* manoeuvres, they are not less bad for having been pressed by individuals rather than by institutional agents. Still less do Aristotle's gang of thirteen sophistical refutations all make the cut. In the *Prologue* we briefly discussed Aristotle's highly peculiar analysis of the fallacy of many questions. It is the quite technical mistake of constructing would-be syllogisms with the aid of conjunctive statements as premisses or with multiple conclusions. It is not a mistake that is minimized or magnified by agency-type, and is therefore not a fallacy in our present resource-bound sense.

What justifies this deviation from older ways of thinking? Why embrace a conception of fallacy that brings down the boundaries of the gang of eighteen and re-structures Aristotle's original list? If we are going to play fast and loose with these traditional groupings, why not go all the way and declare, what pragma-dialecticians have not had the stomach for, that fallacies are, and are nothing but, violations of the rules of critical discussion, and that this is a conception in which the old fallacies (the eighteen and the thirteen) have no *conceptual* purchase whatever?

My answer is that this is not the way to achieve what I think a theory of fallacies should achieve. It was not achieved by Aristotle, it was not achieved by writers of the Standard Treatment, it is not achieved in the Woods-Walton Approach, and it is achieved still less in the pragma-dialectical approach. It is, or ought to be, a central problem for any serious theory of fallacies. It is the problem of giving theoretical articulation the property of "appearing to be good", which is embedded in the historically most dominant conception of fallacy.

As our discussion makes clear, there is utility in recognizing a distinction between the traditional *conception* of fallacies and traditional *lists* of them, whether Aristotle's thirteen, or the more modern gang of eighteen. The distinction at hand is in all essentials the difference between the intension and extension of the predicate "is a fallacy". On the traditional conception, a fallacy is a bad argument that looks good, where looking good is a delusional property. I shall say that a list of would-be fallacies is *properly determined* with regard to a given conception of fallacy if and only if every member of the list satisfies that conception. Alternatively, a list of would-be fallacies is properly determined with regard to a conception K, if and only if the list in question is the extension of the predicate for which K is the intension.

Aristotle tells us very little about what he takes to be the conditions under which an argument looks (delusionally) good. But, again, it is hard to see Many Questions on Aristotle's own account of it as satisfying the delusional standard. There is reason to think, therefore, that Aristotle's list of thirteen is improperly determined with regard to Aristotle's own conception of fallacies. The same would appear to be true of the gang of eighteen in relation to its version of traditional conception of fallacy. If so, they too are improperly determined.

Perhaps it is possible to resist these attributions of improper determination, but I shall not press the matter here. More to the point is that both Aristotle's thirteen and the more modern eighteen are *indubitably* improperly determined with reference to the new conception of fallacy that depends on agency-type. (It also puts paid to any idea of unifying those lists under the new conception.) It is well to emphasize the radical sweep of the proposed innovation. The new conception is conveyable in the old words, and this may blunt one's awareness that the new conception gives the old words ("bad, but looks good") a new meaning.

On the new conception, a fallacy is an argument or a piece of reasoning that meets standards relative to a given agency-type but fails standards relative to agency-type placed higher in the hierarchy of agency. Thus a fallacy is a good argument that fails a certain standard of good argument. This is the crux of the new conception. Given its central importance, it is advisable that we understand it with appropriate clarity.

Here, for example, is a confusion to stay clear of. Every inductively good argument is a bad deductive argument. Indeed every non-demonstrative or ampliative argument is a bad deductive argument. Shall we then say that every good ampliative argument is a fallacy, in as much as it is a good argument of that type, yet an argument that fails the standards of deductive validity? Similarly, shall we say that any good abductive argument (argument from the best explanation), is a fallacy because of its failure to meet the standards imposed by a standard inductive logic? No, this is not what the new conception of fallacy intends. It is true that the conception requires the presence of *standards-asymmetry*: A fallacy is an argument, good by one standard and bad by another. But the standards-asymmetry property is only a necessary condition on fallacies. The further thing that is required is that the standards that the argument in question *fails* themselves fulfill some conditions. Here is one of the most important. Let G be the argument presented by a practical agent A. Suppose that relative to A's agency-type, G is a passable argument. Let S be a standard which G fails, yet whose failure does not vitiate G's goodness in the sense here intended. It is a condition on S that it would wreck G as presented by A were S a standard that A could reasonably *afford*. It is not that S does not apply (and apply correctly) to G, irrespective of the agency-type of the agency whose argument G is. What matters is whether to invoke it, given that A's agency-type is of such low assay. If the costs of S are in doubt, then A would not press forward with argument G. The key constraint on S is that it be a condition that would knock-out G if A could afford to invoke it. This suffices to cut loose our troublesome examples. A good ampliative argument is not a fallacy even though it fails the conditions on deductive validity, because it is not the case (except for fanatical Rationalists) that the agent whose argument we are discussing would apply those deductive criteria simply if he could afford them. Their affordability is not here at issue; it is their applicability to a cognitive task that aims at ampliative expansion of the agent's knowledge-base. Similarly, a good abduction won't qualify as a fallacy just because it is also a bad induction (just because it violates the stands of confirmation theory). Here, too, there are cases in which the question is not whether confirmation theoretic standards can be afforded but rather whether they are properly invoked for the cognitive task in question. When Planck abduced the existence of quanta in his quest to unify the laws of black body radiation, his choice of abduction was not governed by the high cost of invoking confirmation-theoretic standards, but rather their general inapplicability to the conjecture that quanta exist.

I am not certain that the research programme sketched in this Epilogue will bear fruit. But I am minded to think that it will be worthwhile to find out. One of the reasons that people of pragma-dialectical bent have wanted to give up on the traditional idea of fallacy is that they think, or tend to think, that the old conception has not attracted adequate theories. If it is correct to say that a fundamental task for such a theory is to give a disciplined account of the property of appearing good, then it must be said that the traditional idea has not yet attracted adequate complete theories. The pragma-dialectical option is to give up entirely and start anew. But there is another option which attracts me more. It is to solve the Looking Good Problem. How this might be done is only sketched here. As I say, its further development is something for the ongoing research programme.

References

Adams, Ernest (1988). *Modus Tollens* revisited. *Analysis*, 48:122–128.

Agassi, Joseph (1988). Analogies hard and soft. In Helman, David H., editor, *Analogical Reasoning: Perspectives of Aritificial Intelligence*, pages 401–419. Dordrecht: Kluwer Academic Publishers.

Alchourrón, C.E., Gardenfors, P., and Makinson, D. (1985). On the logic of theory change: Partial meet functions for contraction and revision. *Journal of Symbolic Logic*, 50:510–530.

Altheide, D. (1976). *Creating Reality: How TV News Distorts Events.* Newbury Park, CA: Sage.

Anscombe, G.E.M. (1957). *Intention.* New York: Cornell University Press.

Aquist, Lennert (1984). Deontic logic. In Gabbay, Dov M. and Guenthner, F., editors, *Handbook of Philosophical Logic*, volume II, pages 604–714. Dordrecht and Boston: Reidel.

Arnauld, Antoine (1964). *The Art of Thinking [=La Logique, Ou L'Art de Penser, 1662].* Indianapolis, IN: Bobbs-Merrill. Originally published in 1662.

Arnauld, Antoine and Nicole, Pierre (1996). *Logic or the Art of Thinking.* Cambridge: Cambridge University Press. Originally published in 1662. This volume edited by Jill Vance Buroker.

Arrow, Kenneth (1963). *Social Choice and Individual Values.* New York: Wiley, 2nd edition.

Axelrod, Robert (1984). *The Evolution of Cooperation.* New York: Basic Books.

Barnes, Jonathan, editor (1984). *The Complete Works of Aristotle.* Princeton, NJ: Princeton University Press. In 2 volumes. Translated Jonathan Barnes.

Barth, E.M. and Krabbe, E.C.W. (1982). *From Axiom to Dialogue: A Philosophical Study of Logic and Argumentation.* Berlin and New York: de Gruyter.

Barth, E.M. and Martens, J.L. (1977). *Argumentum ad Hominem*: From chaos to formal dialectic. *Logique et Analyse*, 20:76–96.

Bartlett, F.C. (1932). *Remembering.* Cambridge: Cambridge University Press.

Barwise, Jon and Etchemendy, John (1984). *The Liar: An Essay on Truth and Circularity.* Oxford: Oxford University Press.

Batens, Diderik, Mortensen, Chris, Priest, Graham, and Bendegem, Jean-Paul Van, editors (2000). *Frontiers of Paraconsistent Logic.* Baldock, Hertfordshire England: Research Studies Press.

Baum, Robert (1975). *Logic.* New York: Holt Rinehart.

Berlin, Isaiah (1978). *Concepts and Categories.* Harmondsworth, Middlesex: Penguin Books. Edited by H. Hardy.

Black, Max (1946). *Critical Thinking.* New York: Prentice-Hall.

Blair, J.A. and Johnson, R.H. (1987). Argumentation as dialectical. *Argumentation,* 1:41–56.

Blais, Michael (1987). Epistemic TIT FOR TAT. *Journal of Philosophy,* LXXXIV:363–375.

Bolton, R. (1993). Arisotle's account of the *elenchus.* In *Oxford Studies in Ancient Philosophy,* volume XI, pages 121–152. Oxford: Clarendon Press.

Boole, George (1847). *The Mathematical Analysis of Logic Being and Essay Towards a Calculus of Deductive Reasoning.* Cambridge: Macmillan, Barclay and Macmillan, and London: Bell.

Bosanquet, Bernard (1911). *Logic: or, The Morphology of Knowledge.* Oxford: Oxford University Press, 2nd edition.

Bross, Irwin (1971). Critical levels, statistical language, and scientific inference. In Godambe, V. and Sprott, D., editors, *Foundations of Statistical Inference,* pages 500–551. Toronto: Holt Rienhart Winston.

Bunge, Mario (1974). *Treatise on Basic Philosophy,* volume I. Dordrecht: Reidel.

Cantril, H. (1940). *The Invasion from Mars: A Study in the Psychology of Panic.* Princeton: Princeton University Press.

Carney, James D. and Scheer, Richard K. (1980). *Fundamentals of Logic.* New York: Macmillan, 3rd edition.

Carp, Richard (1972). Reducibility among combinatorial problems. In Miller, R. and Thatcher, J., editors, *Complexity of Computer Computations.* New York: Plenum Press.

Cartwright, Nancy (1983). How approximations take us away from theory and towards the truth. *Pacific Philosophical Quarterly,* 64:273–280.

Cartwright, Nancy (1983a). *How the Laws of Physics Lie.* Oxford: The Clarendon Press.

Cauchy, A.L. (1821). *Cours d'Analyse d'Ecole Royale Polytechnique.* Paris: De Bure.

Cherniak, Christopher (1981). Feasible inferences. *Philosophy of Science,* 48:248–268.

Cherniak, Christopher (1986). *Minimal Rationality.* Cambridge, MA: MIT Press.

Cohen, L. Jonathan (1981). Can human irrationality be experimentally demonstrated? *The Behavioral and Brain Sciences,* 4:317–331.

Cohen, L. Jonathan (1989). *An Introduction to the Philosophy of Induction and Probability.* Oxford: Clarendon Press.

Cohen, L. Jonathan (1992). *An Essay on Belief and Acceptance.* Oxford: Clarendon Press.

Cook, Stephen A. (1971). The complexity of theorem-proving procedures. In *Proceedings of the 3rd Annual ACM Symposium on Theory of Computing,* pages 151–158. New York: Association for Computing Machinery.

Copi, Irving and Cohen, Carl (1990). *Introduction to Logic.* New York: Macmillan, 8th edition.

Copi, Irving M. (1986). *Introduction to Logic.* New York: MacMillan, 7th edition.

Copi, Irving M. and Cohen, Carl (1994). *Introduction to Logic.* Toronto and New York: Macmillan Co., 9th edition.

Cormac, Earl Mac (1983). Scientific metaphors as necessary conceptual limitations of science. In Rescher, Nicholas, editor, *The Limits of Lawfulness,* pages 61–68. Lanham, MD: University Press of America.

Darwin, Charles (1964). *On the Origin of Species.* Cambridge MA: Harvard University Press. Originally published in 1859.

Davidson, Donald (1984). *Inquiries into Truth and Interpretation.* New York: Oxford University Press.

de Sousa, Ronald (1987). *The Rationality of Emotions.* Cambridge, MA: MIT Press.

DeMorgan, Augustus (1926). *Formal Logic.* London: Open Court, 2nd edition.

Dewey, John (1976). *Human Nature and Conduct, The Middle Works of John Dewey 1899–1924*, volume 14. Edwardsville, IL: Southern Illinois University Press. Jo Ann Boydston (ed.).

Dewey, John (1981). The theory of enquiry. In Boydston, JoAnn, editor, *John Dewey: The Later Works, 1925–1953*, volume 12. Carbondale, IL: Southern Illinois University Press. Originally published in 1938.

Dewey, John and Tufts, James (1978). *Ethics, The Middle Works of John Dewey*, volume 5. Edwardsville, IL: The Southern Illinois University Press. Originally published from 1899–1924. Edited by Jo Ann Boydston.

Dolan, John M. (1988). Consent as a commodity: A critical study of *Manufacturing Consent. Informal Logic*, X:177–194.

Duhem, Pierre (1969). *To Save the Phenomena.* Chicago: University of Chicago Press.

Economist, The (1990a). Misery in the maternity wards. *The Economist*, 314. 20 January.

Economist, The (1990b). Yes, you are the superpower. *The Economist*, 314. 24 February.

Ellul, Jacques (1965). *Propaganda: The Formation of Men's Attitudes.* New York: Knopf.

Ennis, Robert H. (1969a). *Logic in Teaching.* Englewood Cliffs, NJ: Prentice-Hall.

Ennis, Robert H. (1969b). *Ordinary Logic.* Englewood Cliffs, NJ: Prentice-Hall.

Ennis, Robert H. (1996). *Critical Thinking.* Upper Saddle River, NJ: Prentice Hall.

Fain, Haskell and Griffiths, A. Phillip (1972). On falsely believing that one doesn't know. In Rescher, Nicholas, editor, *Studies in the Philosophy of Mind*, pages 10–23. Oxford: Basil Blackwell.

Finocchiaro, Maurice A. (1987). Six types of fallaciousness: Toward a realistic theory of logical criticism. *Argumentation*, I:263–282.

Fogelin, Robert (1985). The logic of deep disagreements. *Informal Logic*, 7:1–8.

Fogelin, Robert J. and Duggan, Timothy J. (1987). Fallacies. *Argumentation*, 1:255–262.

Frege, Gottlob (1879). *Begriffsschrift, A Formula Language, Modeled upon that of Arithmetic, for Pure Thought.* Cambridge, MA: Harvard University Press. Edited by Jean van Heijenoort.

Gabbay, Dov M. (1996). *Labelled Deductive Systems.* Oxford: Oxford University Press. Volume 1.

Gabbay, Dov M. and Woods, John (1998). *Ad Baculum* is not a fallacy! In van Eemeren, Frans H., Grootendorst, Rob, Blair, J. Anthony, and Willard, Charles A., editors, *Proceedings of the Fourth International Conference in the International Society for the Study of Argumentation*, pages 221–224. Amsterdam: Sic Sat.

Gabbay, Dov M. and Woods, John (2001a). More on non-cooperation in dialogue logic. *Logic Journal of the IGPL*, 9:321–339.

Gabbay, Dov M. and Woods, John (2001b). The new logic. *Logic Journal of the IGPL*, 9:157–190.

Gabbay, Dov M. and Woods, John (2001c). Non-cooperation in dialogue logic. *Synthese*, 127:161–186.

Gabbay, Dov M. and Woods, John (2003). *Agenda Relevance: A Study in Formal Pragmatics*. Studies in Logic and Practical Reasoning. Amsterdam: North-Holland. Volume 1 of *A Practical Logic of Cognitive Systems*.

Gabbay, Dov M. and Woods, John (2004). *The Reach of Abduction*. Studies in Logic and Practical Reasoning. Amsterdam: North-Holland. To appear.

Garcia, John, McGowan, Brenda K., and Green, Kenneth F. (1972). Biological constraints on conditioning. In Black, A. and Prokasy, W., editors, *Classical Conditioning Volume II: Current Research and Theory*, pages 3–27. New York: Appleton-Century Crafts.

Geis, Michael (1982). *The Language of Television Advertising*. New York: Academic Press.

Giere, Ronald (1984). *Understanding Scientific Reasoning*. New York: Holt, Rinehart and Winston, 2nd edition.

Gigerenzer, G. and Selten, R. (2001). Rethinking ratinality. In *Bounded Rationality: The Adaptive Toolbox*, pages 1–12. Cambridge, MA: MIT Press.

Girle, Roderic A. (1993). Dialogue and entrenchment. *Proceedings of the Sixth Florida Artificial Intelligence Research Symposium*, pages 185–189.

Girle, Roderic A. (1996). Knowledge: Organized and disorganized. In *Proceedings of the International Conference on Formal and Applied Practical Reasoning*, pages 246–260. Bonn.

Goodman, Nelson (1965). *Fact, Fiction and Forecast*. Indianapolis, IN: Bobbs-Merrill. Originally published in 1955.

Govier, Trudy (1985). Logical analogies. *Informal Logic*, VII:27–33.

Govier, Trudy (1987). *Problems in Argument Analysis and Evaluation*. Dordrecht and Providence: Foris Publications.

Govier, Trudy (1988). *A Practical Study of Argument*. Belmont, CA: Wadsworth.

Grant, George (1969). *Technology and Empire*. Toronto: House of Anansi.

Grice, H.P. (1967). Logic and conversation. In Grice, Paul, editor, *Studies in the Way of Words*. Cambridge, MA: Harvard University Press.

Grice, H.P. (1975). Logic and conversation. In Cole, Peter and Morgan, Jerry L., editors, *Syntax and Semantics, vol. 3*, pages 41–58. New York: Academic Press.

Grice, Paul (1989). *Studies in the Way of Words*. Cambridge, MA: Harvard University Press.

Groarke, Leo (1991). Critical review of Woods and Walton. *Informal Logic*, XIII:99–112.

Grootendorst, Rob (1987). Some fallacies about fallacies. In van Eemeren, Frans H., Grootendorst, Rob, Blair, J. Anthony, and Willard, Charles A., editors, *Argumentation: Across the Lines of Discipline*, pages 331–342. Dordrecht and Providence: Foris Publications.

Gunther, Arnold (1985). Some ways of representing dialogues. In Vaina, Lucia and Hintikka, Jaakko, editors, *Cognitive Constraints on Communication*, pages 241–250. Dordrecht, Boston and Lancaster: D. Reidel Publishing Company.

Habermas, Jurgen (1973). *Theory and Practice*. Boston: Beacon Press. Translated by John Viertel.

Hamblin, C.L. (1970). *Fallacies*. London: Methuen.

Hamblin, C.L. (1971). Mathematical models of dialogue. *Theoria*, 37:130–155.

Hamblin, C.L. (1978). Quandaries and the logic of rules. *Journal of Philosophical Logic*, 1:74–85.

Hansen, Hans V. and Pinto, Robert (1994). *Fallacies: The Dark Side of Reasoning*. University Park, PA: The Pennsylvania State University Press.

Hansen, Hans V. and Pinto, Robert C. (1995). *Fallacies: Classical and Contemporary Readings*. University Park: Pennslyvania State University Press.

Hardwig, John (1985). Epistemic dependence. *Journal of Philosophy*, LXXXII:335–349.

Harman, Gilbert (1970). Induction: A discussion of the relevance of the theory of knowledge to the theory of induction. In Swain, Marshall, editor, *Induction, Acceptance and Rational Belief*. Dordrecht: Reidel Publishing Company.

Harman, Gilbert (1975). Logical form. In Davidson, Donald and Harman, Gilbert, editors, *The Logic of Grammar*, pages 289–307. Encino, CA: Dickenson.

Harman, Gilbert (1986). *Change in View: Principles of Reasoning*. Cambridge, MA: MIT Press.

Harman, Gilbert H. (1965). The inference to the best explanation. *Philosophical Review*, 74:88–95.

Harper, C.W. (1980). Relative age inference in paleontology. *Lethia*, 13:239–248.

Hempel, C. (1965). *Aspects of Scientific Explanation*. New York: Free Press.

Hempel, Carl (1937). Le problème de la verité. *Theoria*, 3:206–246.

Hempel, C.G. (1943). A purely syntactic definition of confirmation. *Journal of Symbolic Logic*, 8:122–143.

Hempel, C.G. (1945). Studies in the logic of confirmation. *Mind*, 54:1–26 and 97–121. Reprinted in C.G. Hempel *Aspects of Scientific Explanation* by the Free Press of New York in 1965.

Herman, Edward S. and Chomsky, Noam (1988). *Manufacturing Consent: The Political Economy of the Mass Media*. New York: Pantheon Books.

Hesse, Mary (1966). *Models and Analogies in Science*. Notre Dame, IN: University of Notre Dame Press.

Hesse, Mary (1988). Theories, family resemblances and analogy. In Helman, D.H., editor, *Analogical Reasoning: Perspectives of Artificial Intelligence, Cognitive Science, and Philosophy*, pages 317–340. Dordrecht: Kluwer Academic Publishers.

Hintikka, Jaakko (1979). Information seeking dialogues: A model. *Erkenntnis*, 38:355–368.

Hintikka, Jaakko (1987). The fallacy of fallacies. *Argumentation*, 1:211–238.

Hintikka, Jaakko (1989). The role of logic in argumentation. *The Monist*, 72.

Hintikka, Jaakko and Saarinen, E. (1979). Information seeking dialogues: Some of their logical properties. *Studia Logica*, 38:355–363.

Hitchcock, David (1983). *Critical Thinking*. New York: Methuen.

Hitchcock, David (1985). Enythemematic arguments. *Informal Logic*, 7:83–97.

Hitchcock, David (1987). Enthymematic arguments. In van Eemeren, Frans H., Grootendorst, Rob, Blair, J. Anthony, and Willard, Charles A., editors, *Argument: Across the Lines of Discipline*, pages 289–298. Dordrecht and Providence: Foris Publications.

Hitchcock, David (1994). Validity in conductive arguments. In *New Essays in Informal Logic*, pages 58–66. Windsor, ON: Informal Logic.

Hofstadter, Douglas (1988). Letter to John Woods.

Hofstadter, Douglas R. (1986). Analogies and roles in human thinking and machine thinking. In Hofstadter, Douglas R., editor, *Metamagical Themas: A Questing for the Essence of Mind and Pattern*, pages 547–602. New York: Basic Books.

Holland, John H., Holyoak, Keith J., Nisbett, Richard E., and Thagard, Paul R. (1986). *Induction: Processes of Inference, Learning and Discovery*. Cambridge, MA: MIT Press.

Holyoak, Keith J. and Thagard, Paul (1995). *Mental Leaps: Analogy in Creative Thought.* Cambridge, MA: MIT Press.

Hooker, C.A. (1985). Surface dazzle, ghostly depths: An exposition and critical evaluation of van Fraassen's vindication of empiricism against realism. In Churchland, Paul M. and Hooker, C.A., editors, *Images of Science*, pages 153–196. Chicago and London: University of Chicago Press.

Hume, David (1888). *Treatise.* Oxford: Clarendon Press. L.A. Selby-Bigge (ed.).

Hume, David (1947). *Dialogues Concerning Natural Religion.* London and New York: Thomas Nelson and Sons. Originally published in 1779.

Hume, David (1964). A treatise of human nature. In Kemp-Smith, Norman, editor, *The Philosophy of David Hume*, pages 205–228. London: Macmillan and Company Ltd.

Innis, Harold (1951). *The Bias of Communication.* Toronto: University of Toronto Press.

Jacobs, Scott (1993). Failures in higher-order conditions in the organization of witnessing and heckling episodes. In van Eemeren, Frans, Grootendorst, Rob, Jackson, Sally, and Jacobs, Scott, editors, *Reconstructing Argumentative Discourse*, pages 142–169. Tuscaloosa, Alabama: University of Alabama Press.

Jacquette, Dale (1999). *Logic Methodology and Philosophy of Science Abstracts.*

Johnson, Ralph H. (1981). Charity begins at home. *Informal Logic Newsletter.*

Johnson, Ralph H. (1987a). The blaze of her splendors. *Argumentation*, 1:239–253. *Fallacies* a Special Edition edited by John Woods.

Johnson, Ralph H. (1987b). Logic naturalized: Recovering a tradition. In van Eemeren, Frans H., Grootendorst, Rob, Blair, J. Anthony, and Willard, Charles A., editors, *Argumentation: Across the Lines of Discipline*, pages 47–56. Dordrecht and Providence: Foris Publications.

Johnson, Ralph H. (1996). *The Rise of Informal Logic.* Newport News, VA: Vale Press.

Johnson, Ralph H. and Blair; J. Anthony (1983). *Logical Self Defence.* Toronto: McGraw Hill, 2nd edition.

Johnson, R.H. and Blair, J.A. (1994). Sunny prospects for relevance? In *New Essays in Informal Logic*, pages 82–92. Windsor, ON: Informal Logic Press.

Johnson-Laird, P.N. (1988). How is meaning mentally represented? In Eco, Umberto, Santambrogio, Marco, and Violi, Patrizia, editors, *Meaning and Mental Representation.* Bloomington and Indianapolis, IN: Indiana University Press.

Johnstone, H.W. Jr. (1978). *Validity and Rhetoric in Philosophical Argument.* University Park, PA: The Dialogue Press of Man and World.

Joseph, H.W.B. (1916). *An Introduction to Logic.* Oxford: Clarendon Press, 2nd edition.

Kahane, Howard (1969). *Logic and Philosophy: A Modern Introduction.* Belmont, CA: Wadsworth Publishing Co.

Kaufmann, Walter (1906). Über die Konstitution des Elektrons. *Ann. Phys.*, 19:487–553.

Keynes, John Maynard (1963). *A Treatise on Probability.* London: Macmillan.

Krabbe, E.C.W. (1986). Formal systems of dialogue rules. *Synthese*, 63:191–217.

Kuipers, T.A.G. (1987). *What is closer-to-the-truth? A Parade of Approaches to Truthlikeness.* Amsterdam: Rodopi. *Pozan Studies*, Volume 10.

Lakatos, Imre (1976). *Proofs and Refutations: The Logic of Mathematical Discovery.* Cambridge: Cambridge University. Edited by John Worrell and Elie Zahar.

Lakoff, George (1970). Linguistics and natural logic. *Synthese*, 22:151–271.

Lakoff, George and Johnson, Mark (1980). *Metaphors We Live By*. Chicago: University of Chicago Press.

Lambert, Karel and Ulrich, William (1980). *The Nature of Argument*. New York: Macmillan.

Langsdorf, Lenore (1988). Dialogue, distanciation, and engagement: Toward a logic of televisual communication. *Informal Logic*, X:151–168.

Laplace, Pierre (1951). *Philosophical Essay on Probabilities*. New York: Dover. Translation by Frederick Wilson Truscott and Frederick Lincoln Emory. Originally published in 1795.

Laudan, Larry (1981). A confutation of convergent realism. *Philosophy of Science*, 48.

Lewis, C.I. (1912). Implication and the algebra of logic. *Mind*, 21:522–531.

Lewis, David (1969). *Convention: A Philosophical Study*. Cambridge, MA: Harvard University Press.

Lewis, David (1978). Truth in fiction. *American Philosophical Quarterly*, 15:37–46. Reprinted with postscripts in [Lewis, 1983b].

Lewis, David (1983a). General semantics. In *Philosophical Papers*, volume I, pages 189–229. Oxford: Oxford University Press.

Lewis, David (1983b). Truth in fiction. In *Philosophical Papers*, volume 1, pages 261–268. Oxford: Oxford University Press. Reprinted with postscripts from [Lewis, 1978].

Lichter, Tilman (1995). Bill Clinton is the First Lady of the USA: Making and unmaking analogies. *Synthese*, 104:285–297.

Locke, John (1961). *An Essay Concerning Human Understanding*. London: Dent. Originally publisher in 1690. Edited in two volumes with an introduction by John W. Yolton.

Locke, John (1975). *An Essay Concerning Human Understanding*. Oxford: Clarendon Press. This version edited by Peter H. Nidditch. Originally published in 1689.

Lorenzen, Paul and Lorenz, Kuno (1978). *Dialogische Logik*. Darmstadt: Wissenschaftliche Buchgesellschaft.

Lycan, William G. (1984). *Logical Form in Natural Language*. Cambridge, MA: MIT Press.

Mac Cormac, Earl (1985). *A Cognitive Theory of Metaphor*. Cambridge, MA: MIT Press.

MacKenzie, J. (1990). Four dialogue systems. *Studia Logica*, XLIX:567–583.

Manktelow, Ken (1999). *Reasoning and Thinking*. Hove, UK: Psychology Press.

Manor, Ruth (1981). Dialogues and the logics of questions and of answers. *Linguistische Berichte*, 73:1–28.

Massey, Gerald J. (1975a). Are there any good arguments that bad arguments are bad? *Philosophy in Context*, 4:61–77.

Massey, Gerald J. (1975b). In defense of the asymmetry. *Philosophy in Context*, 4:44–45. Supplementary Volume.

Massey, Gerald J. (1981). The fallacy behind fallacies. *Midwest Studies in Philosophy*, 6:489–500.

Maudlin, Tim (1994). *Quantum Non-Locality and Relativity*. Oxford: Blackwell.

May, John D. (1988). Reportage as compound suggestion. *Informal Logic*, X:113–131.

McCarthy, J. (1980). Circumscription — a form of non-monotonic reasoning. *Artificial Intelligence*, 13:27–39.

McGee, Vann (1985). A counterexample of *modus ponens*. *Journal of Philosophy*, 82:462–471.

McKenzie, J. (1981). The dialectics of logic. *Logique et Analyse*, 94:159–177.

McLuhan, Marshall (1964). *Understanding Media: The Extension of Man.* New York: McGraw-Hill.

McMurtry, John (1988). The unspeakable: Understanding the system of fallacy in the media. *Informal Logic*, X:133–150.

McPeck, John (1981). *Critical Thinking and Education.* Oxford and New York: Martin Robertson and Co. and St. Martin's Press.

Mill, J.S. (1865). *An Examination of Sir William Hamilton's Philosophy.* London:Longman's Green.

Mill, J.S. (1959). *A System of Logic.* London: Longman's Green. Originally published in 1843.

Minsky, Marvin (1977). Frame-system theory. In Johnson-Laird, P.N. and Watson, P.C., editors, *Thinking: Readings in Cognitive Science*, pages 355–376. Cambridge: Cambridge University Press.

Moore, R. (1985). Semantical considerations on non-monotonic logics. *Artificial Intelligence*, 25:75–94.

Nash, John (1997). Non-cooperative games. In Kuhn, Harold W., editor, *Classics in Game Theory*, pages 14–26. Princeton, NJ: Princeton University Press.

Niiniluoto, Ilka (1984). *Is Science Progressive?* Dordrecht: Reidel.

Niiniluoto, Ilkka (1988). Analogy and similarity in scientific reasoning. In Helman, David H., editor, *Analogical Reasoning: Perspectives of Aritificial Intelligence*, pages 272–298. Dordrecht: Kluwer Academic Publishers.

Nisbett, R.E. and Ross, L. (1980). *Human Inference: Strategies and Shortcomings of Social Judgement.* Englewood Cliffs, NJ: Prentice-Hall.

Nisbett, Richard E., Krantz, David H., Jepson, Christopher, and Kunda, Ziva (1983). The use of statistical heuristics in everyday inductive reasoning. *Psychological Review*, 90:339–363.

Oddie, Graham (1986). *Likeness to Truth.* Dordrecht: Reidel.

Paley, William (1822). *Natural Theology, or, Evidences of the Existence and Attributes of the Deity, Collected from the Appearances of Nature.* London: F.C. and J. Rivington. First published in 1802.

Parret, Herman (1978). A note on pragmatic universals of language. In *Language Unviersals*, pages 125–140. Tubingen: Narr.

Parsons, Terence (1980). *Non-Existent Objects.* New Haven, CT: Yale University Press.

Paul, Richard (1985). McPeck's mistakes. *Informal Logic*, 7:35–43.

Peirce, C.S. (1931–1958). *Collected Works.* Cambridge, Mass: Harvard University Press. A series of volumes, the first appearing in 1931.

Peirce, C.S. (1955). The fixation of belief. In Buchler, Justus, editor, *Philosophical Writings of Peirce.* New York: Dover Publications.

Peirce, C.S. (1992). *Reasoning and the Logic of Things: The 1898 Cambridge Conferences Lectures by Charles Sanders Peirce.* Cambridge, MA: Harvard University Press. Edited by Kenneth Laine Ketner.

Penelhum, Terence (1971). *Problems of Religious Knowledge.* New York: Herder and Herder.

Perelman, Ch. and Olbrechts-Tyteca, L. (1971). *The New Rhetoric.* South Bend, IN: University of Notre Dame Press.

Perrin, Porter G. and Ebbitt, Wilma R. (1972). *Writer's Guide and Index to English.* Glenview, IL: Scott Foreman.

Polya, George (1954). *Induction and Analogy in Mathematics*, volume Vol 1 of *Mathematics and Plausible Reasoning.* Princeton, NJ: Princeton University Press.

Popper, Karl (1972). *Objective Knowledge.* Oxford: Oxford University Press.

Priest, Graham (1987). *In Contradiction: A Study of the Transconsistent.* Dordrecht: Kluwer Academic.

Putnam, Hilary (1975). The meaning of meaning. In *Mind, Language and Reality: Philosophical Papers, Volume 2*, pages 215–271. Cambridge: Cambridge University Press.

Putnam, Hilary (1988). *Representation and Reality.* Cambridge, MA: MIT Press.

Quine, W.V. (1975). Empirically equivalent systems of the world. *Erkenntnis*, 9:313–328.

Quine, W.V. (1981). *Theories and Things.* Cambridge, MA: Harvard University Press.

Quine, W.V. and Ullian, J.S. (1970). *The Web of Belief.* New York: Random House.

Quine, W.V.O. (1960). *Word and Object.* Cambridge, MA and New York: MIT Press and John Wiley.

Quine, W.V.O. (1970). *Philosophy of Logic.* Englewood Cliffs, NJ: Prentice-Hall.

Quine, W.V.O. (1977). Natural kinds. In Schwartz, Stephen P., editor, *Naming, Necessity and Natural Kinds*, pages 155–175. Ithica, NY: Cornell University Press.

Ramsey, Frank (1931). Truth and probability. In Ramsey, Frank, editor, *The Foundations of Mathematics and Other Essays*, pages 156–198. London: Routledge and Kegan Paul. Originally published in 1926.

Reiter, Raymond (1980). A logic for default reasoning. *Artificial Intelligence*, 12:81–132.

Rescher, Nicholas (1964). *Hypothetical Reasoning.* Amsterdam: North-Holland.

Rescher, Nicholas (1978). *Peirce's Philosophy of Science.* Notre Dame, IN: University of Notre Dame Press.

Rescher, Nicholas (1985). *Pascal's Wager: A Study of Practical Reasoning in Philosophical Theology.* Notre Dame, IN: University of Notre Dame Press.

Robinson, Abraham (1966). *Non-standard Analysis.* Amsterdam: North-Holland.

Ross, James F. (1981). *Portraying Analogy.* New York: Cambridge University Press.

Ross, W.D. (1984). *Aristotle's Metaphysics.* Princeton: Princeton University Press. In 2 volumes.

Ruse, Michael (1974). Cultural evolution. *Theory and Decision*, 5:413–440.

Sandewall, E. (1972). An approach to the frame problem and its implementation. In Meltzer, B. and Michie, D., editors, *Machine Intelligence 7.* Edinburgh: Edinburgh University Press.

Schank, Roger and Abelson, Robert (1977). *Scripts, Plans, Goals and Understanding: An Inquiry into Human Knowledge Structures.* Hillsdale, NJ: Lawrence Erlbaum Associates.

Schipper, Edith W. and Schuh, Edward (1959). *A First Course in Modern Logic.* New York: Henry Holt.

Schwartz, Stephen P., editor (1977). *Naming, Necessity and Natural Kinds.* Ithica, NY: Cornell University Press.

Scriven, Michael (1976). *Reasoning.* New York: McGraw-Hill.

Scriven, Michael (1987a). Fallacies of statistical substitution. *Argumentation*, 1:333–349.

Scriven, Michael (1987b). Proactive logic: Review and preview. In van Eemeren, Frans H., Grootendorst, Rob, Blair, J. Anthony, and Willard, Charles A., editors, *Argumentation: Across the Lines of Discipline*, pages 7–32. Dordrecht-Holland: Foris Publications.

Scriven, Michael (1989). The philosophy of ordinary logic. A lecture to the Third International Symposium on Informal Logic held at the University of Windsor in June.

Searle, J. (1959). On determinables and resemblance. *Aristotelian Society*, Supplemental 33:141–158.

Shaw, William H. and Ashley, L.R. (1983). Analogy and inference. *Dialogue*, XXII:415–432.

Sherwood, William of (1963). *De Obligationibis*: An introduction. In Green, Ronald, editor, *The Logical Treatise "De Obligationibis" with Critical Texts of William of Sherwood and Walter Burley*. Ph.D. Thesis, University of Louvain, 1963.

Sidgwick, Alfred (1910). *The Application of Logic*. London: Macmillan.

Skyrms, Brian (1986). *Choice and Chance*. Belmont, CA: Wadsworth, 3rd edition.

Smith, Edward E. and Medin, Douglas L. (1981). *Categories and Concepts*. Cambridge, MA: Harvard University Press.

Sorensen, Roy (1988). *Blindspots*. Oxford: Clarendon Press.

Sperber, Dan and Wilson, Deirdre (1986). *Relevance*. Oxford: Basil Blackwell. 1st edition.

Stebbing, L. Susan (1961). *A Modern Introduction to Logic*. New York: Harper Torchbooks.

Stich, Stephen P. (1980). Justification and the psychology of human reasoning. *Philosophy of Science*, 47:188–202.

Stich, Stephen P. (1985). Could man be an irrational animal? Some notes on the epistemology of rationality. *Synthese*, 64:115–134.

Stingl, Michael (2004). Engineered death and the (il)logic of social change. In Peacock, Kent and Irvine, Andrew, editors, *Mistakes of Reason*. Toronto: University of Toronto Press. To appear.

Stone, W.H. (1937). Applications of the theory of Boolean rings to general topology. *Transactions of the American Mathematical Society*, 41:375–481.

Stove, David (1986). *The Rationality of Induction*. Oxford: Clarendon.

Sumner, L.W. (1981). *Abortion and Moral Theory*. Princeton: Princeton University Press.

Thagard, Paul, Holyoak, Keith J., Nelson, G., and Gochfield, D. (1990). Retrieval by constraint satisfaction. *Artificial Intelligence*, 46:259–310.

Thom, René (1975). *Structural Stability and Morphogenesis*. Reading, MA: W.A. Benjamin.

Thomas, Stephen N. (1977). *Practical Reasoning in Natural Language*. Englewood Cliffs, NJ: Prentice-Hall.

Thomson, Judith Jarvis (1971). A defence of abortion. *Philosophy and Public Affairs*, 1:47–66.

Toulmin, Stephen (1969). *The Uses of Argument*. Cambridge: Cambridge University Press.

Tversky, A. (1977). Feature of similarity. *Psychological Review*, 84:327–352.

Tversky, A. and Kahneman, D. (1983). Extensional versus intensional reasoning: The conjunction fallacy in probability judgements. *Psychological Review*, 90:293–315.

Ulam, S.M. (1983). *Adventures of a Mathematician*. New York: Charles Scribner and Sons, 2nd edition.

van Benthem, Johan (1994). General dynamic logic. In Gabbay, D.M., editor, *What Is A Logical System?*, pages 107–140. Oxford: Oxford University Press.

van Eemeren, Frans H. and Grootendorst, Rob (1983). *Speech Acts in Argumentative Discourse*. Dordrecht: Foris Publication.

van Eemeren, Frans H. and Grootendorst, Rob (1984). *Speech Acts in Argumentative Discussions: A Theoretical Model for the Analysis of Discussions Directed Towards Solving Conflicts of Opinion.* Dordrecht: Foris/Berlin.

van Eemeren, Frans H. and Grootendorst, Rob (1987). Fallacies in pragma-dialectical perspective. *Argumentation*, 1:283–302.

van Eemeren, Frans H. and Grootendorst, Rob (1988). Rationale for a pragma-dialectical perspective. *Argumentation*, 2:271–292.

van Eemeren, Frans H. and Grootendorst, Rob (1989). A transition stage in the theory of fallacies. *Journal of Pragmatics*, 13:99–109.

van Eemeren, Frans H. and Grootendorst, Rob (1992). *Argumentation, Communication, and Fallacies: A Pragma-Dialectical Perspective.* Hillsdale, NJ and London: Lawrence Erlbaum Associates.

van Eemeren, Frans H. and Grootendorst, Rob (1994). *Studies in Pragma-Dialectics.* Amsterdam: Sic Sat.

van Eemeren, Frans H., Grootendorst, Rob, Henkemans, Francisca Snoeck, Blair, J. Anthony, Johnson, Ralph H., Krabbe, Erik C.W., Plantin, Christian, Walton, Douglas N., Willard, Charles A., Woods, John, and Zarefsky, David (1996). *Fundamentals of Argumentation Theory: A Handbook of Historical Backgrounds and Contemporary Developments.* Mahwah, NJ: Erlbaum.

van Eemeren, Frans H., Grootendorst, Rob, and Kruiger, Tjark (1987). *Handbook of Argumentation Theory: A Critical Survey of Classical Backgrounds and Modern Studies.* Dordrecht: Foris.

van Fraassen, Bas C. (1973). Values and the heart's command. *Journal of Philosophy*, 70:5–19.

van Fraassen, Bas C. (1980). *The Scientific Image.* Oxford: Clarendon Press.

von Neumann, John (1958). *The Computer and the Brain.* New Haven, CT: Yale University Press.

Walton, Douglas (1992a). *The Place of Emotion in Argument.* University Park: The Pennsylvania State University Press.

Walton, Douglas and Krabbe, E.C.W. (1995). *Commitment in Dialogue.* Albany, NY: SUNY Press.

Walton, Douglas N. (1984). *Logical Dialogue-Games and Fallacies.* Lanham, MD: Univeristy Press of America.

Walton, Douglas N. (1985a). The logic of dialogue. *Synthese*, 63:259–388.

Walton, Douglas N. (1987). *Informal Fallacies.* Amsterdam: John Benjamin.

Walton, Douglas N. (1988). Argumentation in dialogues. *Argumentation*, 2:393–538.

Walton, Douglas N. (1991). *Begging the Question: Circular Reasoning as a Tactic of Argumentation.* New York: Greenwood Press.

Walton, Douglas N. (1992). *Slippery Slope Arguments.* Oxford: Clarendon Press.

Weddle, Perry (1978). *'Comparison', in Argument: A Guide to Critical Thinking.* New York: McGraw-Hill.

Weinstein, Mark (1990). Towards an account of argumentation in science. *Argumentation*, 4:269–298.

Whately, Richard (1848). *Elements of Logic, 1826; 9th edition.* London: Longmans.

Williams, Bernard (1981). *Moral Luck: Philosophical Papers, 1973-1980.* Cambridge: Cambridge University Press.

Wilson, N.L. (1959). Substances without substrata. *Review of Metaphysics*, 12:521–539.

Woods, John (1965). Was Achilles' Achilles' heel Achilles' heel? *Analysis*, XXV:142–146.

Woods, John (1974a). *The Logic of Fiction: Philosophical Soundings of Deviant Logic.* The Hague and Paris: Mouton.

Woods, John (1974b). Semantic kinds. *Philosophia*, 3:117–151.

Woods, John (1978). *Engineered Death: Abortion, Suicide, Euthanasia and Senecide.* Ottawa: University of Ottawa Press.

Woods, John (1980). What is informal logic? In *Informal Logic: The First International Conference*, pages 57–68. San Francisco: Edgepress Publishers. Reprinted in Woods and Walton *Fallacies: Selected Papers 1972-1982*, Berlin and New York: Foris de Gruyter, 1989.

Woods, John (1985). Utilitarian abortion. *Dialogue: The Canadian Philosophical Review*, XXIV:671–682.

Woods, John (1987). *Ad baculum*, self-interest and Pascal's Wager. In van Eemeren, F.H., Grootendorst, Rob, Blair, J. Anthony, and Willard, Charles A, editors, *Argumentation: Across the Lines of Discipline*, pages 343–349. Dordrecht and Providence: Foris Publications.

Woods, John (1988). Ideals of rationality in dialogic. *Argumentation*, 2:395–408.

Woods, John (1989a). The maladroitness of epistemic TIT FOR TAT. *Journal of Philosophy*, LXXXVI:324–331.

Woods, John (1989b). The relevance of relevant logic. In Norman, J. and Sylvan, R., editors, *Directions in Relevant Logics*, pages 77–86. Dordrecht: Kluwer Academic Publisher.

Woods, John (1991). Pragma-dialectics: A radical departure in fallacy theory. *Communication and Cognition*, 24:43–53.

Woods, John (1992a). Public policy and standoffs of force five. In Barth, E.M. and Krabbe, E.C.W., editors, *Logic and Political Culture*, pages 97–108. Amsterdam: North-Holland.

Woods, John (1992b). Who cares about the fallacies? In van Eemeren, F.H., Grootendorst, Rob, Blair, J. Anthony, and Willard, Charles A., editors, *Argumentation Illuminated*, pages 22–48. Amsterdam: Sic Sat Press.

Woods, John (1993a). Charity: Can we find it a home? *Speech Communication and Argumentation*, pages 46–59. Precis in Russian, 60–61.

Woods, John (1993b). Dialectical blindspots. *Philosophy and Rhetoric*, 26:251–265.

Woods, John (1993c). Missing premisses in pragma-dialectics. *Logique et Analyse*, 129-130:155–168.

Woods, John (1995). Fearful symmetry. In Hansen, Hans V. and Pinto, Robert C., editors, *Classical and Contemporary Readings.* University Park, PA: The Pennsylvania State University Press.

Woods, John (1996). Deep disagreements and public demoralization. In Gabbay, Dov M. and Jürgen Ohlbach, Hans, editors, *Practical Reasoning: Springer Notes on Artificial Intelligence*, pages 650–662. Berlin: Springer.

Woods, John (1999a). Antoine Arnauld. *Argumentation*, 13:631–374.

Woods, John (1999b). Aristotle. *Argumentation*, 13:203–220.

Woods, John (1999c). Augustus DeMorgan. *Argumentation*, 13:393–397.

Woods, John (1999d). John Stuart Mill. *Argumentation*, 13:317–334.

Woods, John (2000a). How philosophical is informal logic? *Informal Logic*, 20:139–167.

Woods, John (2000b). Privatizing death: Metaphysical discouragements of ethical thinking. In French, Peter A. and Wettstein, Howard K., editors, *Midwest Studies in Philosophy*, volume XXIV, pages 199–217. Boston: Blackwell.

Woods, John (2001). *Aristotle's Earlier Logic.* Oxford: Hermes Science.

Woods, John (2002a). John Locke. *Argumentation*. To appear.
Woods, John (2002b). *Paradox and Paraconsistency: Conflict Resolution in the Abstract Sciences*. Cambridge and New York: Cambridge University Press.
Woods, John (2002c). Speaking your mind: Inarticulacies constitutional and circumstantial. *Argumentation*, 16:59–78.
Woods, John (2004a). Life, death and slippery slopes. In Stingl, Michael, editor, *Rodrigues and Latimer: Euthanasia in Canada*. To appear.
Woods, John (2004b). Pluralism's embrace: Life, death and sex. To appear.
Woods, John and Hansen, Hans V. (2001). The subtleties of Aristotle's non-cause. *Logique et Analyse*, 43.
Woods, John and Hudak, Brent (1989). By parity of reasoning. *Informal Logic*, XI:125–140.
Woods, John and Hudak, Brent (1992). Verdi is the Puccini of music. *Synthese*, 92:189–220.
Woods, John, Irvine, Andrew, and Walton, Douglas (2000). *Argument: Critical Thinking, Logic and the Fallacies*. Toronto: Prentice-Hall.
Woods, John, Johnson, Ralph H., Gabbay, Dov M., and Ohlbach, Hans Jürgen (2002). Logic and the practical turn. In Gabbay, Dov M., Johnson, Ralph H., Ohlbach, Hans Jürgen, and Woods, John, editors, *Handbook of the Logic of Argument and Inference: The Turn Toward the Practical*, volume one of *Studies in Logic and Practical Reasoning*. Amsterdam: North-Holland.
Woods, John and Walton, Douglas (1976). *Ad Baculum*. *Grazer Philosophische Studien*, 2:133–140.
Woods, John and Walton, Douglas (1982). *Argument: The Logic of the Fallacies*. Toronto: McGraw-Hill.
Woods, John and Walton, Douglas (1989). *Fallacies: Selected Papers 1972-1982*. Berlin and New York: Foris de Gruyter.
Wreen, Michael (1995). Knockdown arguments. *Informal Logic*, 17:316–336.
Zbynek, Zeman (1973). *Nazi Propaganda*. New York: Oxford University Press.
Zwart, S.D. (1998). *Approach to Truth: Verisimilitude and Truthlikeness*. Amsterdam: Institute for Logic, Language and Computation.

Index

AB-hypothesis, 134
Abstract, 20, 36, 189–190, 206, 261–262, 267–268, 270, 278, 282
Action-matrix, 87–89
Ad baculic, 31
Ad baculum, 4, 75–76, 79, 355
Ad hominem, 4, 11, 13, xxi–xxii, xxv, 31, 34, 40–41, 95, 97–99, 102–107, 109–112, 118, 122–123, 156, 161, 163, 178–179, 185, 301, 320, 355
Ad ignorantiam, 17–18, xxii
Ad misericordiam, 4, 156
Ad populum, 4, 11, 31, 75, 125–126, 131, 135, 138, 154
Ad verecundiam, 4, 11, 18, xxii, xxv, 101, 108, 125, 138, 154, 172, 178, 355
Affirming the consequent, 4, 16, 355
Alethic argument, 83, 324
Alethic rationality, xxvii, 75
Ambiguation Rule, 58
Amphiboly, 4
Analogical prediction, 276
Analogy, 5, 36, 153, 203, 251, 253–261, 263–265, 267–271, 274–276, 278–283, 285–290, 292, 295, 297, 341–342
Analytical hypothesis, 223–224
Argumentation theory, xix–xxi, 25–26, 28, 31–34, 36, 38–39, 55, 151, 161, 219–221, 231
Aristotelian-Lockean, 97
Arrow Barrier Theorem, 194
Artificial language, 44–46
Attitude-modification, 78
Autoepistemic logic, xx
Baconian probability, 14
Begging the question, 5, xxi–xxii, 143, 174–175, 178, 181, 309, 346
Bias, 95, 98, 125–131, 135–137, 139, 141–143, 145–148
Bias Rule, 129
BiasRule, 145
Boolean algebra, 180
Boundary question, 4–5, 7, 12, 21, 164
Can Do Principle, 144, 230
ChallRule, 140
Charity Principle, 138, 219
Charity principle, 219
Charity Principle, 222, 226, 233, 235, 240–241
CKRule, 141
Clutter Avoidance fallacy, 129
Clutter Avoidance Principle, 128–130, 240
Cognitive psychology, xxi, 175
Cognitive rationality, 316
Common knowledge, 11, 131–141, 143–146, 231, 296
Compactness Metatheorem, 26, 39
Complex question, 5, xxii–xxiii, 163, 178
Conceptual realism, 161, 169
Conflict resolution, 8, 82, 152, 158, 226–227
Connective Rule, 49, 54–55, 59
Convention T, 46–47
Cost-benefit, 8, 12, 68, 78–79, 87–88, 91, 103, 129, 147
Decision theory, 32, 66, 68–69, 207
Deep structure, 33, 256–259, 261–262, 269–273, 293
Default logic, xx
Demoralization rule, 205–206
Denying the antecedent, 4, 12, 16, 19, 355
Deontic argument, 83–85, 324
Deontic belief, 201–203
Deontic-prudential argument, 84
Dialectic, 110, 168, 186, 188
Dialectical argument, xxii, 112
Dialectical minimum, 117
Dialectical proposition, xxii
Dialogical Charity, 225–226, 231, 233, 238
Dialogical Principle, 226

Dialogue logic, xiii, xxi, xxvi
Disambiguation Rule, 50–52, 55
Disguised threat, 83
Doctrine of entrenchment, 328
Dominant élites, 195, 197–199, 208, 212, 214
Doxastic surrender, 69–70, 72
Easy-Easy Principle, 22–23
Empirical adequacy, 20, 36
Epistemic community, 131–132, 139, 147
Equivocation, 5, 156, 158, 190, 319
Ex falso quodlibet, 27–28
Exemplar theory, 174–177
Expert knowledge, 120–121, 140
Explanation condition, 81–82
FactRule, 145
Fallacies of relevance, 11, 68
Fallacy Dilemma, 6, 8
Fallacy of many questions, xxii, 355
Fallacy of misrepresentation, 84
Fallacy of relevance, 67, 76
Fallacy theory, 1, 10, xix–23, xxv–xxvi, 33–35, 126, 149, 151, 161, 163, 165, 167, 169–176, 185, 239, 251, 301, 307, 350
False analogy, 154, 253, 273, 294–295
FDL – formal deductive logic, 25–26, 28, 30, 33–35, 38–41, 91, 173, 243
First order logic, 26, 28–29
Folk psychology, 171–172
Formal logic, 26–28, 43–44, 46, 52, 55, 60
Formalization, 48–55, 57, 59–60, 351
Frame-system, 136, 233, 247, 313
Gang of eighteen, 4, 10, 15–16, 18, 20, xxv, 125–126, 164, 178, 239, 311, 355–356
Generalization, 9–10, 14–15, 17, 19, 31, 97, 174–176, 245, 248–249, 265, 274, 301–302, 305, 308, 311–314, 320–323, 325–327, 330, 332, 334, 340, 354
Generic fallacy, 41
Good faith condition, 81–82
Goody Two-Shoes Model, xxvi, 185
Gricean Charity, 231
Grice's Quantity Maxim, 83
Grice's Relation Maxim, 221
Grue Paradox, 325, 327–328, 331
Habermas' Consensus Theory, 185
Hasty conclusion, 314
Hasty generalization, 5, 13–15, 126, 174–175, 178, 181, 299, 303, 305, 307–308, 310–312, 314, 333–335, 354
Ignoratio elenchi, 5, 123, 172
Inconsistency, 40, 45, 51, 97, 99–101, 103–104, 109, 138, 141, 147, 201, 205, 245–246, 255–256, 322
Indoctrination, 214–216
Induction, 10, 12, 16–17, 19, 174–176, 234–237, 254, 264, 274, 299, 304–305, 311, 314–315, 317, 320, 324–325, 328, 330–331, 333, 335–347, 357
Inference to the best explanation, 17–19, 99, 144, 337, 345
Informal logic, 1, xiii, xix, 21, 25, 43, 53, 55, 60, 149
Intuitionistic logic, xxiii, 32, 54–55, 181
Isomorphism, 285
KK-hypothesis, 132
Kripke-semantics, xxi
Kripke-structures, 12
Labelled deductive systems, xx
Law of Distributivity, 53
Lewis' Charity, 232
Limitation theorem, 21, 130, 135, 186, 199, 346
Logical Inertia Rule, 51–52, 59
Lowenheim-Skolem theorem, 173, 181
Manifestation Principle, 232
Mathematical logic, xx, 45
Maxim of Moderation, 190–191
Mechanical Rule, 229
Meta-arguments, 257, 264–265, 267, 273
Metaphor, 16, xxv, 36, 78, 109, 195, 275, 279, 283, 287
Minimax, 77–78, 80, 193, 315, 317
Moduality hypothesis, 231
Modus ponens, 56–57, 262, 267, 270
Monotonic, 113–114, 220
Moorean sentence, 109
Moore's paradox, 108–109
Natural logic, 33, 52, 272
Negative-proof, 122
Nomic systematicity, 26, 30
Non-cognitivism, 111–112, 198
Non-cooperation arguments, 63, 65, 75, 89, 219
Non-falsifying refutation, 123
Nonmonotonic, xx–xxi, 113
No-retraction rule, 115
Normative model, 220–221
One-way unification, 181
Ordo cognescendi, 169, 329
Ordo essendi, 169, 329
Paraconsistent logic, xx–xxi, xxiii
Pascalian probability, 14
Pascal's Wager, 65, 68, 70, 77, 90, 127
Peirasis, 121
Post hoc
ergo propter hoc, 11, 31, 313
Practical syllogism, 84–86, 88–93, 318, 320
Pragmatic inconsistency, 102, 201, 204–205, 245–246, 248
Pragmatic Optimum Rule, 245
PreCondRule, 145
Predicational analogues, 268–269
Premissorily practical syllogisms, 85–86, 88

Premiss-selection rule, 115
Primitive inductions, 8–10, 13, 274, 314
Principle of Charity, 18, 138, 217, 219, 221–222, 240
Principle of Generativity, 232
Principle of Informedness, 128–129
Principle of Logical Inertia, 58
Principle of Rationality, 185, 232
Principle of Truthfulness, 232
Probability theory, xxii, 312
ProbDogRule, 148
Problem of induction, 16, 311, 320, 339
Problem validity, 158–159
ProPopRule, 140
Propositional calculus, 11–12, 28, 45
Propositional simplification, xxiv
Prototype, 88, 136–137, 233, 247
ProtRule, 136–137
Prudential argument, 66, 73, 76, 79, 81, 83–84
Quantity Maxim, 100, 229
Quantum logic, 53–55, 60
Quantum theory, 35–36
Radical Charity, 225–226, 231–233
Radical translation, 18, 222–223, 225, 232
Ramsey's Maxim, 189–191, 198
Raven Paradox, 312, 322
Real possibility, 71–72, 77–78, 102, 127–128, 193–194, 207
Received opinion, 213–214
Reductio ad absurdum, 294, 322
Reductio ad impossible, 117
Refutation, xxiii–xxv, 98, 111–115, 117–123, 263–264, 301
Relevance, 11, 26, 28, 30, 60, 98–99, 102, 105, 128–129, 221, 227–228, 231, 253–254, 266–268, 280, 322
Representation Theorem, 180
Rhetoric, 197, 210
Risk aversion strategies, 77
Risk-aversion strategy, 176, 315
Rules of inference, 27, 220
Sample-induction, 336, 339–341, 343, 345
Satisfiability, 29, 39
Secundum quid, 5, 17, 153–154, 172, 299, 301–312
Sheffer stroke, 26, 39
Shifty Utilitarianism, 196
Single-argumentation, 226, 229
Skepticism, 25–26, 35, 68, 70–72, 272, 317, 337, 342, 344–347
Skolem functions, 16, 234
Slippery slope, 154
Standard syllogism, 84, 86
Standard Treatment, 6, 15–16, xix–xx, 23, 65–67, 151–152, 154, 156, 159, 161, 168–170, 172, 178–179, 356
Standoff, 143, 186–189, 191–199, 201–204, 206–207, 210, 212, 259–260, 317
Stimulus meaning, 223, 236
Stipulation, 80, 165–166, 177–178, 242, 246, 326, 351
Strategic rationality, 3, xxvii, 65, 316
Strategies of intervention, 206
Structural re-description, 261, 270
Subset sum problem, 29
Technology of persuasion, 197–198, 203, 207–208
The Violinist, 258–259, 261, 263, 270–271
Theoretical entity, 157, 162
Theory-dependency, 162–163, 167, 169–170
Topology, 179–180, 306, 319
Transitivity, 111, 269
Triangle Principle, 232
Trickle-down property, 23
Truth conditional, 46, 48, 273, 292
Tu quoque, 105–106, 123, 355
Turing machine, 29
Two-way unification principle, 180
Validity, 33–35, 44, 47, 50–55, 57, 59–60, 113–114, 126–127, 157–158, 227, 242–243, 249, 256–257, 263, 269–270, 272, 357
VerRule, 138
Who-cares question, 4, 7, 21
Woods-Walton Approach, xiii, 86, 149, 161, 171, 349

APPLIED LOGIC SERIES

1. D. Walton: *Fallacies Arising from Ambiguity.* 1996 ISBN 0-7923-4100-7
2. H. Wansing (ed.): *Proof Theory of Modal Logic.* 1996 ISBN 0-7923-4120-1
3. F. Baader and K.U. Schulz (eds.): *Frontiers of Combining Systems.* First International Workshop, Munich, March 1996. 1996 ISBN 0-7923-4271-2
4. M. Marx and Y. Venema: *Multi-Dimensional Modal Logic.* 1996 ISBN 0-7923-4345-X
5. S. Akama (ed.): *Logic, Language and Computation.* 1997 ISBN 0-7923-4376-X
6. J. Goubault-Larrecq and I. Mackie: *Proof Theory and Automated Deduction.* 1997 ISBN 0-7923-4593-2
7. M. de Rijke (ed.): *Advances in Intensional Logic.* 1997 ISBN 0-7923-4711-0
8. W. Bibel and P.H. Schmitt (eds.): *Automated Deduction - A Basis for Applications.* Volume I. Foundations - Calculi and Methods. 1998 ISBN 0-7923-5129-0
9. W. Bibel and P.H. Schmitt (eds.): *Automated Deduction - A Basis for Applications.* Volume II. Systems and Implementation Techniques. 1998 ISBN 0-7923-5130-4
10. W. Bibel and P.H. Schmitt (eds.): *Automated Deduction - A Basis for Applications.* Volume III. Applications. 1998 ISBN 0-7923-5131-2 (Set vols. I-III: ISBN 0-7923-5132-0)
11. S.O. Hansson: *A Textbook of Belief Dynamics.* Theory Change and Database Updating. 1999 Hb: ISBN 0-7923-5324-2; Pb: ISBN 0-7923-5327-7
    Solutions to exercises. 1999. Pb: ISBN 0-7923-5328-5
    Set: (Hb): ISBN 0-7923-5326-9; (Pb): ISBN 0-7923-5329-3
12. R. Pareschi and B. Fronhöfer (eds.): *Dynamic Worlds from the Frame Problem to Knowledge Management.* 1999 ISBN 0-7923-5535-0
13. D.M. Gabbay and H. Wansing (eds.): *What is Negation?* 1999 ISBN 0-7923-5569-5
14. M. Wooldridge and A. Rao (eds.): *Foundations of Rational Agency.* 1999 ISBN 0-7923-5601-2
15. D. Dubois, H. Prade and E.P. Klement (eds.): *Fuzzy Sets, Logics and Reasoning about Knowledge.* 1999 ISBN 0-7923-5911-1
16. H. Barringer, M. Fisher, D. Gabbay and G. Gough (eds.): *Advances in Temporal Logic.* 2000 ISBN 0-7923-6149-0
17. D. Basin, M.D. Agostino, D.M. Gabbay, S. Matthews and L. Viganò (eds.): *Labelled Deduction.* 2000 ISBN 0-7923-6237-3
18. P.A. Flach and A.C. Kakas (eds.): *Abduction and Induction.* Essays on their Relation and Integration. 2000 ISBN 0-7923-6250-0
19. S. Hölldobler (ed.): *Intellectics and Computational Logic.* Papers in Honor of Wolfgang Bibel. 2000 ISBN 0-7923-6261-6

APPLIED LOGIC SERIES

20. P. Bonzon, M. Cavalcanti and Rolf Nossum (eds.): *Formal Aspects of Context.* 2000 ISBN 0-7923-6350-7
21. D.M. Gabbay and N. Olivetti: *Goal-Directed Proof Theory.* 2000 ISBN 0-7923-6473-2
22. M.-A. Williams and H. Rott (eds.): *Frontiers in Belief Revision.* 2001 ISBN 0-7923-7021-X
23. E. Morscher and A. Hieke (eds.): *New Essays in Free Logic.* In Honour of Karel Lambert. 2001 ISBN 1-4020-0216-5
24. D. Corfield and J. Williamson (eds.): *Foundations of Bayesianism.* 2001 ISBN 1-4020-0223-8
25. L. Magnani, N.J. Nersessian and C. Pizzi (eds.): *Logical and Computational Aspects of Model-Based Reasoning.* 2002 Hb: ISBN 1-4020-0712-4; Pb: ISBN 1-4020-0791-4
26. D.J. Pym: *The Semantics and Proof Theory of the Logic of Bunched Implications.* 2002 ISBN 1-4020-0745-0
27. P.B. Andrews: *An Introduction to Mathematical Logic and Type Theory: To Truth Through Proof.* Second edition. 2002 ISBN 1-4020-0763-9
28. F.D. Kamareddine: *Thirty Five Years of Automating Mathematics.* 2003 ISBN 1-4020-1656-5
29. F. Kamareddine, T. Laan and R. Nederpelt: *A Modern Perspective on Type Theory.* From its Origins until Today. 2004 ISBN 1-4020-2334-0
30. H. Leitgeb: *Inference on the Low Level.* An Investigation into Deduction, Nonmonotonic Reasoning, and the Philosophy of Cognition. 2004 ISBN 1-4020-2492-4
31. R. Caferra, A. Leitsch and P. Nicholas: *Automated Model Building.* 2004 ISBN 1-4020-2652-8
32. J.H. Woods: *The Death of Argument.* Fallacies in Agent-Based Reasoning. 2004 ISBN 1-4020-2663-3

KLUWER ACADEMIC PUBLISHERS – DORDRECHT / BOSTON / LONDON

MIX
Papier aus verantwortungsvollen Quellen
Paper from responsible sources
FSC® C105338

If you have any concerns about our products,
you can contact us on
ProductSafety@springernature.com

In case Publisher is established outside the EU,
the EU authorized representative is:
Springer Nature Customer Service Center GmbH
Europaplatz 3, 69115 Heidelberg, Germany

Printed by Libri Plureos GmbH
in Hamburg, Germany